Motors, Generators, Transformers, and Energy

Motors, Generators, Transformers, and Energy

Pericles Emanuel

Queensborough Community College
of the City University of New York

PRENTICE-HALL, INC., Englewood Cliffs, N.J. 07632

Library of Congress Cataloging in Publication Data

EMANUEL, PERICLES JOHN. date
   Motors, generators, transformers, and energy.

   Includes index.
   1. Electric machinery. I. Title.
TK2000.E4 1985   621.31′3   84-6964
ISBN 0-13-604026-8

Editorial/production supervision and
   interior design: Tom Aloisi
Cover design: Photo Plus Art, Celine Brandes
Manufacturing buyer: Anthony Caruso
Cover photo: Courtesy of Motorola, Inc.

Printed in the United States of America

10 9 8 7 6 5 4 3 2 1

ISBN: 0-13-604026-8

Prentice-Hall International, Inc., *London*
Prentice-Hall of Australia Pty. Limited, *Sydney*
Editora Prentice-Hall do Brasil, Ltda., *Rio de Janeiro*
Prentice-Hall Canada Inc., *Toronto*
Prentice-Hall of India Private Limited, *New Delhi*
Prentice-Hall of Japan, Inc., *Tokyo*
Prentice-Hall of Southeast Asia Pte. Ltd., *Singapore*
Whitehall Books Limited, *Wellington, New Zealand*

FOR:

My Daughter, Melanie—the prettiest and best 10-year-old pitcher in the world.

My Son, Michael—a 6-year-old possessing the confidence to do whatever he wants on this
   earth.

and My Wife, Sandra—who has not let her love falter during these two agonizing years.

Contents

3   DC MACHINE CONSTRUCTION   56

4   DC GENERATOR CHARACTERISTICS   75

5   DC MOTOR CHARACTERISTICS   121

6   CONTROL OF DIRECT-CURRENT MOTORS   169

9   THE THREE-PHASE INDUCTION MOTOR   297

Preface

This book, although written with an up-to-date slant, is in effect a classical power technology textbook. It has been written for both two- and four-year technology programs and contains sufficient material for a one- or possibly two-semester course. This would depend on both the level of the student and the depth to which the material is covered.

The text is different from the other books available today in many ways:

1. The language has been kept simple, making it a very readable text. Wherever possible, sentences were kept short and "multisyllabic" words were avoided.

2. It makes use of both the English and International System of Units (SI) in such a way that either one or both could be used without any degree of difficulty. Equations are given for both systems. The numerous worked-out examples are presented on a one-for-one basis. When one is done in the English system it is immediately repeated in SI with slightly different numbers, making them similar yet distinct. The end-of-chapter problems are arranged in three groups: those which can be done by the student using the English system, those which can be done by the student using SI, and those that can be done by both.

3. The mathematics have been kept simple. Calculus is avoided throughout the text.

4. Many experimental sets of data are used in examples and problems. In each instant the student is asked to graph the data and calculated results. This should be quite beneficial to those schools that do not have laboratory work to accompany the theory.

5. Computer control of motors, in particular microprocessor control, is presented in Chapters 6 and 13. The treatment is not superficial but rather quite complete.

Not only is the motor control circuitry presented and explained, but the computer input/output circuitry as well as actual programs are explained in detail. The programs are written in machine code and assembly language for the Motorola 6800 microprocessor. To fully understand the sections on computer control, the student should have some background in microprocessors (both hardware and software).

6. This classical subject is given a refreshing shot in the arm with the inclusion of two topics which have been neglected in other books. Solar and wind energy conversion systems are discussed in Chapters 14 and 15. The conversion from these alternative energy sources to usable electrical power is simply and completely presented. In this way the student and handyperson could give serious thought to supplying their homes with electrical energy someday from the sun and/or wind if ever desired.

7. The induction generator, long a forgotten device, is presented in Chapter 11. The coverage is complete, yet easy to understand.

8. All of Chapter 1 deals with magnetic principles and magnetic circuits. Students are not expected to understand electromagnetic devices without an understanding of electric circuits. It therefore makes sense that the student should also have a basic understanding of magnetic circuits.

As is the case in any scientific textbook, there are many symbols used throughout. To prevent confusion, a table, listing the symbols introduced in each chapter together with their definitions and units, is given at the end of each chapter. In addition to the end-of-chapter problems, there is also a set of questions included at each chapter's end.

There are two appendices that contain the conversion formulas from the English system to SI, and vice versa. The conversions given are those that would be helpful in solving the problems given in this text. Also included are the numerical answers to all of the odd numbered end-of-chapter problems.

I would like to thank the various companies who made available much of the technical information, in particular, the Reliance Electric Company who supplied the photographs used in the book. I must also thank Gaetano Giudice, Leon Katz, Jackson Lum, and Peter Stark, whose valuable comments throughout the writing of this text helped make it what it is. Above all, my thanks to Joseph Aidala, who read and edited the initial chapters for me. His valuable comments helped shape the format and content of the finished product.

Finally, I must thank my family for enduring throughout this project. My daughter Melanie and my son Michael now realize the sacrifice that other children have made when their parents have written a book. Most important there is that person who for two years has encouraged me to complete the project, shared me with the four walls of my office, and impeccably typed the entire manuscript. I'm sure every author knows that I can only be referrring to my loving and caring wife, Sandra.

Pericles Emanuel

Motors,
Generators,
Transformers,
and Energy

Chapter 1

Introduction to Magnetism and Magnetic Circuits

The world we live in today has been described as an electronic world: a world in which humankind has become so dependent on electricity that without it we could not live a normal life as we know it. There is, however, another phenomenon that is quite often overlooked in the way it affects our lives. This invisible, magical phenomenon is **magnetism.** Without magnetism, everyday electrical devices such as radios, televisions, refrigerators, air conditioners, heating equipment, and power tools could not work. In fact, without magnetism, the electricity that we have become so dependent on would not be present in our homes or places of work. Thus we can draw the same conclusion as above: Without magnetism humankind could not live a normal life as we know it.

1-1 SIMILARITIES BETWEEN MAGNETISM AND ELECTRICITY

The devices covered in this text fall into a category called **electromagnetic** components. In other words, their operation is a function of their electric and magnetic properties. In addition, their operation depends on the way in which their electric and magnetic properties affect each other. It is felt that an introduction to simple magnetic circuits at this point will enable us better to understand the operation of electromagnetic devices. It will be seen in the following paragraphs that magnetic circuits are very similar to electric circuits.

1-1.1 Flux (ϕ)

The flux (ϕ) in a magnetic circuit is analogous to the current in an electric circuit. There is a general law frequently used in technical descriptions (effect = cause/

1

opposition). With reference to this law, flux and current are both "effects." The flux is represented by imaginary lines having a direction and varying paths in a magnetic circuit. The magnetic circuit is made up of sections of materials that easily permit lines of flux to be formed. These materials are called **ferromagnetic.** Ferromagnetic materials in a magnetic circuit are comparable to "good conductors" in an electric circuit.

The unit for flux in the English system is the **line** (or equivalently the **maxwell**). In SI (International System of Units) the unit for flux is the **weber** (Wb).

$$1 \text{ weber} = 10^8 \text{ lines (or } 10^8 \text{ maxwells)} \tag{1-1}$$

A quantity of importance in magnetic circuits is the **flux density (B).** It is a measure of the amount of flux perpendicular to a cross-sectional area (A) in the magnetic circuit.

(English)

$$B = \frac{\phi}{A} \frac{\text{lines}}{\text{in}^2} \quad \left(\text{or } \frac{\text{maxwells}}{\text{in}^2} \right) \tag{1-2a}$$

(SI)

$$B = \frac{\phi}{A} \frac{\text{Wb}}{\text{m}^2} \quad [\text{or tesla (T)}] \tag{1-2b}$$

Example 1-1 (English)

What is the flux in a core whose cross-sectional area is 3 in²? The flux density is 80,000 lines/in².

Solution

Using Eq. 1-2a, we have

$$B = \frac{\phi}{A}$$

and solving for the unknown quantity (ϕ), we get

$$\phi = BA$$
$$= 80{,}000 \text{ lines/in}^2 \times 3 \text{ in}^2$$
$$= 240{,}000 \text{ lines} \quad (\text{or maxwells})$$

Example 1-2 (SI)

What is the flux in a core whose cross-sectional area is 0.04 m²? The flux density is 1.5 T (tesla, or webers/m²).

Solution

Using Eq. 1-2b, we have

$$B = \frac{\phi}{A}$$

and solving for the unknown quantity (ϕ), we get

$$\phi = B\mathbf{A}$$
$$= 1.5 \text{ T} \times 0.04 \text{ m}^2$$
$$= 0.06 \text{ Wb}$$

1-1.2 Magnetomotive Force, MMF(U)

The magnetomotive force (MMF) in a magnetic circuit is analogous to a voltage source (EMF) in an electric circuit. Just as an EMF causes a current in an electric circuit, an MMF causes flux in a magnetic circuit. Again in relation to the general law, EMF and MMF are both "causes."

A difference in the comparison between electric and magnetic circuits being made should be noted here. In a simple electric circuit a current cannot exist without a source (such as an EMF). In a magnetic circuit flux can exist without a source (MMF). This occurs in the case of a permanent magnet. An example of a permanent magnet is the simple bar or U-shaped magnet that everyone reading this book has had experience with at one time or another.

The simplest type of EMF considered in electricity is the battery. Besides the permanent magnet, the simplest MMF would be formed by a wire, carrying a direct current, wrapped around a ferromagnetic material as shown in Figure 1-1.

The MMF (U) is numerically equal to the product of the number of turns of wire (N) and the current (I) in amperes.

(English)

$$U = NI \qquad \text{ampere-turns} \qquad \text{(1-3a)}$$

(SI)

$$U = NI \qquad \text{amperes} \qquad \text{(1-3b)}$$

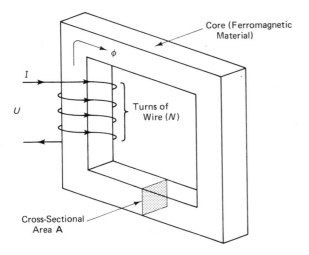

Figure 1-1 Simple magnetomotive force.

It is interesting and useful to note that the applied MMF equals the sum of the MMF drops around a loop in a magnetic circuit. This rule, which is analogous to Kirchhoff's voltage law, will be used later in this chapter to solve simple magnetic circuits.

The direction of the flux produced can be found by wrapping the right hand around the core with the fingers going around in the direction of the current in the coil. The thumb will then be pointing in the direction of the flux.

Another quantity of importance in magnetic circuits is the magnetic field **intensity** (H). It is given numerically by

(English)

$$H = \frac{NI}{l} \qquad \frac{\text{ampere-turns}}{\text{inch}} \qquad (1\text{-}4a)$$

(SI)

$$H = \frac{NI}{l} \qquad \frac{\text{amperes}}{\text{meter}} \qquad (1\text{-}4b)$$

In Eqs. 1–4, l represents the mean (average) length of the magnetic circuit in inches and meters for the English and SI systems of units, respectively.

Example 1-3 (English)

The ferromagnetic core shown in Figure 1-2 has 1000 turns of wire with 2 amperes (A) of current. Find the magnetomotive force and the magnetic field intensity. Assume that the entire core is made of the same material and that it has a uniform cross-sectional area.

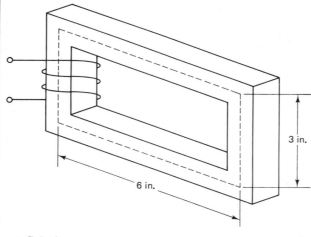

3 in.

6 in.

Figure 1-2 Core for Example 1-3.

Solution

The MMF is calculated using Eq. 1-3a.

$$U = NI$$

$$= 1000(2) = 2000 \text{ A-turns}$$

The magnetic field intensity is calculated using Eq. 1-4a. First, however, the mean length (*l*) must be calculated. The mean length is the average distance around the core and is shown by the dashed line.

$$l = 3 \text{ in.} + 6 \text{ in.} + 3 \text{ in.} + 6 \text{ in.}$$

$$= 18 \text{ in.}$$

$$H = \frac{NI}{l}$$

$$= 2000 \text{ A-turns}/18 \text{ in.}$$

$$= 111.11 \text{ A-turns/in.}$$

Example 1-4 (SI)

The ferromagnetic core shown in Figure 1-3 has 1000 turns of wire with 2 A of current. Find the magnetomotive force and the magnetic field intensity. Assume that the entire core is made of the same material and that it has a uniform cross-sectional area.

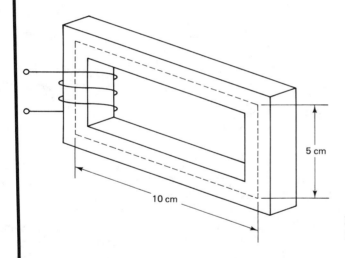

Figure 1-3 Core for Example 1-4.

Solution

The MMF is calculated using Eq. 1-3b.

$$U = NI$$

$$= 1000(2) = 2000 \text{ A}$$

The magnetic field intensity is calculated using Eq. 1-4b. First, however, the mean length (*l*) must be calculated. The mean length is the average distance around the core and is shown by the dashed line.

$$l = 5 \text{ cm} + 10 \text{ cm} + 5 \text{ cm} + 10 \text{ cm}$$

$$= 30 \text{ cm}$$

This length must now be converted to meters (1 cm = 0.01 m):

$$l = 30(0.01) = 0.3 \text{ m}$$

and

$$H = \frac{NI}{l} = \frac{2000 \text{ A}}{0.3 \text{ m}} = 6666.67 \text{ A/m}$$

1-1.3 Reluctance ($\mathcal{R}$)

The reluctance ($\mathcal{R}$) in a magnetic circuit is analogous to resistance in electric circuits. Again referring to the general law, reluctance and resistance are both "oppositions." Recall the formula for the resistance of a copper wire and examine it here.

$$R = \frac{\rho l}{\mathbf{A}} \tag{1-5}$$

Also, if we make the substitution for resistivity (ρ), which is the reciprocal of the conductivity (σ), we obtain

$$R = \frac{l}{\sigma \mathbf{A}} \tag{1-6}$$

Equation 1-6 states that the resistance of a copper wire is proportional to the length (l) of the wire and inversely proportional to the conductivity (σ) and cross-sectional area ($\mathbf{A}$) of the wire.

In magnetic circuits, the reluctance is given by

$$\mathcal{R} = \frac{l}{\mu \mathbf{A}} \tag{1-7}$$

In Eq. 1-7, l is the mean (average) length of the magnetic circuit or core section under examination, $\mathbf{A}$ is the effective cross-sectional area of the circuit (in this text, only uniform cross-sectional areas are considered), and μ is the **permeability** of the portion of the circuit under consideration. Permeability is a measure of a material's ability to permit a magnetic field to exist in it. The higher the permeability, the more flux can be obtained for a given MMF. The lower the permeability, the less flux can be obtained for a given MMF. The permeability of air is denoted μ_0 and is given by

(English)

$$\mu_0 = 3.19 \times 10^{-3} \frac{\text{kilolines/in}^2}{\text{A-turns/in.}} \tag{1-8a}$$

(SI)

$$\mu_0 = 4\pi \times 10^{-7} \frac{\text{T}}{\text{A/m}} \tag{1-8b}$$

Although the permeability of air is constant, the permeability of ferromagnetic materials is not. It is a nonlinear quantity that decreases as the flux density (B) increases. This nonlinear trait is due to a phenomenon to be considered and defined shortly, called **saturation.**

The permeability of a material also relates the flux density (B) to the magnetic field intensity (H). The relationship is given by

$$B = \mu H \qquad \textit{magnetic material} \qquad (1\text{-}9)$$

If μ is constant, as it is for air and other nonferromagnetic materials, a plot of Eq. 1-9 would be a straight line, as shown in Figure 1-4. The slope of the straight line is the permeability of the material:

$$\text{slope} = \mu = \frac{\Delta B}{\Delta H} = \frac{B}{H} \qquad (1\text{-}10)$$

Notice in Figure 1-4 that as H increases, B increases proportionately; hence the term "linear" is used. It is also important to note that when dealing with nonferromagnetic materials, if H is known, Eq. 1-9 can be used to solve for B (and vice versa).

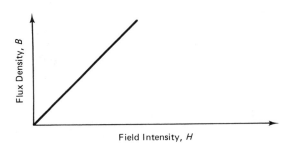

Field Intensity, H

Flux Density, B

Figure 1-4 Plot of B versus H for nonferromagnetic materials (air, wood, plastic, etc.).

Example 1-5 (English)

In a given air space the field intensity is 30 A-turns/in. Find the flux density in the air space.

Solution

Equation 1-9 is used, and since the material is air, $\mu = \mu_0$ and is given by Eq. 1-8a.

$$B = \mu_0 H \qquad \textit{NON magnetic material}$$

$$= 3.19 \times 10^{-3} \frac{\text{kilolines/in}^2}{\text{A-turns/in.}} (30 \text{ A-turns/in.})$$

$$= 95.7 \text{ lines/in}^2$$

Example 1-6 (SI)

In a given air space the field intensity is 900 A/m. Find the flux density in the air space.

Solution

Equation 1-9 is used, and since the material is air, $\mu = \mu_0$ and is given by Eq. 1-8b.

$$B = \mu_0 H$$

$$= 4\pi \times 10^{-7} \frac{T}{A/m} (900 \text{ A/m})$$

$$= 0.113 \times 10^{-2} \text{ T}$$

Example 1-7 (English)

An air space having a cross-sectional area of 4 in² has 200,000 lines of flux. Find the magnetic field intensity in the air space.

Solution

This is a two-step problem. First, Eq. 1-2a will be used to find the flux density, B. Then, Eq. 1-9 will be used to find the field intensity, H.

$$B = \frac{\phi}{A}$$

$$= 200,000 \text{ lines/4 in}^2$$

$$= 50,000 \text{ lines/in}^2 = 50 \text{ kilolines/in}^2$$

$$B = \mu_0 H$$

Rearranging yields

$$H = \frac{B}{\mu_0}$$

$$= \frac{50 \text{ kilolines/in}^2}{3.19 \times 10^{-3} \frac{\text{kilolines/in}^2}{\text{A-turns/in.}}} = 15,670 \text{ A-turns/in.}$$

Example 1-8 (SI)

An air space having a cross-sectional area of 0.1 m² has 0.002 Wb. Find the magnetic field intensity in the air space.

Solution

This is a two-step problem. First, Eq. 1-2b will be used to find the flux density, B. Then, Eq. 1-9 will be used to find the field intensity, H.

$$B = \frac{\phi}{A}$$

$$B = \frac{0.002 \text{ Wb}}{0.1 \text{ m}^2} = 0.02 \text{ Wb/m}^2 = 0.02 \text{ T}$$

$$B = \mu_0 H$$

Rearranging yields

$$H = \frac{B}{\mu_0}$$

$$= \frac{0.02 \text{ T}}{4\pi \times 10^{-7} \dfrac{\text{T}}{\text{A/m}}}$$

$$= 15.9 \times 10^3 \text{ A/m} = 15,900 \text{ A/m}$$

1-2 NONLINEAR EFFECTS OF FERROMAGNETIC MATERIALS

When dealing with nonferromagnetic materials such as air, analyzing and problem solving of magnetic circuits is fairly simple, as seen in the preceding examples. However, when dealing with ferromagnetic materials, the situation is made a little bit more difficult by the fact that the permeability of these materials is not constant. As the flux density of a ferromagnetic material increases, the permeability decreases. This is due to a phenomenon called **saturation.** As a ferromagnetic core becomes magnetized by a magnetomotive force, the flux in the core increases until the core fills up. At this point, further increases in the field intensity will cause the flux to increase; however, the increase will not be as great as before. The core is said to be **saturated.**

Saturation can be compared to boarding a subway car during the rush hour. When the car is empty, people can board it quite easily. As it is filling up, boarding it remains relatively easy. Finally, when the car is filled up with passengers and there is no longer any room for people to enter it, we can say the car is saturated just like a ferromagnetic core. In reality with a little pushing and shoving, there is always room for one more person to squeeze into the car. The situation is the same in the ferromagnetic core. Even though the core is saturated, if the field intensity is increased, the flux in the core will increase very slightly. Figure 1-5 is a plot of flux density versus field intensity for a typical ferromagnetic core. It is commonly referred to as the B–H curve or **magnetization curve** * of a particular material.

Referring to Figure 1-5, we see three distinct regions on the curve. Region 1 indicates that portion of the curve where increases in field intensity cause corresponding increases in flux density. As the flux density increases, region 2 is reached. This is commonly called the **knee** of the curve. At this point the core is beginning to saturate. In region 3 the core is saturated. Notice that increases in field intensity now cause only slight increases in flux density.

Figures 1-6 and 1-7 are magnetization curves for cast iron and cast steel. Figure

*The terms "magnetization curve" and "saturation curve" are used interchangeably in practice.

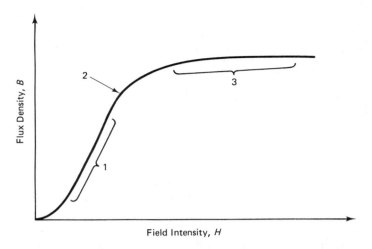

Figure 1-5 Magnetization curve of a typical ferromagnetic material.

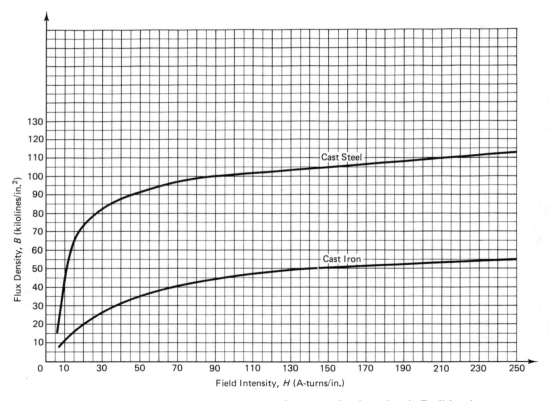

Figure 1-6 Magnetization curves for cast steel and cast iron in English units.

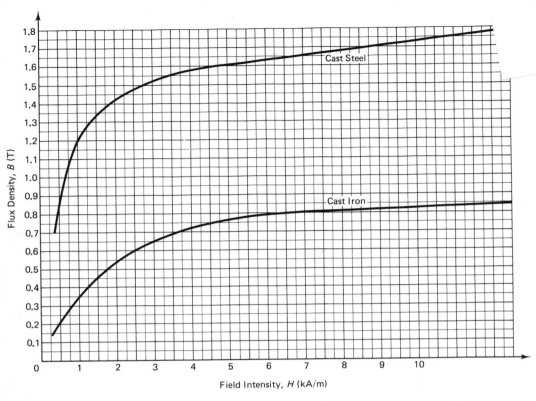

Figure 1-7 Magnetization curves for cast steel and cast iron in SI units.

1-6 is plotted in English units while Figure 1-7 is plotted in SI units. Care should be taken to use the proper curve, depending on the system of units being used.

Example 1-9 (English)

For the cast-iron core shown in Figure 1-8, find the flux density. It has uniform cross-sectional area.

Solution

$$l = 3 \text{ in.} + 1.5 \text{ in.} + 3 \text{ in.} + 1.5 \text{ in.}$$

$$= 9 \text{ in.}$$

Using Eq. 1-4a gives us

$$H = \frac{NI}{l}$$

$$= \frac{270(4)}{9} = 120 \text{ A-turns/in.}$$

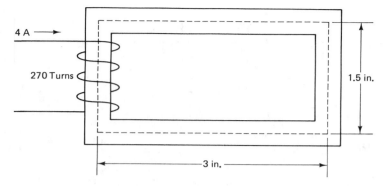

Figure 1-8 Core for Example 1-9.

Now using Figure 1-6, knowing H we can find B. From the curve for cast iron when H is 120 A-turns/in., B is found to be approximately 49 kilolines/in². Note that Eq. 1-9 was not used in this example since μ was not known.

Example 1-10 (SI)

For the cast-iron core shown in Figure 1-9, find the flux density. It has uniform cross-sectional area.

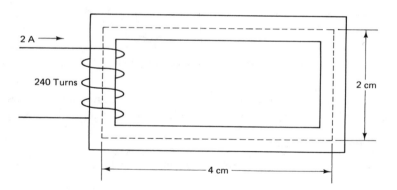

Figure 1-9 Core for Example 1-10.

Solution

$$l = 4 \text{ cm} + 2 \text{ cm} + 4 \text{ cm} + 2 \text{ cm}$$

$$= 12 \text{ cm}$$

$$= 12(0.01) = 0.12 \text{ m}$$

Using Eq. 1-4b, we have

$$H = \frac{NI}{l}$$

$$H = \frac{240(2)}{0.12}$$

$$= 4000 \text{ A/m} = 4 \text{ kA/m}$$

Now using Figure 1-7, knowing H we can find B. From the curve for cast iron when H is 4 kA/m, B is found to be approximately 0.71 T. Note that Eq. 1-9 was not used in this example since μ was not known.

1-2.1 Hysteresis

Another interesting and important phenomenon occurs in ferromagnetic materials. It is best described by referring to a typical B–H curve such as that in Figure 1-10. When a current flows through a coil wrapped around a ferromagnetic core, it creates an MMF. As the MMF increases, so does the field intensity H, until the core saturates (point 1 to point 2 on the curve). If the current is now decreased to zero, the MMF and hence the field intensity H will also go to zero. However, the flux density B will not go to zero (point 2 to point 3 on the curve). Even though the current and field intensity have gone to zero, the core remains magnetized. This magnetism that remains in the core is called **residual magnetism.** It is this effect that enables us to make permanent magnets.

If the current is now reversed and slowly increases in the negative direction, the flux density will be driven to zero. The negative field intensity needed to drive B to

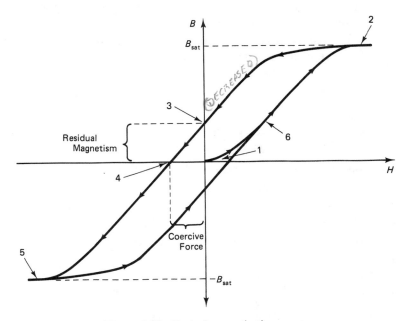

Figure 1-10   Typical magnetization curve.

zero is called the **coercive force** (point 4 on the curve). As the current is made more negative, the core will eventually saturate. The flux density will have a polarity opposite to that in the original case (point 5 on the curve).

As the current is reduced to zero and then made positive again, the curve will join up with the original curve at point 6. The closed loop joining points 2, 3, 4, 5, 6, and 2 again is called a **hysteresis loop** (This phenomenon whereby B lags H (that is, does not return to zero at the same time H does) is called **hysteresis.**)

The following items should be noted:

1. The hysteresis loop has polar symmetry. That is, if B_1 is the flux density for field intensity H_1, $-B_1$ will be the flux density for a field intensity $-H_1$.

2. For a given field intensity H, there are two possible values for B depending on whether H is increasing or decreasing. This property will often cause discrepancies in data taken in the laboratory by students. For example, the field current in a direct-current (dc) generator has been increased to a large value, and then decreased to an intermediate value to check some data. Because of hysteresis, the flux in the machine will not be the same as the first setting and therefore some data will not be verified.

3. Once a ferromagnetic material has been magnetized, the only way to de-magnetize it (make B and H both zero) is to put it through magnetization cycles while gradually decreasing the intensity.

4. The maximum flux density in a ferromagnetic material is a function of temperature. As the temperature increases, the saturation flux density decreases. In addition, beyond a certain temperature (called the **Curie point**) the material loses its ferromagnetic properties.

5. Different materials have different saturation flux densities. This factor will affect the size of the core needed in an electromagnetic device. The following examples will illustrate this.

Example 1-11 (English)

A maximum flux (ϕ_{max}) of 350,000 lines is desired in a core of uniform cross-sectional area. Find the minimum area needed if the material is:

(a) Cast iron with a maximum flux density of 50 kilolines/in^2

(b) Cast steel with a maximum flux density of 100 kilolines/in^2

Solution

(a) Equation 1-2a is rewritten with the following subscripts:

$$B_{max} = \frac{\phi_{max}}{A_{min}}$$

Solving for A_{min} yields

$$A_{min} = \frac{\phi_{max}}{B_{max}}$$

$$\mathbf{A}_{min} = \frac{350 \text{ kilolines}}{50 \text{ kilolines/in}^2}$$

$$= 7 \text{ in}^2$$

(b)

$$\mathbf{A}_{min} = \frac{\phi_{max}}{B_{max}}$$

$$= \frac{350 \text{ kilolines}}{100 \text{ kilolines/in}^2}$$

$$= 3.5 \text{ in}^2$$

Example 1-12 (SI)

A maximum flux (ϕ_{max}) of 3.5×10^{-3} Wb is desired in a core of uniform cross-sectional area. Find the minimum area needed if the material is:

(a) Cast iron with a maximum flux density of 0.8 T

(b) Cast steel with a maximum flux density of 1.6 T

Solution

(a) Equation 1-2b is rewritten with the following subscripts:

$$B_{max} = \frac{\phi_{max}}{\mathbf{A}_{min}}$$

Solving for $\mathbf{A}_{min}$ gives us

$$\mathbf{A}_{min} = \frac{\phi_{max}}{B_{max}}$$

$$= \frac{3.5 \times 10^{-3} \text{ Wb}}{0.8 \text{ T}}$$

$$= 4.375 \times 10^{-3} \text{ m}^2$$

(b)

$$\mathbf{A}_{min} = \frac{\phi_{max}}{B_{max}}$$

$$= \frac{3.5 \times 10^{-3} \text{ Wb}}{1.6 \text{ T}}$$

$$= 2.1875 \times 10^{-3} \text{ m}^2$$

1-3 MAGNETIC CIRCUITS

Some elementary series and parallel magnetic circuits are described in this section. The concepts are introduced with numerical examples. A typical problem can be presented in two ways. One way, which will be discussed here, is the following: Given a magnetic circuit, find the MMF required to produce a given flux. This type of problem is analogous to a series electric circuit where we are asked to find the voltage (EMF) required to produce a given current. The solution is simple. By adding up all of the $I \times R$ drops around the loop, we find the applied voltage. In the magnetic circuit the solution is almost as easy.

1. Knowing ϕ and the cross-sectional area, find B.
2. Knowing B and the core material, find H using the magnetization curves. If the material is air, Eq. 1-9 will be used to find H, where μ is given by Eqs. 1-8.
3. Now add up all the $H \times l$ (MMF) drops around the loop to find the total MMF required.

The second way that a problem can be presented is the following: Given an applied MMF, find the flux in a magnetic circuit. This problem, although it can be solved by trial and error, will not be covered here because it will only tend to confuse the reader. Because of the nonlinear nature of ferromagnetic materials, we do not know how a total MMF distributes itself around a loop. We would have to make an initial guess, solve for ϕ, and then repeat the procedure, each time improving the prior guess. Eventually, we reach a guess that produces a flux constant in all parts of a series loop.

Example 1-13 (English)

The core in Figure 1-11 has a uniform cross-sectional area of 1.25 in². To get a flux of 50 kilolines, find:

(a) The coil MMF required

(b) The current in the coil

Solution

A systematic way to solve the problem is to set up Table 1-1. The table will have a row for each part of the circuit. In this example there are two parts: one made of cast iron,

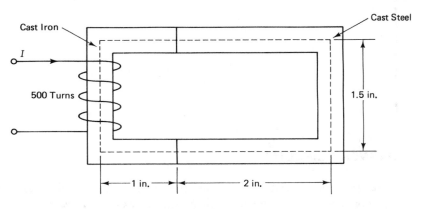

Figure 1-11 Core for Example 1-13.

the other made of cast steel. The table is filled in from the left to right as the calculations are made.

Table 1-1

Part	ϕ	A	B	H	l	Hl
Cast iron	50 k	1.25	40 k	68	3.5	238
Cast steel	50 k	1.25	40 k	10	5.5	55
Total MMF						293

For both parts of the circuit,

$$B = \frac{\phi}{A} = \frac{50 \text{ k}}{1.25} = 40 \text{ kilolines/in}^2$$

Using Figure 1-6, find H.
For cast iron,

$$l = 1 \text{ in.} + 1.5 \text{ in.} + 1 \text{ in.}$$
$$= 3.5 \text{ in.}$$

For cast steel,

$$l = 2 \text{ in.} + 1.5 \text{ in.} + 2 \text{ in.}$$
$$= 5.5 \text{ in.}$$

Now multiply $H \times l$ and enter it in the last column.

(a) The total MMF is the sum of the Hl drops:

$$\text{total MMF} = 238 + 55 = 293 \text{ A-turns}$$

(b) Total MMF = NI

$$293 = 500I$$
$$I = \frac{293}{500} = 0.586 \text{ A}$$

Example 1-14 (SI)

The core in Figure 1-12 has a uniform cross-sectional area of 9×10^{-4} m^2. To get a flux of 0.75×10^{-3} Wb, find:

(a) The coil MMF required

(b) The current in the coil

Solution

A systematic way to solve the problem is to set up Table 1-2. The table will have a row for each part of the circuit. In this example there are two parts; one made of cast iron, the other made of cast steel. The table is filled in from left to right as the calculations are made.

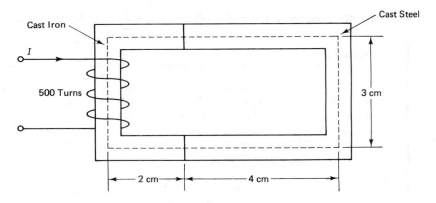

Figure 1-12 Core for Example 1-14.

Table 1-2

Part	ϕ	A	B	H	l	Hl
Cast iron	0.75×10^{-3}	9×10^{-4}	0.83	9.4 k	0.07	658
Cast steel	0.75×10^{-3}	9×10^{-4}	0.83	0.5 k	0.11	55
Total MMF						713

For both parts of the circuit,

$$B = \frac{\phi}{A} = \frac{0.75 \times 10^{-3}}{9 \times 10^{-4}}$$

$$= 0.83 \text{ T}$$

Using Figure 1-7, find H.
For cast iron,

$$l = 2 \text{ cm} + 3 \text{ cm} + 2 \text{ cm}$$

$$= 7 \text{ cm} = 0.07 \text{ m}$$

For cast steel,

$$l = 4 \text{ cm} + 3 \text{ cm} + 4 \text{ cm}$$

$$= 11 \text{ cm} = 0.11 \text{ m}$$

Now multiply $H \times l$ and enter it in the last column.

(a) The total MMF is the sum of the Hl drops:

$$\text{total MMF} = 658 + 55 = 713 \text{ A}$$

(b) Total MMF $= NI$

$$713 = 500I$$

$$I = \frac{713}{500} = 1.43 \text{ A}$$

Example 1-15 (English)

The core in Figure 1-13 is made of cast iron. Find the coil MMF necessary to get a flux of 150-kilolines. Assume that the cross-sectional area for the core as well as the air gap is uniform and equals 3 in^2.

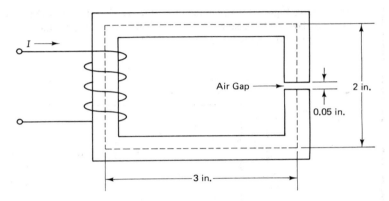

Figure 1-13   Core for Example 1-15.

Solution

There are two parts to this circuit, the cast iron core and the air gap (see Table 1-3).

Table 1-3

Part	ϕ	A	B	H	l	Hl
Cast iron	150k	3	50k	130	10	1300
Air gap	150k	3	50k	15.67×10^3	0.05	783.5
Total MMF						2083.5

$$B = \frac{\phi}{A} = \frac{150k}{3} = 50k$$

For the air gap, H is found using Eqs. 1-9 and 1-8a.

$$B = \mu_0 H$$

$$H = \frac{B}{\mu_0} = \frac{50 \text{ kilolines/in}^2}{3.19 \times 10^{-3} \dfrac{\text{kilolines/in}^2}{\text{A-turns/in}}}$$

$$= 15.67 \times 10^3 \text{ A-turns/in.}$$

For the cast-iron part, neglecting the air gap,

$$l = 3 \text{ in.} + 2 \text{ in.} + 3 \text{ in.} + 2 \text{ in.}$$

$$= 10 \text{ in.}$$

From Figure 1-6, H is found to be 130 A-turns/in. The total MMF is the sum of the Hl drops:

$$\text{total MMF} = 1300 + 783.5 = 2083.5 \text{ A-turns}$$

Example 1-16 (SI)

The core in Figure 1-14 is made of cast iron. Find the coil MMF necessary to get a flux of 15×10^{-4} Wb. Assume that the cross-sectional area for the core as well as the air gap is uniform and equals 20×10^{-4} m².

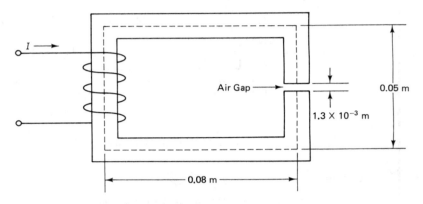

Figure 1-14 Core for Example 1-16.

Solution

There are two parts to this circuit, the cast-iron core and the air gap (see Table 1-4).

Table 1-4

Part	ϕ	A	B	H	l	Hl
Cast iron	15×10^{-4}	20×10^{-4}	0.75	4.5×10^{3}	0.26	1170
Air gap	15×10^{-4}	20×10^{-4}	0.75	5.97×10^{5}	1.3×10^{-3}	776.1
Total MMF						1946.1

$$B = \frac{\phi}{A} = \frac{15 \times 10^{-4}}{20 \times 10^{-4}} = 0.75 \text{ T}$$

For the air gap, H is found using Eqs. 1-9 and 1-8b.

$$B = \mu_0 H$$

$$H = \frac{B}{\mu_0} = \frac{0.75 \text{ T}}{4\pi \times 10^{-7} \dfrac{\text{T}}{\text{A/m}}}$$

$$= 5.97 \times 10^5 \text{ A/m}$$

For the cast-iron part, neglecting the air gap,

$$l = 0.08 \text{ m} + 0.05 \text{ m} + 0.08 \text{ m} + 0.05 \text{ m}$$

$$= 0.26 \text{ m}$$

From Figure 1-7, H is found to be 4.5 kA/m. The total MMF is the sum of the Hl drops:

$$\text{total MMF} = 1170 + 776.1 = 1946.1 \text{ A}$$

It is interesting to note the following with reference to Examples 1-15 and 1-16. In both examples the length of the cast iron is 200 times the length of the air gap. However, in both cases approximately 40% of the total MMF is dropped across the air gap. For this reason air gaps are made as small as possible in electromagnetic devices where they are needed (motors and generators). By making the air gap very small, less total MMF is needed to produce the required flux. This means that less current and/or fewer turns are needed for the MMF producing coil. However, some air gap is necessary to eliminate friction between moving parts. Typical air gaps are $\frac{1}{20}$ in. or 1 to 2 mm.

Example 1-17 (English)

The core in Figure 1-15 is made of cast steel. Find the coil MMF necessary to get an air gap flux (ϕ_1) of 240 kilolines. Assume that the cross-sectional area of the center section (a to b) and air gaps is 4 in². The cross-sectional areas of the remaining sections are 8 in² for the left and 6 in² for the right. Each air gap is 0.05 in.

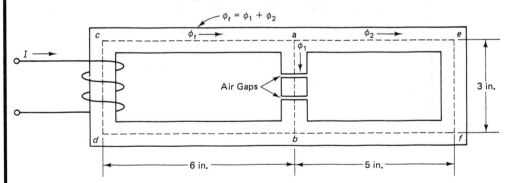

Figure 1-15 Core for Example 1-17.

Solution

There are three sections to the core: a center section ab (which includes two air gaps), a left section $acdb$, and a right section $aefb$. Table 1-5 is first filled in for the center section and air gaps, going from left to right. The total MMF for section ab is found by adding the two MMFs. This MMF is entered as the Hl drop for the right section $aefb$ since it is in parallel with the center section. The table is now filled in for the right section ($aefb$) going from right to left. When the right section flux (ϕ_2) is found, it is added to the center section flux (ϕ_1) to obtain the flux (ϕ_t) in the left section. The table is now filled in for the left section ($acdb$) going from left to right. Finally, the MMF of the coil is obtained by adding the *Hl* drops around either one of the two closed loops (*acdb* + *ab*, or *acdb* + *aefb*).

First fill in all the lengths, cross-sectional areas, and the flux in the center section.

Table 1-5

Part	ϕ	A	B	H	l	Hl	
Left *acdb*	870k ($\phi_1 + \phi_2$) $--------\to$	8	109k	190 $--------------\to$	15	2850	
Center *ab*	240k (ϕ_1) $-----------\to$	4	60k	12 $--------------\to$	3	36	sum of
Two air gaps	240k (ϕ_1) $-----------\to$	4	60k	18,810 $--------------\to$	0.1	1881	MMFs
Right *aefb*	630k $\phi_2 \leftarrow-----------$	6	105k	147.46 $\leftarrow--------------$	13	1917	$\leftarrow$

$$B = \frac{\phi}{A} = 60 \text{ kilolines/in}^2$$

From Figure 1-6,

$$H_{ab} = 12 \text{ A-turns/in.}$$

$$H_{\text{air gap}} = \frac{B}{\mu_0} = \frac{60}{3.19 \times 10^{-3}} = 18.81 \times 10^3 \text{ A-turns/in.}$$

Multiply $H \times l$ for the two parts. For *aefb*,

$$Hl = 36 + 1881 = 1917 \text{ A-turns}$$

$$H = \frac{Hl}{l} = \frac{1917}{13} = 147.46$$

From Figure 1-6,

$$B = 105 \text{ kilolines/in}^2$$

$$\phi_2 = B \times A = 105k(6) = 630 \text{ kilolines}$$

For *acdb*,

$$\phi_t = \phi_1 + \phi_2$$

$$= 240k + 630k = 870 \text{ kilolines}$$

$$B = \frac{870k}{8} \approx 109 \text{ kilolines/in}^2$$

From Figure 1-6,

$$H = 190 \text{ A-turns/in.}$$

$$Hl = 190 \times 15 = 2850 \text{ A-turns}$$

$$\text{total coil MMF} = 2850 + 1917$$

$$= 4767 \text{ A-turns}$$

or 2850 + 36 + 1881 = 4767 A-turns.

Example 1-18 (SI)

The core in Figure 1-16 is made of cast steel. Find the coil MMF necessary to get an air gap flux (ϕ_1) of 24×10^{-4} Wb. Assume that the cross-sectional area of the center section (*a* to *b*) and air gaps is 28×10^{-4} m^2. The cross-sectional areas of the remaining sections are 50×10^{-4} m^2 for the left and 35×10^{-4} m^2 for the right. Each air gap is 0.001 m.

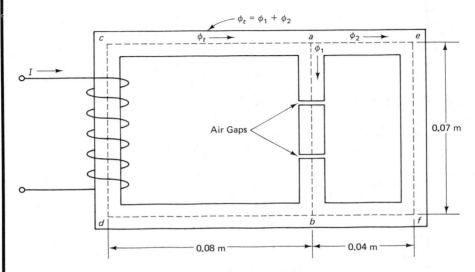

Figure 1-16 Core for Example 1-18.

Solution

There are three sections to the core: a center section *ab* (which includes two air gaps), a left section *acdb*, and a right section *aefb*. Table 1-6 is first filled in for the center section and air gaps, going from left to right. The total MMF for section *ab* is found by adding the two MMFs. This MMF is entered as the *Hl* drop for the right section since it is in parallel with the center section. The table is now filled in for the right section (*aefb*) going from right to left. When the right section flux (ϕ_2) is found, it is added to the center section flux (ϕ_1) to obtain the flux (ϕ_t) in the left section. The table is now filled in for the left section (*acdb*) going from left to right. Finally, the MMF of the coil is

Table 1-6

Part	ϕ	A	B	H	l	Hl
Left *acdb*	84.2×10^{-4} ($\phi_1 + \phi_2$)	50×10^{-4}	1.68	7.75k	0.23	1782.5
Center *ab*	24×10^{-4} (ϕ_1)	28×10^{-4}	0.86	500	0.07	35
Two air gaps	24×10^{-4} (ϕ_1)	28×10^{-4}	0.86	684k	0.002	1368
Right *aefb*	60.2×10^{-4} (ϕ_2)	35×10^{-4}	1.72	9.35k	0.15	1403

sum of MMFs

obtained by adding the Hl drops around either one of the two closed loops ($acdb + ab$, or $acdb + aefb$).

First fill in all the lengths, cross-sectional areas, and the flux in the center section.

$$B = \frac{\phi}{A} = \frac{24 \times 10^{-4}}{28 \times 10^{-4}}$$

$$= 0.86 \text{ T}$$

From Figure 1-7,

$$H_{ab} = 0.5 \text{ kA/m} = 500 \text{ A/m}$$

$$H_{\text{air gap}} = \frac{B}{\mu_0} = \frac{0.86}{4\pi \times 10^{-7}} = 684 \text{ kA/m}$$

Multiply $H \times l$ for each part. For $aefb$,

$$Hl = 35 + 1368 = 1403 \text{ A}$$

$$H = \frac{Hl}{l} = \frac{1403}{0.15} = 9.353 \text{ kA/m}$$

From Figure 1-7,

$$B = 1.72 \text{ T}$$

$$\phi_2 = B \times A = 1.72 \times 35 \times 10^{-4} = 60.2 \times 10^{-4}$$

For $acdb$,

$$\phi_t = \phi_1 + \phi_2$$

$$= 24 \times 10^{-4} + 60.2 \times 10^{-4} = 84.2 \times 10^{-4}$$

$$B = \frac{84.2 \times 10^{-4}}{50 \times 10^{-4}} \approx 1.68 \text{ T}$$

From Figure 1-7,

$$H = 7.75 \text{ kA/m}$$

$$Hl = 7.75 \text{ k} \times 0.23 = 1782.5 \text{ A}$$

$$\text{Total coil MMF} = 1782.5 + 1403$$

$$= 3185.5 \text{ A}$$

or $1782.5 + 35 + 1368 = 3185.5$ A.

SYMBOLS INTRODUCED IN CHAPTER 1

Symbol	Definition	Units English	SI
ϕ	Magnetic flux	lines (or maxwells)	webers (Wb)
B	Flux density	lines/in^2	Wb/m^2 [or tesla (T)]
A	Cross-sectional area	in^2	m^2
U	Magnetomotive force	ampere-turns	amperes (A)
H	Magnetic field intensity	A-turns/in.	A/m
l	Mean length of magnetic circuit	inches	meters
N	Number of turns of wire	turns	turns
$\mathscr{R}$	Reluctance	$\dfrac{\text{A-turns}}{\text{lines/in}^2}$	A/T
μ	Permeability of magnetic material	$\dfrac{\text{lines/in}^2}{\text{A-turns/in.}}$	$\dfrac{\text{T}}{\text{A/m}}$
μ_0	Permeability of air	$3.19 \times 10^{-3} \dfrac{\text{kilolines/in}^2}{\text{A-turns/in.}}$	$4\pi \times 10^{-7} \dfrac{\text{T}}{\text{A/m}}$

QUESTIONS

1. What magnetic quantities are analogous to the following electrical quantities: current; voltage; resistance; good conductors?
2. Define permeability. To what electrical quantity is it analogous?
3. What is a magnetization curve?
4. Why do we need magnetization curves?
5. What is residual magnetism? By what phenomenon is it caused?
6. Define the following terms: coercive force; Curie point.

PROBLEMS

(English)

1. The flux density in a core is 100 kilolines/in^2. Find the flux if the cross-sectional area is:
(a) 4 in^2
(b) Circular with a diameter of 4 in.
(c) 0.05 ft^2
2. A magnetic core has a 500-turn coil. The current in the coil is 5 A. The core has a mean length of 14 in. Find the magnetomotive force and the magnetic field intensity.
3. A core with a mean length of 12 in. has a coil of 250 turns wrapped around it. Find the current necessary to produce a field intensity of 90 A-turns/in.
4. If the field intensity in an air space is 50 A-turns/in., find the flux density.
5. The flux density in a given air space is 20 kilolines/in^2. Find the field intensity.

6. The flux in an air space is 300 kilolines. If the cross-sectional area is 8 in^2, find the field intensity.

7. The cast-steel core in Figure 1-17 has the following values: $\mathbf{A} = 8$ in^2, $l = 20$ in., $N = 80$, and $\phi = 720$ kilolines. Find the current flowing in the coil.

8. The cast-iron core in Figure 1-17 has 10 A of current in the coil. If $\mathbf{A} = 6$ in^2, $N = 100$, and $l = 12$ in., find the flux present in the core.

9. The core in Figure 1-17 is made of cast steel. If $\mathbf{A} = 2$ in^2, $l = 10$ in., $N = 200$, and $I = 8$ A, find the flux density and the flux.

10. A maximum flux of 400 kilolines is desired in a core. Find the minimum area needed if the core is:

(a) Cast iron with a maximum flux density of 65 kilolines/in^2.

(b) Cast steel with a maximum flux density of 120 kilolines/in^2.

11. The doughnut-shaped core (toroid) shown in Figure 1-18 has $r_1 = 2$ in., $r_2 = 3$ in., $N = 100$, and $I = 4.5$ A. It is made of cast steel. Find the flux in the core.

12. The core in Figure 1-19 has a cross-sectional area of 2 in^2. Part A is cast steel, part B is cast iron, $l_1 = 3$ in., $l_2 = 4$ in., and $N = 200$ turns. To get a flux of 60 kilolines, find:

(a) The coil MMF required

(b) The current in the coil

13. Repeat problem 12 if part A is cast iron and part B is cast steel.

14. The cast-iron core in Figure 1-20 has a flux of 100 kilolines. The mean length is 14 in., the air gap is 0.04 in., and the cross-sectional area is 2.5 in^2. Find the coil MMF required.

15. Repeat Problem 14 if the core is made of cast steel.

16. The core in Figure 1-21 is made of cast iron. Find the coil MMF required to get an air gap flux of 60 kilolines. The cross-sectional area of the center section and air gaps is 4 in^2. The remaining sections have a cross-sectional area of 6 in^2. The air gaps are both 0.06 in. The lengths are $l_1 = 7$ in. and $l_2 = 4$ in.

17. Repeat Problem 16, but this time the core is made of cast steel.

(SI)

18. The flux density in a core is 1.6 T. Find the flux if the cross-sectional area is:

(a) 0.003 m^2

(b) Circular with a diameter of 0.1 m

(c) 25 cm^2

19. A magnetic core has a 400-turn coil. The current in the coil is 6 A. The core has a mean length of 0.4 m. Find the magnetomotive force and the magnetic field intensity.

20. A core with a mean length of 0.3 m has a coil of 300 turns wrapped around it. Find the current necessary to produce a field intensity of 4000 A/m.

21. If the field intensity in an air space is 2000 A/m, find the flux density.

22. The flux density in a given air space is 0.3 T. Find the field intensity.

23. The flux in an air space is 0.003 Wb. If the cross-sectional area is 0.005 m^2, find the field intensity.

24. The cast-steel core in Figure 1-17 has the following values: $\mathbf{A} = 0.005$ m^2, $l = 0.5$ m, $N = 100$, and $\phi = 0.008$ Wb. Find the current flowing in the coil.

25. The cast-iron core in Figure 1-17 has 8 A of current in the coil. If $\mathbf{A} = 0.004$ m^2, $N = 100$, and $l = 0.3$ m, find the flux present in the core.

26. The core in Figure 1-17 is made of cast steel. If $\mathbf{A} = 0.002$ m^2, $l = 0.25$ m, $N = 200$, and $I = 7.5$ A, find the flux density and the flux.

27. A maximum flux of 0.004 Wb is desired in a core. Find the minimum area needed if the core is:

(a) Cast iron with a maximum flux density of 1 T.

(b) Cast steel with a maximum flux density of 1.6 T.

28. The doughnut-shaped core (toroid) shown in Figure 1-18 has $r_1 = 0.05$ m, $r_2 = 0.075$ m, $N = 120$, and $I = 4.5$ A. It is made of cast steel. Find the flux in the core.

29. The core in Figure 1-19 has a cross-sectional area of 0.0015 m². Part A is cast steel, part B is cast iron, $l_1 = 0.08$ m, $l_2 = 0.1$ m, and $N = 180$ turns. To get a flux of 0.8×10^{-3} Wb, find

(a) The coil MMF required

(b) The current in the coil

30. Repeat Problem 29 if part A is cast iron and part B is cast steel.

31. The cast-iron core in Figure 1-20 has a flux of 0.001 Wb. The mean length is 0.35 m, the air gap is 1×10^{-3} m, and the cross-sectional area is 1.25×10^{-3} m². Find the coil MMF required.

32. Repeat Problem 31 if the core is made of cast steel.

33. The core in Figure 1-21 is made of cast iron. Find the coil MMF required to get an air gap flux of 0.0005 Wb. The cross-sectional area of the center section and air gaps is 0.0027 m². The remaining sections have a cross-sectional area of 0.004 m². The air gaps are both 1.5×10^{-3} m. The lengths are $l_1 = 0.2$ m and $l_2 = 0.1$ m.

34. Repeat problem 33, but this time the core is made of cast steel and the desired air gap flux is 0.8×10^{-3} Wb.

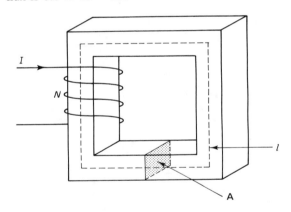

Figure 1-17

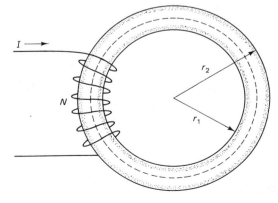

Figure 1-18

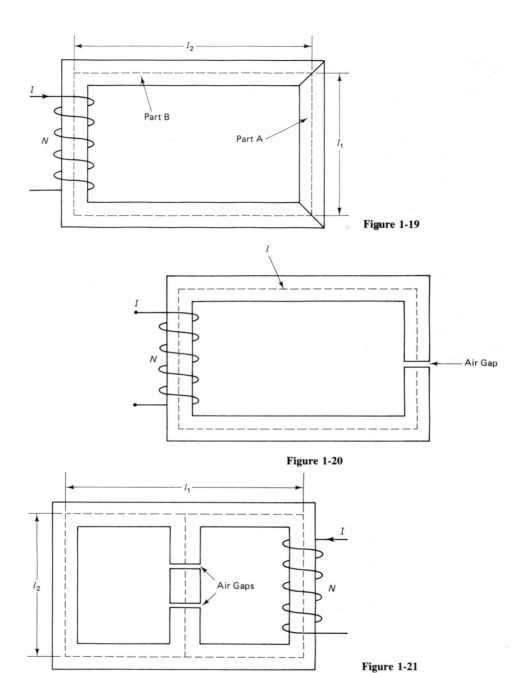

Figure 1-19

Figure 1-20

Figure 1-21

Chapter 2

Principles of Voltage and Torque Generation

A dynamo (a specific form of transducer) is a device that converts mechanical energy into electrical energy (a generator) or electrical energy into mechanical energy (a motor). In Chapter 1 you were introduced to the concepts of flux, flux density, air gaps, and how a magnetic field is produced. In this chapter the basic principles on which generators and motors work are presented. You will be introduced to some very important laws that explain how magnetism interacts with electricity.

2-1   VOLTAGE INDUCED IN A CONDUCTOR

About 150 years ago a very important discovery was made. It was found that if a piece of wire (conductor) moved through a magnetic field in a direction such that it cut the lines of flux, a voltage would be induced (generated) in the wire. The scientist who made this discovery was Michael Faraday. In particular, he found that if the wire cut 100 million lines (or 1 weber) in 1 second, then 1 volt would be generated. As a result of this fact, he was able to derive an equation so that calculations and predictions could be made. The following equations describe Faraday's law.

(English)

$$E = Blv(\sin \theta) \times 10^{-8} \qquad (2\text{-}1\text{a})$$

(SI)

$$E = Blv(\sin \theta) \qquad (2\text{-}1\text{b})$$

The units shown in Table 2-1 should be used with Eqs. 2-1.

Table 2-1

English (Eq. 2-1a)		SI (Eq. 2-1b)	
B	lines/in^2	B	tesla or Wb/m^2
l	inches	l	meters
v	inches/s	v	meters/s

In both equations, E represents the instantaneous voltage induced in the wire, B is the flux density, l is the length of the wire in the magnetic field, v is the velocity of the wire, and θ is the angle the wire makes with the lines of flux. If the wire moves parallel to the field, θ equals 0°; and if the wire moves perpendicular to the field, θ equals 90°.

Figure 2-1 will help clarify the preceding discussion. Referring to the figure, the following items should be noted:

1. In the case when the conductor is moving in the direction of arrow a, no lines of flux will be cut and the induced voltage will be zero. This represents the condition when $\theta = 0°$.

2. In the case when the conductor is moving in the direction of arrow b, a maximum number of lines will be cut and the induced voltage will be a maximum. This represents the condition when $\theta = 90°$, and is the normal condition in dynamos.

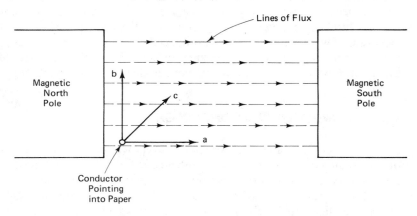

Figure 2-1 One conductor moving in a magnetic field.

3. In the case when the conductor is moving in the direction of arrow c, some lines will be cut and the induced voltage will be greater than zero but less than the maximum of direction b.

4. Obviously, if the conductor moves in and out of the paper, no lines will be cut and no voltage will be induced.

Example 2-1 (English)

The conductor in Figure 2-2 is 10 in. long. The flux density is 100,000 lines/in^2 and the velocity of the conductor is 40 in./s. Find the voltage induced in the conductor when:

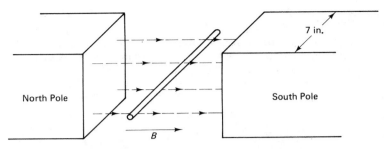

Figure 2-2 Diagram for Example 2-1.

(a) $\theta = 35°$

(b) $\theta = 90°$ (maximum voltage)

Solution

Equation 2-1a will be used to solve for the voltages. It should first be noted that although the conductor is 10 in. long, its length in the magnetic field is only 7 in. This is because the pole face is only 7 in. long.

(a) $E = Blv(\sin\theta) \times 10^{-8}$

$\qquad = 100 \times 10^3(7)(40)(\sin 35°) \times 10^{-8}$

$\qquad = 16{,}060.2 \times 10^{-5}$

$\qquad = 0.16 \text{ V}$

(b) $E = 100 \times 10^3(7)(40)(\sin 90°) \times 10^{-8}$

$\qquad = 28{,}000 \times 10^{-5}$

$\qquad = 0.28 \text{ V}$

Example 2-2 (SI)

The conductor in Figure 2-3 is 0.5 m long. The flux density is 1.5 T (Wb/m^2) and the velocity of the conductor is 1.3 m/s. Find the voltage induced in the conductor when:

(a) $\theta = 35°$

(b) $\theta = 90°$ (maximum voltage)

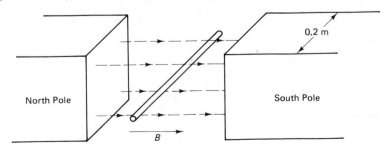

Figure 2-3 Diagram for Example 2-2.

Solution

Equation 2-1b will be used to solve for the voltages. It should first be noted that although the conductor is 0.5 m long, its length in the magnetic field is only 0.2 m. This is because the pole face is only 0.2 m long.

(a) $E = Blv(\sin \theta)$

$$= 1.5(0.2)(1.3)(\sin 35°)$$

$$= 0.22 \text{ V}$$

(b) $E = 1.5(0.2)(1.3)(\sin 90°)$

$$= 0.39 \text{ V}$$

2-1.1 Polarity of the Induced Voltage

From the preceding discussion, we know how to calculate the voltage generated in a single conductor. There is, however, one other important note to be made regarding the induced voltage. If a closed circuit were connected to the conductor, a current would flow as long as lines of flux were being cut. In addition, depending on the direction of the flux and motion of the conductor, the current would flow in either of two directions. There is a simple rule used to determine the direction of the current. It is called **Fleming's right-hand rule** because we use the right hand and a scientist named Fleming devised it.

Refer to Figure 2-4. If the index finger is pointed in the direction of magnetic flux (north to south) and the thumb is pointed in the direction that the conductor is

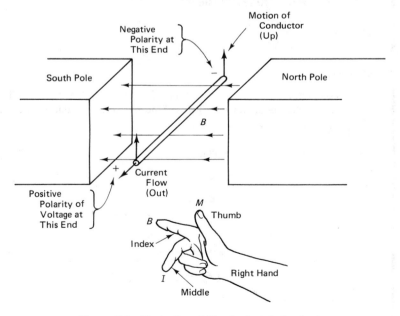

Figure 2-4 Illustration of Fleming's right-hand rule.

moving (up in Figure 2-4), the middle finger will point in the direction that current would flow if the circuit was closed. The middle finger also points toward the end of the conductor where the polarity of the induced voltage is positive. Note that reversing the direction of flux, or the motion of the conductor, will reverse the polarity of the induced voltage. However, reversing both will not change it.

Example 2-3

In each case shown in Figure 2-5, verify for yourself the polarity of the induced voltage by using the right-hand rule.

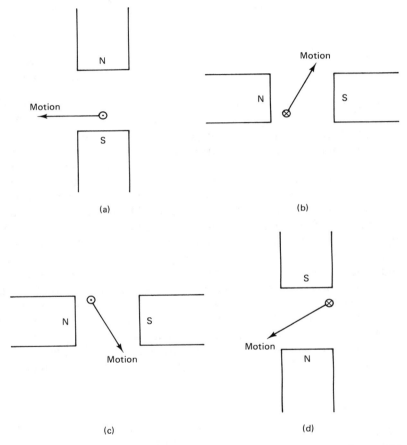

Figure 2-5 Four cases for Example 2-3: (a) and (c) current flow out of paper; (b) and (d) current flow into paper.

2-1.2 Single Conductor in a Circular Path

Although the examples just considered are valid, they must be modified so that they represent a more useful and realistic case. To derive a continuous source of electrical energy from a generator, the conductor would have to continuously move in the

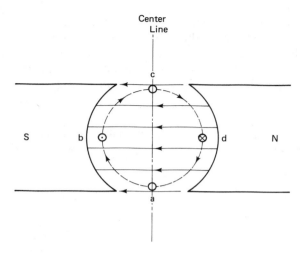

Figure 2-6 One conductor rotating clockwise in a magnetic field.

magnetic field. The simplest way to do this would be to have it rotate between the pole faces as shown in Figure 2-6. As the conductor rotates a full 360°, it will complete one full cycle of a changing voltage. At point a the conductor is moving parallel to the field, no lines of flux are being cut, and the induced voltage is zero. As the conductor moves away from a toward b, it begins to cut lines of flux at an increasing rate until at point b it is cutting lines at a maximum rate. At this point the voltage induced is a maximum and the direction is out of the paper. This direction is obtained by using the right-hand rule, and is indicated by the dot in the conductor at point b. As the conductor moves away from b toward c, it cuts fewer and fewer lines until it reaches c. At that point it is moving parallel to the field, no lines are being cut, and the induced voltage is zero again. As it moves toward d the conductor begins to cut flux again; however, the polarity is opposite from before since it is cutting the lines in a downward direction. At d the voltage is a maximum and the direction is into the paper. This, too, can be verified using the right-hand rule and is indicated by an X in the conductor at point d. As the conductor completes its 360° revolution, the induced voltage decreases until it reaches zero again at point a.

A plot of the induced voltage versus time is shown in Figure 2-7. Note that when the conductor is moving from a to c (to the left of the center line in Figure 2-6), the

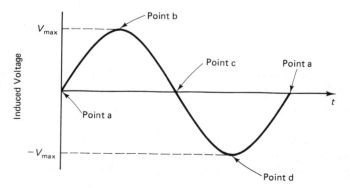

Figure 2-7 Voltage versus time for the single conductor in Figure 2-6.

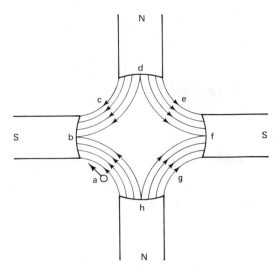

Figure 2-8 One conductor rotating clockwise in a four-pole machine.

voltage is positive, and when it moves from c to a (to the right of the center line), the voltage is negative.

If instead of two poles there were four poles surrounding the rotating conductor, the picture would look as shown in Figure 2-8. In this case the voltage would be zero at four points (a, c, e, and g) and a maximum at each of the pole faces (b, d, f, and h). A plot of the induced voltage is shown in Figure 2-9. Notice that in this case for one 360° revolution of the conductor there will be two full cycles of induced voltage. If the conductor rotated at the same speed in each case, doubling the number of poles would double the frequency of the induced voltage. This important point will be recalled in a later chapter.

2-1.3 Average Voltage Induced by a Single Conductor

Equations 2-1 are useful in determining the instantaneous voltage induced in a conductor. However, a more important quantity to be calculated is the **average** voltage (E_g) induced by a rotating conductor. Referring to Figure 2-6, the average voltage gener-

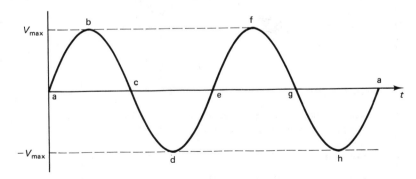

Figure 2-9 Voltage versus time for the single conductor in Figure 2-8.

ated by a single conductor as it moves from point a to point b (zero flux cut to maximum flux cut) is given by

(English)

$$E_g = \frac{1}{2}\frac{\phi}{t} \times 10^{-8} \qquad (2\text{-}2a)$$

(SI)

$$E_g = \frac{1}{2}\frac{\phi}{t} \qquad (2\text{-}2b)$$

In Eqs. 2-2, ϕ is the flux in lines and webers for the English and SI systems, respectively. The quantity t is the time it takes for the conductor to go from zero flux cut to maximum flux cut in seconds.

Example 2-4 (English)

The conductor in Figure 2-6 is rotating at 1200 rev/min. If the flux per pole is 100 kilolines, find the average voltage induced in the conductor.

Solution

Equation 2-2a will be used. The only problem is calculating the time t. If the speed is 1200 rev/min, it can be expressed as

$$1200 \frac{\text{rev}}{\cancel{\text{min}}} \times \frac{1}{60 \dfrac{\text{s}}{\cancel{\text{min}}}} = 20 \text{ rev/s}$$

If the conductor makes 20 rev in 1 s, it takes $\frac{1}{20}$ s for one revolution. For two poles, t is the time for $\frac{1}{4}$ of a revolution. Therefore,

$$t = \frac{1}{4} \times \frac{1}{20} = \frac{1}{80} = 0.0125 \text{ s}$$

Solving for E_g gives us

$$E_g = \frac{1}{2}\frac{\phi}{t} \times 10^{-8}$$

$$= \frac{1}{2}\left(\frac{100 \times 10^3}{0.0125}\right) \times 10^{-8}$$

$$= 0.04 \text{ V}$$

Example 2-5 (SI)

The conductor in Figure 2-6 is rotating at 125 rad/s. If the flux/pole is 1×10^{-3} Wb, find the average voltage induced in the conductor.

Solution

Equation 2-2b will be used. The only problem is calculating the time t. If the speed is

125 rad/s, the time to rotate 1 rad would be $1/125$ s. For two poles, t is the time for $\frac{1}{4}$ of a revolution (90° or $\pi/2$ rad); therefore,

$$t = \frac{\pi}{2} \times \frac{1}{125} = 0.0126 \text{ s}$$

Solving for E_g, we have

$$E_g = \frac{1}{2} \frac{\phi}{t}$$

$$= \frac{1}{2} \left(\frac{1 \times 10^{-3}}{0.0126} \right)$$

$$= 0.397 \text{ V}$$

As can be seen from the preceding examples, the solution is difficult only because it is hard to calculate the quantity t. To simplify calculations, Eqs. 2-2 can be expressed in terms of the number of poles and the speed of the conductor and are rewritten as

(English)

$$E_g = \frac{\phi \mathbf{P} S \times 10^{-8}}{60} \qquad (2\text{-}3a)$$

(SI)

$$E_g = \frac{\phi \mathbf{P} \omega}{2\pi} \qquad (2\text{-}3b)$$

In both equations, $\mathbf{P}$ represents the number of poles (in Figure 2-8, $\mathbf{P} = 4$). In Eq. 2-3a, ϕ is the flux per pole in lines, and S is the angular speed of the conductor in rev/min. In Eq. 2-3b, ϕ is the flux per pole in webers, and ω is the angular speed of the conductor in rad/s.

Example 2-6 (English)

Given a single conductor rotating at 4000 rev/min between two poles. Find:

(a) The average voltage induced if the flux per pole is 300 kilolines

(b) The flux per pole needed for it to generate 1 V

Solution

(a) Equation 2-3a will be used to solve for E_g.

$$E_g = \frac{300\text{k}(2)(4000) \times 10^{-8}}{60}$$

$$= 0.4 \text{ V}$$

(b) Equation 2-3a will be used again; however, it will first be rearranged to solve for the unknown quantity, ϕ.

$$E_g = \frac{\phi PS \times 10^{-8}}{60}$$

Solving for ϕ, we obtain

$$\phi = \frac{60 E_g}{PS \times 10^{-8}}$$

Substituting the known quantities gives us

$$\phi = \frac{60(1)}{2(4000) \times 10^{-8}}$$

$$= 750{,}000 \text{ lines} = 750 \text{ kilolines}$$

Example 2-7 (SI)

Given a single conductor rotating at 400 rad/s between two poles. Find:

(a) The average voltage induced if the flux per pole is 3×10^{-3} Wb

(b) The flux per pole needed for it to generate 1 V

Solution

(a) Equation 2-3b will be used to solve for E_g.

$$E_g = \frac{3 \times 10^{-3} \times 2 \times 400}{2\pi}$$

$$= 0.38 \text{ V}$$

(b) Equation 2-3b will again be used; however, it will first be rearranged to solve for the unknown quantity, ϕ.

$$E_g = \frac{\phi P \omega}{2\pi}$$

Solving for ϕ, we obtain

$$\phi = \frac{2\pi E_g}{P \omega}$$

Substituting the known quantities gives us

$$\phi = \frac{2\pi(1)}{2(400)}$$

$$= 7.9 \times 10^{-3} \text{ Wb}$$

2-2 VOLTAGE INDUCED BY A COIL

Having a good understanding of the voltage induced by a single conductor, we can now examine the effects of several conductors. Figure 2-10 shows a coil of one turn rotating clockwise between two poles. As the coil rotates, notice that conductors c and

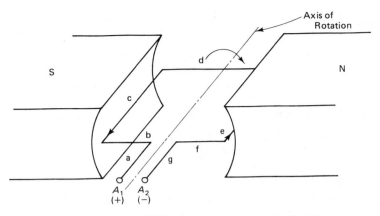

Figure 2-10 One-turn coil rotating in a magnetic field.

e are the only ones cutting lines of flux. Conductors b, d, and f are all moving in a plane parallel to the field and therefore are not cutting the lines of flux. Conductors a and g are outside the magnetic field; therefore, they too will not cut lines of flux. Thus we see that for a coil having one turn, there are two conductors generating a voltage. In general, if a coil has N turns, the number of conductors (z) will be equal to 2 times N:

$$z = 2N \tag{2-4}$$

Again referring to Figure 2-10 and applying the right-hand rule, we see that the voltage generated by conductor c is out of the paper and that generated by conductor e is into the paper. Since the two conductors are in series and the polarities of their induced voltages are aiding each other, the voltage between the ends of the coil (A_1A_2) would be the sum of the two instantaneous voltages. The fact that the two induced voltages are always aiding each other enables us to conclude that the average voltage generated by a one-turn coil would be 2 times the value obtained for one conductor. Equations 2-3 can now be written in a more general way.

(English)

$$E_g = \frac{z\phi PS \times 10^{-8}}{60} \tag{2-5a}$$

(SI)

$$E_g = \frac{z\phi P\omega}{2\pi} \tag{2-5b}$$

In Eqs. 2-5, z represents the number of conductors in a coil.

(English)

$$E_g = \frac{N\phi PS \times 10^{-8}}{30} \tag{2-6a}$$

(SI)

$$E_g = \frac{N\phi P \omega}{\pi} \qquad (2\text{-}6b)$$

In Eqs. 2-6, N represents the number of turns in a coil.

Example 2-8 (English)

A coil of 200 turns rotates at 1800 rev/min between four poles. If the flux/pole is 300 kilolines, find the average voltage induced in the coil.

Solution

Either Eq. 2-5a or 2-6a can be used. The former will be used here. Since there are 200 turns on the coil, the number of conductors z will be $2 \times 200 = 400$ (see Eq. 2-4). Substituting into Eq. 2-5a, we obtain

$$E_g = \frac{400(300\text{k})(4)(1800) \times 10^{-8}}{60}$$

$$= 144 \text{ V}$$

Example 2-9 (SI)

A coil of 200 turns rotates at 190 rad/s between four poles. If the flux/pole is 3×10^{-3} Wb, find the average voltage induced in the coil.

Solution

Either Eq. 2-5b or 2-6b can be used. The former will be used here. Since there are 200 turns on the coil, the number of conductors z will be $2 \times 200 = 400$ (see Eq. 2-4). Substituting into Eq. 2-5b, we obtain

$$E_g = \frac{400(3 \times 10^{-3})(4)(190)}{2\pi}$$

$$= 145.1 \text{ V}$$

Example 2-10 (English)

A coil of 50 turns is wrapped around a solid cast-steel cylinder that rotates at 1800 rev/min. The air gaps are each 0.05 in. Assume that the effective cross-sectional area of the core as well as the cylinder is 4 in². Find the current necessary in the field excitation coil (I) to produce an average generated voltage in the rotating coil of 25 V. Refer to Figure 2-11.

Solution

This is a two-part problem involving the concepts learned in Chapter 1 and earlier in this chapter. The first part will be to calculate the flux ϕ necessary to produce the desired voltage. This will be done using Eq. 2-6a. The second part will be similar to Example 1-15. That is, knowing the desired flux, find the coil MMF and hence the current necessary to produce it.

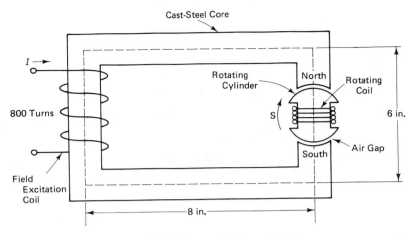

Figure 2-11 Diagram for Example 2-10.

Part 1

Solving Eq. 2-6a for ϕ, we obtain

$$\phi = \frac{30E_g}{NPS \times 10^{-8}}$$

Substituting the known quantities (note that there are only two poles in the diagram), we have

$$\phi = \frac{30(25)}{50(2)(1800)(10^{-8})}$$

$$= 416{,}667 \text{ lines} = 416.67 \text{ kilolines}$$

Part 2

Knowing ϕ, we can fill in Table 2-2 to solve for the total MMF required. Note that there are in effect two parts to the magnetic circuit: the cast-steel core, including the cylinder

Table 2-2

Part	ϕ	A	B	H	l	$H \times l$
Cast steel	416.67k	4	104.2k	135	28	3780
Two air gaps	416.67k	4	104.2k	32,665	0.1	3267
Total MMF						7047

and the two air gaps. The calculations that follow are used to fill in the table.

$$B = \frac{\phi}{A} = \frac{416.67k}{4} = 104.2 \text{ kilolines/in}^2$$

Using Figure 1-6, H is found to be about 135 A-turns/in. for the cast-steel part. For the two air gaps, using Eqs. 1-9 and 1-8a, we get

$$H = \frac{B}{\mu_0}$$

$$= \frac{104.2}{3.19 \times 10^{-3}}$$

$$= 32{,}665 \text{ A-turns/in.}$$

Now compute the $H \times l$ drops:

$$\text{total MMF} = 3780 + 3267 = 7047 \text{ A-turns}$$

Using Eq. 1-3a gives us

$$7047 = NI = 800I$$

$$I = \frac{7047}{800} = 8.8 \text{ A}$$

Example 2-11 (SI)

A coil of 50 turns is wrapped around a solid cast-steel cylinder that rotates at 188 rad/s. The air gaps are each 1.3×10^{-3} m. Assume that the effective cross-sectional area of the core as well as the cylinder is 2.5×10^{-3} m^2. Find the current necessary in the field excitation coil (I) to produce an average generated voltage in the rotating coil of 25 V. Refer to Figure 2-12.

Solution

This is a two-part problem involving the concepts learned in Chapter 1 and earlier in this chapter. The first part will be to calculate the flux ϕ necessary to produce the desired voltage. This will be done using Eq. 2-6b. The second part will be similar to Example 1-16. That is, knowing the desired flux, find the coil MMF and hence the current necessary to produce it.

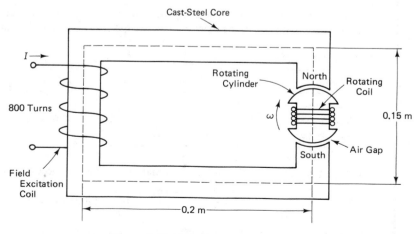

Figure 2-12 Diagram for Example 2-11.

Part 1

Solving Eq. 2-6b for ϕ, we obtain

$$\phi = \frac{E_g \pi}{NP\omega}$$

Substituting the known quantities (note that there are only two poles in the diagram), we have

$$\phi = \frac{25\pi}{50(2)(188)}$$

$$= 4.178 \times 10^{-3} \text{ Wb}$$

Part 2

Knowing ϕ, we can fill in Table 2-3 to solve for the total MMF required. Note that there are in effect two parts to the magnetic circuit: the cast-steel core including the cylinder,

Table 2-3

Part	ϕ	A	B	H	l	$H \times l$
Cast steel	4.18×10^{-3}	2.5×10^{-3}	1.67	7.3k	0.7	5110
Two air gaps	4.18×10^{-3}	2.5×10^{-3}	1.67	1329k	2.6×10^{-3}	3455
Total MMF						8565

and the two air gaps. The calculations that follow are used to fill in the table.

$$B = \frac{\phi}{A} = \frac{4.18 \times 10^{-3}}{2.5 \times 10^{-3}}$$

$$= 1.67 \text{ Wb/m}^2(\text{T})$$

Using Figure 1-7, H is found to be about 7.3 kA/m for the cast-steel part. For the two air gaps, using Eqs. 1-9 and 1-8b, we get

$$H = \frac{B}{\mu_0}$$

$$= \frac{1.67}{4\pi \times 10^{-7}}$$

$$= 1329 \text{ kA/m}$$

Now compute the $H \times l$ drops:

$$\text{total MMF} = 5110 + 3455 = 8565$$

Using Eq. 1-3b, we have

$$8565 = NI = 800I$$

$$I = \frac{8565}{800} = 10.7 \text{ A}$$

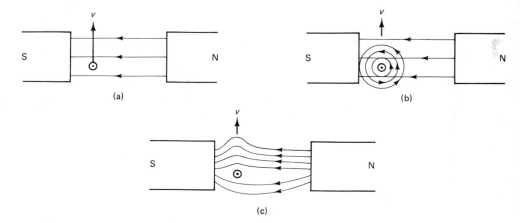

Figure 2-13 (a) Conductor with only field due to poles shown; (b) conductor with both fields shown; (c) conductor with resultant field shown.

2-3   LENZ'S LAW

If one were to try to rotate the coil shown in Figure 2-10, it would be fairly easy. The only opposition to rotation would be a frictional force that is present in any moving part, and wind resistance as the coil turns. If, however, a closed circuit were connected to the coil and current began to flow, there would be an additional force opposing the rotation. This additional opposition is a magnetic force that increases as the current in the coil increases. It is this phenomenon that makes it difficult to turn a generator as current is drawn from it. It is explained by the following effect, which is known as Lenz's law. Figure 2-13a shows a conductor moving up through a magnetic field. Fleming's right-hand rule tells us that if a closed circuit were connected to the conductor, current would flow out of the paper (indicated by the dot in the conductor). Figure 2-13b shows the magnetic field formed by the current in the conductor. **Lenz's law** states that this magnetic field will always oppose the motion that induced the voltage in the conductor to begin with. Figure 2-13c shows the resultant field formed by the vector addition of the two fields. Notice how much denser the field is above the conductor than below it. It is the imbalance in fields that creates the opposing force to the motion of the conductor. As the current in the conductor increases, the imbalance increases, causing a proportional increase in the opposing force.

2-4   FORCE PRODUCED BY A CONDUCTOR

In Sections 2-1 through 2-3 the principles of voltage generation were discussed. That is, when an external mechanical force causes a conductor to rotate in a magnetic field, a voltage will be generated in the conductor. In the remaining sections of this chapter, the principles of torque generation (motor action) are discussed. That is, when an external voltage is applied to a conductor causing a current in it, a force will be developed that will make the conductor move.

2-4.1 Biot–Savart Law

The **Biot–Savart law** given by Eqs. 2-7 enables us to calculate the force F produced by a current-carrying conductor in a magnetic field.

(English)

$$F = 0.885BIl \times 10^{-7} \qquad (2\text{-}7a)$$

(SI)

$$F = BIl \qquad (2\text{-}7b)$$

The units in Table 2-4 should be used with Eqs. 2-7. As in Eqs. 2-1, l is the length of the conductor in the magnetic field.

Table 2-4

English (Eq. 2-7a)		SI (Eq. 2-7b)	
F	pounds	F	newtons
B	lines/in^2	B	webers/m^2 (tesla)
l	inches	l	meters
I	amperes	I	amperes

2-4.2 Direction of Force

Just as we have a rule for finding the polarity of the voltage induced in a conductor, we have a rule for finding the direction of the force produced by a current-carrying conductor. It is called the **left-hand rule**, and we use the left hand, as shown in Figure 2-14. With the index finger pointing in the direction of the flux (north to south) and

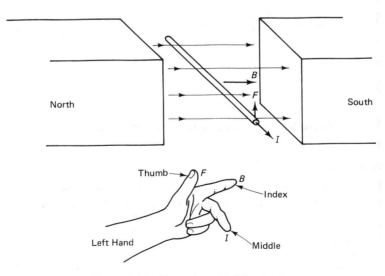

Figure 2-14 Illustration of left-hand rule.

the middle finger pointing in the direction of current flow, the thumb will point in the direction of the force on the conductor.

Example 2-12

In each case shown in Figure 2-15, verify for yourself the direction of the force (F) produced.

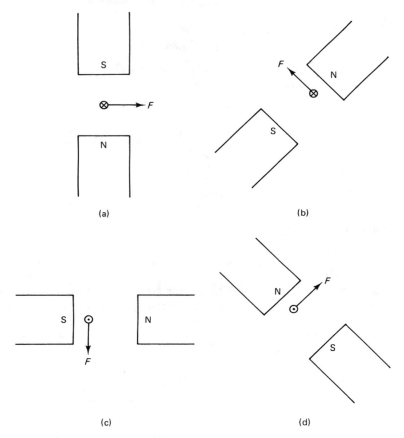

(a)

(b)

(c)

(d)

Figure 2-15 Four cases for Example 2-12: (a) and (b) current into paper; (c) and (d) current out of paper.

Example 2-13 (English)

The conductor shown in Figure 2-16 is 10 in. long. The pole face is a square 6 in. on each side; and the flux is 400,000 lines. Find the magnitude and direction of the force produced if the current is 8 A into the paper.

Solution

Equation 2-7a will be used to solve for the force. Since the pole face is 6 in. long, only 6 in. of the conductor will be in the magnetic field (i.e., $l = 6$ in.). In addition, we must

Figure 2-16 Diagram for Example 2-13.

first compute the flux density (B) using Eq. 1-2a. The cross-sectional area must also be computed.

$$A = 6 \times 6 = 36 \text{ in}^2$$

$$B = \frac{\phi}{A} = \frac{400k}{36}$$

$$= 11.11 \text{ kilolines/in}^2$$

$$F = 0.885BIl \times 10^{-7}$$

$$= 0.885 \times 11.11 \times 10^3 \times 8 \times 6 \times 10^{-7}$$

$$= 0.0472 \text{ lb}$$

Using the left-hand rule, the force will be *up*.

Example 2-14 (SI)

The conductor shown in Figure 2-17 is 0.3 m long. The pole face is a square 0.2 m on each side, and the flux is 4×10^{-3} Wb. Find the magnitude and direction of the force produced if the current is 8 A into the paper.

Figure 2-17 Diagram for Example 2-14.

Solution

Equation 2-7b will be used to solve for the force. Since the pole face is 0.2 m long, only 0.2 m of the conductor will be in the magnetic field (i.e., $l = 0.2$ m). In addition, we must first compute the flux density (B) using Eq. 1-2b. The cross-sectional area must also be computed.

$$A = 0.2 \times 0.2 = 0.04 \text{ m}^2$$

$$B = \frac{\phi}{A} = \frac{4 \times 10^{-3}}{0.04}$$

$$= 0.1 \text{ Wb/m}^2 \text{ (T)}$$

$$F = BIl$$

$$= 0.1 \times 8 \times 0.2$$

$$= 0.16 \text{ N (newtons)}$$

Using the left-hand rule, the force will be *up*.

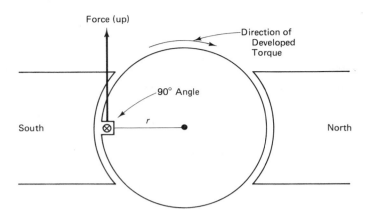

Figure 2-18 Single conductor developing torque in a magnetic field.

2-5 TORQUE DEVELOPED BY A CONDUCTOR

The conductor in Figure 2-18 is mounted in a solid cylinder that is free to rotate. If a current were to flow through the conductor, a force would be produced (Section 2-4). When a force is applied to a body that is free to rotate about an axis, a **torque** will be developed. Torque is numerically equal to a force times its perpendicular distance to the axis of rotation. It is given by

(English)

$$T = F \times r \tag{2-8a}$$

(SI)

$$T = F \times r \tag{2-8b}$$

In Eq. 2-8a, the force F is in pounds, the distance r is in feet, and the torque T will be in lb-ft. In Eq. 2-8b, the force F is in newtons, the distance r is in meters, and the torque T will be in N-m.

The torque developed on the conductor in Figure 2-18 will cause the cylinder to rotate. As the conductor rotates it will assume the positions shown in Figure 2-19. Note that as the conductor moves from its original position, the perpendicular distance between the force and axis of rotation decreases. The distance r_2 is less than r_1. When the conductor is directly above the axis (F_3), the perpendicular distance (not shown in Figure 2-19) is zero, and the torque developed at that point will be zero. Since the field, current, and conductor length (B, I, and l) remain constant during rotation, the magnitude of the force will also remain constant. However, since the distance r is changing as the conductor rotates, the torque will change proportionately. Note that when the conductor is directly in front of the pole face (position 1 in Figure 2-19), r is a maximum (equal to the radius of the cylinder) and the developed torque will also be a maximum. When the conductor is 90° away from the pole face (position 3 in Figure 2-19), r is zero and the developed torque will be zero.

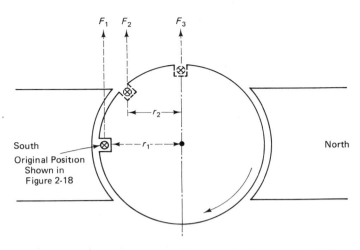

Figure 2-19 Single conductor shown rotating in a magnetic field.

It can thus be seen that the variation in torque is analogous to the variation in the induced voltage (see Section 2-1.2); that is, both are maximum when the conductor is at the center of the pole face, and both are zero when the conductor is at the midpoint between the two poles (for a two-pole configuration).

If Eqs. 2-7 are substituted into Eq. 2-8, the following equations for the maximum torque developed by a conductor are obtained.

(English)

$$T = 0.885BIlr \times 10^{-7} \tag{2-9a}$$

(SI)

$$T = BIlr \tag{2-9b}$$

In Eqs. 2-9, r is the radius of the cylinder. Equations 2-9 can be used to calculate the instantaneous torque developed at any point in rotation. The appropriate r would then have to be calculated using simple trigonometry, as shown in the following examples.

Example 2-15 (English)

For the configuration shown in Figure 2-20, find the magnitude and direction of the developed torque for the two positions shown. The flux density is 150 kilolines/in^2, l is 5 in., and I is 8 A. In addition, if we wish to double the torques in question, what must be the new current I?

Solution

At position 1, using the left-hand rule, the force will be down and the torque will be counterclockwise. Equation 2-9a is used, where r equals the radius $\frac{1}{2}$ ft (6 in.). The

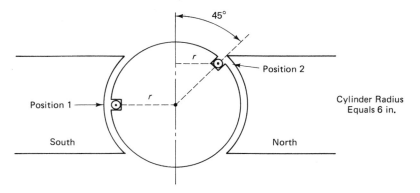

Figure 2-20 Diagram for Example 2-15.

torque at this position also corresponds to the maximum torque developed as the conductor rotates.

$$T = 0.885(150k)(8)(5)(\tfrac{1}{2}) \times 10^{-7}$$

$$= 0.2655 \text{ lb-ft}$$

At position 2, using the left-hand rule, the force will be down again; however, this time the torque will be clockwise. The distance r is less than the radius of the cylinder now and must first be calculated.

$$r = (\tfrac{1}{2} \text{ ft})(\sin 45°)$$

$$= 0.5(0.707) = 0.3535 \text{ ft}$$

$$T = 0.885(150k)(8)(5)(0.3535) \times 10^{-7}$$

$$= 0.1877 \text{ lb-ft}$$

Since the torque is proportional to the current in the conductor, to double the torque the current would have to double. The current would be 16 A.

Example 2-16 (SI)

For the configuration shown in Figure 2-21, find the magnitude and direction of the developed torque for the two positions shown. The flux density is 1.5 T (Wb/m²), l is 0.12 m, and I is 8 A. In addition, if we wish to double the torques in question, what must be the new current I?

Solution

At position 1 using the left-hand rule, the force will be down and the torque will be counterclockwise. Equation 2-9b is used, where r equals the radius 0.15 m. The torque at this position also corresponds to the maximum torque developed as the conductor rotates.

$$T = 1.5 \times 8 \times 0.12 \times 0.15$$

$$= 0.216 \text{ N-m}$$

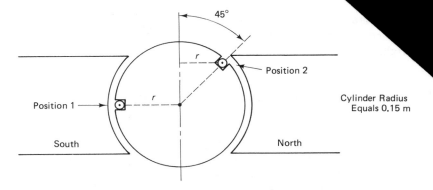

Figure 2-21 Diagram for Example 2-16.

At position 2, using the left-hand rule, the force will be down again; however this time the torque will be clockwise. The distance r is less than the radius of the cylinder now and must first be calculated.

$$r = (0.15 \text{ m})(\sin 45°)$$

$$= (0.15)(0.707) = 0.1061 \text{ m}$$

$$T = 1.5 \times 8 \times 0.12 \times 0.1061$$

$$= 0.1528 \text{ N-m}$$

Since the torque is proportional to the current in the conductor, to double the torque the current would have to double. The new current would be 16 A.

2-5.1 Torque Developed by a Coil

If a one-turn coil is formed by a continuous piece of wire and mounted on a cylinder, it will look as shown in Figure 2-22. Since we have a continuous loop of wire, when current flows into the left-side conductor, it will be flowing out of the right-side conductor. This gives rise to a situation where one force is up and the other is down. However, this is exactly the condition desired, since both forces will produce clock-

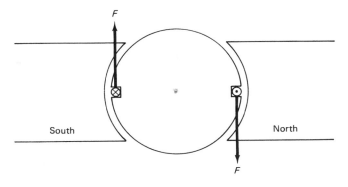

Figure 2-22 One-turn coil mounted on a cylinder.

...id the cylinder will rotate. All that remains now to provide a useful ... e to find a way to have the current flow in on the left side and out on the ... vice versa) all the time. This will be dealt with in Section 3-1.1.

...ROMOTIVE FORCE (BACK EMF)

... ...is rotating as in Figure 2-19 due to a current that is flowing in it (motor action), a voltage will simultaneously be induced in the conductor. This is called **generator action** and was discussed in Sections 2-1 through 2-3. The polarity of the induced voltage will always oppose the current that made the conductor move to begin with. Since the induced voltage opposes the current it is called a **back electromotive force (EMF).** It is also commonly referred to as **counter EMF.** The back EMF is an extremely important effect in motors and is discussed in detail in Chapter 5. The reader may wish to verify that in Figure 2-19 the induced voltage will oppose the current flow. Be sure to use the right-hand rule.

SYMBOLS INTRODUCED IN CHAPTER 2

Symbol	Definition	Units English	SI
v	Velocity of a conductor	in./s	m/s
l	Length of conductor in magnetic field	inches	meters
θ	Angle between wire's motion and lines of flux	degrees	degrees
t	Time it takes for a conductor to go from zero flux to maximum flux cut	seconds	seconds
z	Number of conductors	—	—
E_g	Average induced voltage	volts	volts
r	Radius	feet	meters
P	Number of poles	—	—
S	Rotational speed (English)	rev/min	—
ω	Rotational speed (SI)	—	rad/s
F	Force produced by a conductor	pounds	newtons
T	Torque	lb-ft	N-m

QUESTIONS

1. Define the following terms: dynamo; torque; back EMF.
2. What is Faraday's law?
3. Describe the right-hand and left-hand rules. What are they used for?
4. What is Lenz's law?

PROBLEMS

(English)

1. In Figure 2-23a, the length of the conductor in the magnetic field is 8 in. Its velocity is 50 in./s and the flux density is 200 kilolines/in². Find the induced voltage.

2. In Figure 2-23b, the length of the conductor in the magnetic field is 1.5 ft. Its velocity is 3 ft/s and the flux density is 150 kilolines/in². Find the induced voltage.

3. In Figure 2-23c, the length of the conductor in the magnetic field is 1 ft. Its velocity is 60 in./s and the flux density is 300 kilolines/in². Find the induced voltage.

4. In each case shown in Figure 2-23, determine the polarity of the induced voltage; that is, would current flow into or out of the paper?

5. For each of the conductors (a, b, c, d, and e) shown in Figure 2-24, determine the polarity of the induced voltage; that is, would current flow into or out of the paper?

6. A single conductor is rotating in a two-pole magnetic field at 1800 rev/min (see Figure 2-6). The flux per pole is 300 kilolines. What is the average voltage induced in the conductor?

7. A single conductor is rotating at 1200 rev/min in an eight-pole magnetic field. The flux per pole is 350 kilolines. What is the average voltage induced in the conductor?

8. A conductor is rotating in a six-pole magnetic field at 600 rev/min. Find the flux per pole necessary to induce an average voltage of 1.5 V.

9. Repeat Problem 8, but use 36 poles.

10. A 400-turn coil is rotating at 1650 rev/min between eight poles. If the flux per pole is 200 kilolines, find the average voltage induced in the coil.

11. Repeat Problem 10 if the speed were to drop to 1450 rev/min.

12. What must be the flux per pole in a six-pole machine so that the average coil voltage induced is 85 V? The coil has 300 turns and is rotating at 1600 rev/min.

13. Repeat Problem 12 if the speed increases to 1800 rev/min.

14. Refer to Figure 2-11. The rotating coil has 120 turns, the air gaps are each 0.04 in., and the cross-sectional area of the core as well as the air gaps is 6 in². If the coil is rotating at 1750 rev/min, find the current necessary in the field coil to generate an average voltage of 40 V.

15. For each case shown in Figure 2-25, find the direction of the force produced. The current is always out of the paper.

16. Repeat Problem 15 if the current is into the paper.

17. Referring to Figure 2-24, what current direction is necessary in each of the conductors to produce the indicated rotation? Note that each conductor must be examined separately.

18. A conductor 14 in. long is in a magnetic field whose flux density is 50 kilolines/in². Find the magnitude of the force produced on the conductor if the current in it is 10 A.

19. In Problem 18, find the current necessary to produce a force of 1.5 lb.

20. With reference to Figure 2-20, the length of the conductor is 8 in., the current is 10 A, and the flux density is 100 kilolines/in². Find the magnitude of the torque produced for:
(a) Position 1
(b) Position 2

(SI)

21. In Figure 2-23a, the length of the conductor in the magnetic field is 0.2 m. Its velocity is 1.25 m/s and the flux density is 3 T. Find the induced voltage.

22. In Figure 2-23b, the length of the conductor in the magnetic field is 45 cm. Its velocity is 90 cm/s and the flux density is 2 T. Find the induced voltage.

23. In Figure 2-23c, the length of the conductor in the magnetic field is 30 cm. Its velocity is 1.5 m/s and the flux density is 4.5 T. Find the induced voltage.

24. In each case shown in Figure 2-23, determine the polarity of the induced voltage; that is, would current flow into or out of the paper?

25. For each of the conductors (a, b, c, d, and e) shown in Figure 2-24, determine the polarity of the induced voltage; that is, would current flow into or out of the paper?

26. A single conductor is rotating in a two-pole magnetic field at 180 rad/s (see Figure 2-6). The flux per pole is 0.003 Wb. What is the average voltage induced in the conductor?

27. A single conductor is rotating at 120 rad/s in an eight-pole magnetic field. The flux per pole is 0.004 Wb. What is the average voltage induced in the conductor?

28. A conductor is rotating in a six-pole magnetic field at 60 rad/s. Find the flux per pole necessary to induce an average voltage of 2.4 V.

29. Repeat Problem 28, but use 36 poles.

30. A 400-turn coil is rotating at 160 rad/s between eight poles. If the flux per pole is 0.002 Wb, find the average voltage induced in the coil.

31. Repeat Problem 30 if the speed were to drop to 140 rad/s.

32. What must be the flux per pole in a six-pole machine so that the average coil voltage induced is 90 V? The coil has 240 turns and is rotating at 150 rad/s.

33. Repeat Problem 32 if the speed increases to 175 rad/s.

34. Refer to Figure 2-12. The rotating coil has 100 turns, the air gaps are each 0.001 m, and the cross-sectional area of the core as well as the air gaps is 0.004 m². If the coil is rotating at 170 rad/s, find the current necessary in the field coil to generate an average voltage of 50 V.

35. For each case shown in Figure 2-25, find the direction of the force produced. The current is always out of the paper.

36. Repeat Problem 35 if the current is into the paper.

37. Referring to Figure 2-24, what current direction is necessary in each of the conductors to produce the indicated rotation? Note that each conductor must be examined separately.

38. A conductor 0.35 m long is in a magnetic field whose flux density is 0.8 T. Find the magnitude of the force produced on the conductor if the current in it is 11 A.

39. In Problem 38, find the current necessary to produce a force of 6 N.

40. With reference to Figure 2-21, the length of the conductor is 0.2 m, the current is 7 A, and the flux density is 1.4 T. Find the magnitude of the torque produced for:
(a) Position 1
(b) Position 2

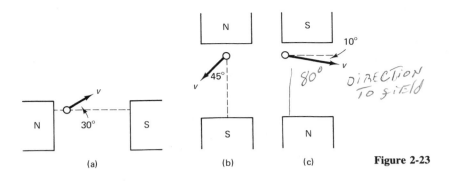

(a)   (b)   (c)   **Figure 2-23**

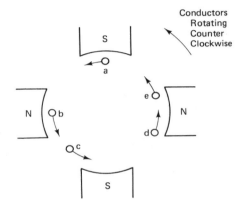

Conductors
Rotating
Counter
Clockwise

Figure 2-24

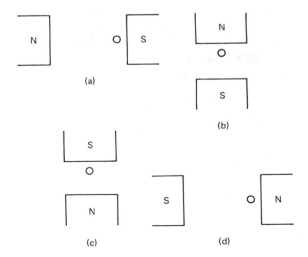

(a)

(b)

(c)   (d)   **Figure 2-25**

Chapter 3

DC Machine Construction

Before going on to examine the characteristics of the dc motor and dc generator in detail, we will look at how these machines are constructed. The purpose of this chapter is merely to acquaint you with the important parts of the dc machine and their purpose.

3-1   ARMATURE

In Chapter 2 you were introduced to the principles of voltage and torque generation. In the case of voltage generation we saw that a coil had to rotate in a magnetic field. In the case of torque generation we saw that a current had to flow through a coil that was free to rotate in a magnetic field. In dc machines, the part on which this coil is mounted is called the **armature.** The rotating part of a machine is also called the **rotor.** Therefore, when we talk about a dc machine, the armature and rotor both refer to the same part.

The armature is made up of a shaft on which a core of laminated disks is mounted. The disks are made of a steel alloy. The purpose of the laminated disk construction is to reduce a phenomenon called **eddy currents.** Eddy currents are discussed in Chapter 4. Figure 3-1 is a picture of a typical armature. The coil of wire making up the armature winding is wrapped around the armature and embedded into slots. The slots can be seen in Figure 3-1. Normally, there are several coils mounted on an armature. The way in which these coils are interconnected is discussed in Section 3-6.

Insulation
between Commutator
Segments

Commutator

Figure 3-1   A typical dc motor armature. (Courtesy of Reliance Electric Company.)

3-1.1 Commutation

The place at which the interconnection of armature coils is made is called a **commutator segment**. Commutator segments are conducting strips placed axially on the shaft of the armature. They are electrically insulated from each other and any other metal parts of the armature. It is through the commutator segments that the armature makes contact with the outside world. Current is either taken from the rotating coil (generator action) or received by the rotating coil (motor action) through the commutator.

In Section 2-1.2 we saw that a conductor rotating in a magnetic field will generate an alternating voltage (shown in Figure 2-7). Another function of the commutator is to provide internal switching which rectifies the generated alternating voltage to provide a dc output. Figure 3-2 shows a one-turn coil rotating in a two-pole magnetic field.

As the armature rotates, coil end A is always connected to commutator segment A and coil end B is always connected to commutator segment B. The induced voltage is picked up by two brushes positioned as shown. The brushes are connected to the frame of the generator and do not rotate. As a result of this, the top brush will always be making contact with the coil side in the right half-plane (i.e., passing the north pole) and the bottom brush will always be making contact with the coil side in the left half-plane (i.e., passing the south pole). Rectification has been accomplished since current flow will always be out of the top brush and into the bottom brush. As conductors A and B pass by the top and bottom brushes, respectively, the switching is taking place. The top brush breaks contact with commutator segment A and simul-

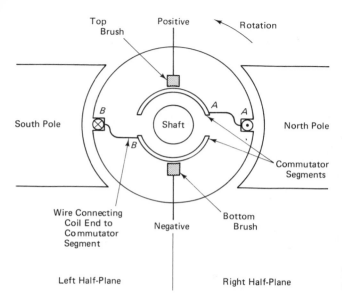

Figure 3-2 Two-pole generator illustrating commutation.

taneously makes contact with commutator segment B. It should be noted here that as the switching is taking place, the voltage being induced in each conductor is zero since at that instant no lines of flux are being cut by the coil.

However, due to an effect called **armature reaction** (see Section 3-1.3), there will be a small voltage induced. It is the switching of this small voltage which accounts for the sparks seen as you observe a motor or generator in operation. The sparking is caused by the brush momentarily shorting the two commutator segments as the switching takes place. At that instant a current will flow (this is the visible spark) from segment A to segment B through the brush. Ideally, the voltage induced at the point of switching is zero. In this case no sparks would occur since a difference in voltage would not exist between the two segments. However, in a real machine this is not the case.

3-1.2 Brushes

The brushes that are necessary for commutation must be good conductors, long wearing, and relatively soft compared to the copper commutating segments against which they are constantly rubbing. For this reason, brushes are made of carbon, graphite, and sometimes a mixture of carbon and copper. After many hours of use, however, brushes will wear down and must be replaced. The brushes are held in place and kept in contact with the commutator segments by brush rigging. The rigging is mounted on but electrically insulated from the machine housing. The rigging includes a spring that forces the brush against the commutator segments to ensure good electrical contact.

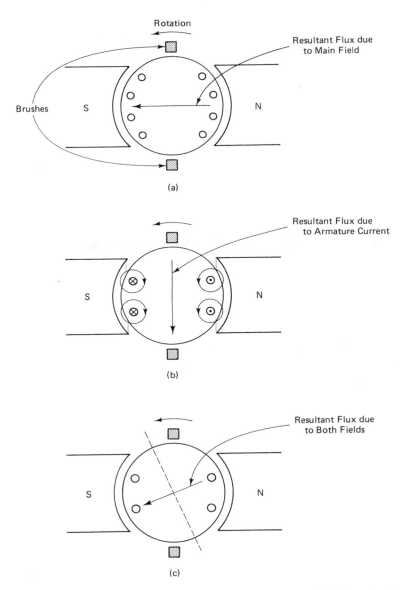

Figure 3-3 Magnetic fields in a two-pole generator: (a) main field flux only; (b) armature current flux only; (c) net flux in armature due to both fields.

3-1.3 Armature Reaction

If we examine the magnetic fields present in the armature of a dc generator while it is rotating and under load, we would see the simplified picture shown in Figure 3-3. When a generator is under load, its armature supplies the current to the load. As a result a current exists in the armature winding which increases as the load increases. The current flowing in the armature winding will create a flux of its own which has a circular path around each conductor (see Figure 3-3b). These small circular paths of flux add up in the armature and form a resultant flux in the downward direction. Since magnetic fields are vector quantities (have magnitude and direction), the two fields add vectorally to form the resultant or net flux shown in Figure 3-3c. As the conductors rotate, the points at which they are moving parallel to the magnetic field would be where they cross the dashed line in Figure 3-3c. At this point no lines of flux are being cut and the induced voltage in a conductor is zero. This would be the desired location of the brushes; however, they have been placed in the positions shown in Figure 3-3. Since the brushes are now out of place, when commutation takes place, sparking will occur.

As you can see from the discussion above, armature reaction makes it look as if the brushes have shifted from their ideal location. For this reason, the phenomenon of armature reaction is also called **brush shifting.** In addition, as the generator supplies more load, the armature current increases and armature reaction will increase as well.

3-2   *INTERPOLES*

One way to eliminate or reduce armature reaction is to introduce a magnetic pole between each of the main field poles (see Figure 3-4). These poles are called **interpoles** and are wired so that their polarity opposes the armature current flux shown in Figure 3-3b. They are wired in series with the armature winding. In this way, as the armature current flux increases due to a rise in armature current, the interpole flux increases proportionately. It should be noted that since the interpoles are wired in series with the armature winding, the total armature circuit resistance would be the sum of the interpole winding resistance and the armature winding resistance.

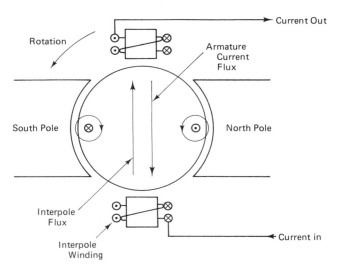

Figure 3-4 Diagram of a two-pole machine showing interpole connection and location.

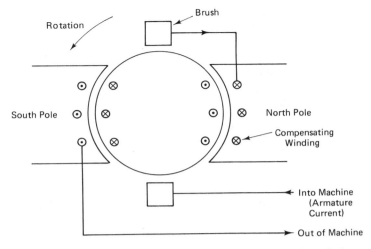

Figure 3-5 Diagram of a two-pole machine showing compensating winding connection and location.

3-3 COMPENSATING WINDINGS

Another technique of reducing armature reaction is the use of compensating windings. These windings are set in the pole faces of the main field poles. Their conductors run parallel to the armature conductors as shown in Figure 3-5. They, too, are wired in series with the armature winding so that an increase in armature current causes an increase in compensating winding flux as well as the armature winding flux. The connection is in such a way that the two fields oppose each other and effectively cancel each other out. Here, again, since the compensating winding is in series with the armature winding, the total armature circuit resistance would be the sum of the resistances of the two windings.

3-4 FIELD POLES

In each of the diagrams in this chapter representing an armature, the main field flux has been produced by a pair of poles labeled north and south. They are more commonly called the **field poles.** Field poles are produced in one of two different ways. In small machines, particularly motors used in toys, the field poles are created by a **permanent magnet.** A permanent magnet is a ferromagnetic material that has been magnetized initially by an external source. This external excitation could be a current flowing in a coil of wire wrapped around the material (see Figure 1-1), or it could be a strong magnetic field into which the material has been placed. In any event, when the external excitation is removed, the material will remain magnetized due to hysteresis (Section 1-2.1). The magnetism that remains in the material was called **residual magnetism** in Chapter 1. The strength of a permanent magnet depends on the strength of the excitation that magnetizes it and the type of material of which it is made. One of the popular materials used for permanent magnets is Alnico (an alloy consisting of aluminum, nickel, cobalt, copper, and iron).

The alternative to a permanent magnet is an electromagnet. It is constructed much as shown in Figure 1-13. The strength of an electromagnet, although dependent on the size and type of material used, can be easily controlled by the number of turns of wire making up the coil and the amount of current in the coil. It should be noted that unlike a permanent magnet, a source of electricity is needed to maintain the current in the coil. If the current is removed, all that remains will be the residual magnetism, which is small for the type of material used in electromagnets.

When electromagnets are used to produce the field poles, they will generally be one of two types. Either a small current will flow through many turns of thin wire, or a large current will flow through a few turns of a thick wire. In Chapter 4 we will see why the former is called a **shunt field** and the latter is called a **series field.** Figure 3-6b is a photograph of a typical field pole. We can see from the picture that it is made of laminations (of a steel alloy); in addition, since there are many turns of a thin wire making up the coil, it would be a shunt-field pole.

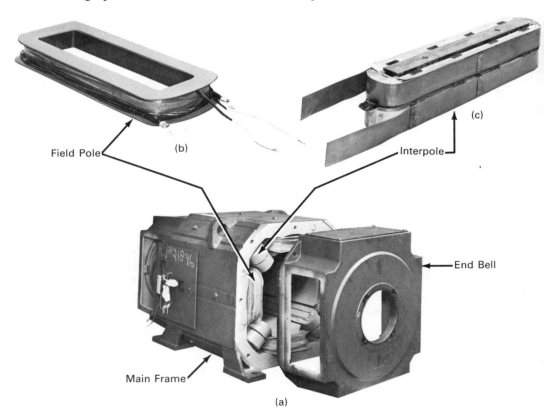

Figure 3-6 (a) dc motor with end bell removed; (b) shunt field pole; (c) interpole. (Courtesy of Reliance Electric Company.)

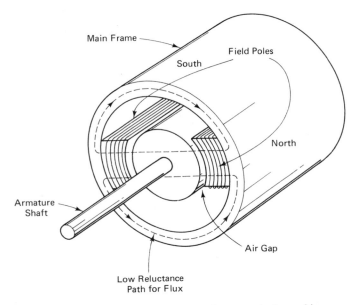

Figure 3-7 Simplified diagram of a two-pole dc machine.

3-5   MECHANICAL STRUCTURE

Since the armature must rotate freely with very little friction, its shaft is mounted in **bearings** on both ends. The bearings are mounted in circular plates called **end bells** at opposite ends of the machine. The end bells fit into a cylindrical steel alloy structure called the **main frame.** Not only does the main frame support the end bells, armature, brushes, and field poles, but it has another important function. It provides a low-reluctance path for the magnetic flux outside the armature and for this reason is generally thicker than needed for structural support (see Figure 3-6 and 3-7).

3-6   ARMATURE WINDINGS

The coils on an armature and how they are interconnected are the primary factors determining a dc machine's characteristics. There are two basic types of windings, the **lap winding** and the **wave winding.** In addition, there are several variations of these windings; however, only two will be discussed here.

To understand how a lap winding behaves, we will examine Figure 3-8. The machine shown in Figure 3-8 has four poles, eight slots, eight commutator segments (S1 to S8), and four brushes (B1 to B4). The diagram shows the end view of the armature, which is rotating counterclockwise. There are eight coils (labeled A to H) wound on the armature, each having N turns. The beginning and end of coil B are labeled b and b', respectively; the others are labeled similarly. For the instant shown, coils A, C, E, and G are cutting lines of flux and are therefore generating a voltage.

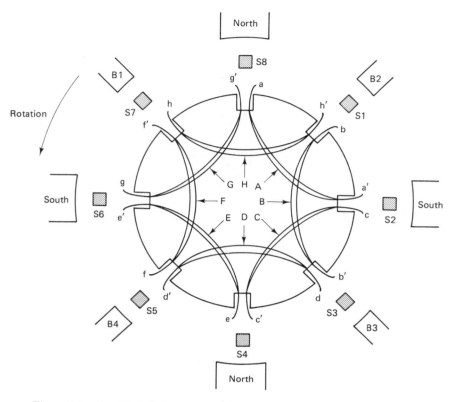

Figure 3-8 Simplified diagram of an eight-coil armature in a four-pole machine.

Coils B, D, F, and H are moving parallel to the flux lines and therefore are not generating a voltage. The winding shown in Figure 3-8, although simplified, will be used to illustrate a lap winding. The difference between a lap and wave winding is in how the coils are interconnected at the commutator segments and the number of segments used.

3-6.1 Simplex Lap Singly Reentrant Winding

The first thing to notice is that for each coil, when one side is under a north pole the other side will be under a south pole. This is in accordance with the theory presented in Section 2-2. In a simplex lap winding the end of the first coil (coil A) gets connected to the beginning of the coil that starts in the slot adjacent to where coil A started. That would be coil B. Therefore, a' is connected to b at commutator segment S1. The end of coil B is then connected to the beginning of the coil that starts in the slot adjacent to where coil B started. That would be coil C. Therefore, b' is connected to c at commutator segment S2. This pattern continues around the armature with the following connections; c' to d at S3, d' to e at S4, e' to f at S5, f' to g at S6, g' to h at S7, and h' to a at S8. In addition, brushes B_1 and B_3 are connected together and brought

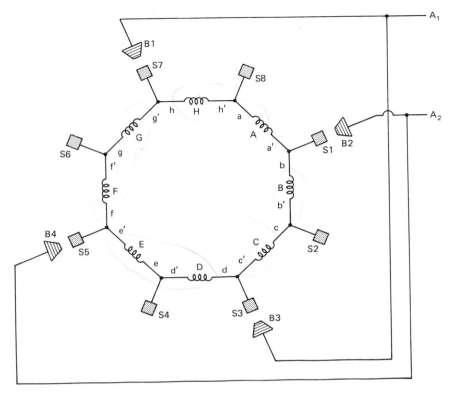

Figure 3-9 Schematic of a lap-wound armature.

out of the machine as one armature terminal A_1. Similarly, B_2 and B_4 are connected and brought out as A_2.

With the connections above made, a schematic diagram of the armature can be drawn. It is shown in Figure 3-9. Note that it is for the position that the armature is in at the instant shown in Figure 3-8.

Figure 3-9 can be redrawn to obtain a clearer picture. It is shown in Figure 3-10. As can be seen, there are four parallel paths. If the picture was redrawn, an instant

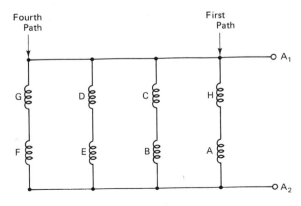

Figure 3-10 Simplified schematic of a lap-wound armature.

later after the armature had rotated 45 degrees, it would look the same except that there would be a different pair of coils in each path. The first path would contain A and B, the second C and D, the third E and F, and the fourth G and H. In any case the number of coils (turns or conductors) would be the same for each path. The following statements can be made for a lap-wound armature:

1. For a simplex lap winding there are as many parallel paths as there are poles.
2. Although not discussed here, a duplex lap winding would have twice as many parallel paths as there were poles, and a triplex lap winding would have three times as many paths as there were poles. Hence a four-pole triplex winding would have 12 parallel paths.
3. The number of coils on an armature is not important. What is important is the total number of conductors (z) on the armature and the number of parallel paths.
4. The voltage generated by a lap-wound armature would be the voltage per path since all paths are in parallel.
5. The rated current of a lap-wound armature would be the rated current per path times the number of paths. The rated current per path is simply a function of the gauge of the wire used for the coils.

Example 3-1

An armature is wound simplex lap in a six-pole generator. The total number of conductors is 600. The resistance per conductor is 0.04 Ω and the average induced voltage per conductor is 0.48 V. If the wire used is rated 15 A, find:

(a) The armature resistance

(b) The armature voltage

(c) The rated armature current

(d) The power output of the armature when supplying rated current

Solution

A six-pole simplex lap armature has six parallel equivalent paths, as shown by Figure 3-11. Since there are a total of 600 conductors, each path must have 100 conductors. As can be seen in Figure 3-10, the conductors in each path are in series. Therefore, the resistance per path would be (100 cond./path) × (0.04 Ω/cond.) = 4 Ω/path, the volt-

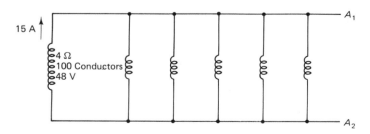

Figure 3-11 Diagram for Example 3-1.

age per path would be (100 cond./path) $\times$ (0.48 V/cond.) = 48 V and each path could supply 15 A.

(a) The armature resistance would be the parallel combination of six 4-Ω resistors.

$$R_a = \frac{4 \; \Omega}{6} = 0.67 \; \Omega$$

(b) Since the paths are in parallel, the armature voltage would be

$$V_a = 48 \; V$$

(c) Since the paths are in parallel, the rated armature current would be

$$I_a = 6 \text{ paths} \times 15 \text{ A/path} = 90 \text{ A}$$

(d) When supplying rated current, the power output would be

$$P = V_a I_a$$
$$= 48 \; V \times 90 \; A = 4320 \; W$$

3-6.2 Simplex Wave Singly Reentrant Winding

An analysis for the wave winding could be carried out in a way similar to that for the lap winding. It is, however, more difficult to visualize. We will therefore not go into the details of the wave-winding construction but rather highlight its characteristics.

1. For a simplex wave winding there will always be two parallel paths regardless of the number of poles.
2. A duplex wave winding will double the number of parallel paths present in the simplex. Thus there will be 2 $\times$ 2, or four parallel paths; the triplex wave winding will have 3 $\times$ 2, or six parallel paths; and so on.
3. Since there are always two parallel paths (simplex wave), the number of conductors per path will be the total number of armature conductors divided by 2.
4. The voltage generated by a simplex wave armature will be that due to half of the armature conductors since there are two paths.
5. Brushes of like polarity are always shorted out by a coil whose conductors are moving parallel to the lines of flux (zero voltage generated). Therefore, all that is needed is one pair of brushes, one positive and the other negative. An obvious advantage of the wave winding is the need for only two brushes. It makes it somewhat easier to construct.
6. The rated current of a simplex wave-wound armature would be twice the rated current per path. The rated current per path is simply a function of the gauge of the wire used for the coils.

Example 3-2

An armature is wound simplex wave in a six-pole generator. The total number of conductors is 600. The resistance per conductor is 0.04 Ω and the average induced voltage per conductor is 0.48 V. If the wire used is rated 15 A, find:

(a) The armature resistance

(b) The armature voltage

(c) The rated armature current

(d) The power output of the armature when supplying rated current

Solution

A simplex wave armature always has two parallel equivalent paths, as shown by Figure 3-12. Since there are a total of 600 conductors, each path must have 300 conductors. As can be seen in Figure 3-12, the conductors in each path are in series. Therefore, the

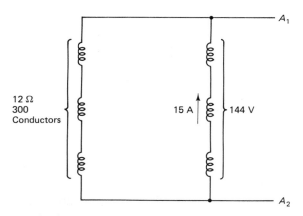

12 Ω
300
Conductors

15 A

144 V

A_1

A_2

Figure 3-12 Diagram for Example 3-2.

resistance per path would be (300 cond./path) × (0.04 Ω/cond.) = 12 Ω/path, the voltage per path would be (300 cond./path) × (0.48 V/cond.) = 144 V/path and each path could supply 15 A.

(a) The armature resistance would be the parallel combination of two 12-Ω resistors.

$$R_a = \frac{12 \ \Omega}{2} = 6 \ \Omega$$

(b) Since the paths are in parallel, the armature voltage would be

$$V_a = 144 \ V$$

(c) Since the paths are in parallel, the rated armature current would be

$$I_a = 2 \text{ paths} \times 15 \text{ A/path} = 30 \text{ A}$$

(d) When supplying rated current, the power output would be

$$P = V_a I_a$$

$$= 144 \ V \times 30 \ A = 4320 \ W$$

3-6.3 Comparison of Lap and Wave Windings

In Sections 3-6.1 and 3-6.2 we examined an armature winding which was first connected simplex lap and then simplex wave. By comparing Examples 3-1 and 3-2, some general conclusions can be drawn.

1. Depending on how the armature is connected (lap or wave), very different characteristics are obtained. A lap-wound armature has a high current rating but generates a low voltage, whereas a wave winding will be used for lower currents and higher voltages. In either case, however, the power rating will be the same.

2. A simplex lap winding will have as many parallel paths as there are poles. The number of paths is further multiplied if the winding is duplex ($\times$ 2) or triplex ($\times$ 3). Therefore, an eight-pole duplex lap winding would have 16 (8 $\times$ 2) parallel paths.

3. A simplex wave winding will always have two parallel paths independent of the number of poles. The number of paths, however, gets multiplied if the winding is duplex (2 $\times$ 2) or triplex (3 $\times$ 2). Therefore, an eight-pole duplex wave winding would have four (2 $\times$ 2) parallel paths.

4. A two-pole armature wound simplex lap would have exactly the same characteristics and look exactly the same as a two-pole armature-wound simplex wave.

5. In the wave winding, since coils that are not generating a voltage electrically connect brushes of similar voltage polarity, only two brushes are required.

At this point, we can develop further a basic equation derived in Chapter 2. Equations 2-5 were used to solve for the average voltage generated by an armature with one coil having a total of z conductors. As we have seen in this section, the number of coils is not important. What is important, in addition to the number of conductors, is the number of parallel paths. Therefore, Eqs. 2-5 will be rewritten here to include the number of parallel paths (**a**). The quantities z, ϕ, **P**, S, and ω have the same meaning and units as stated for Eqs. 2-5.

(English)

$$E_g = \frac{z\phi \mathbf{P}S \times 10^{-8}}{60\mathbf{a}} \tag{3-1a}$$

(SI)

$$E_g = \frac{z\phi \mathbf{P}\omega}{2\pi\mathbf{a}} \tag{3-1b}$$

Example 3-3 (English)

A six-pole generator has a total of 300 conductors. It is driven at 1800 rev/min and the flux per pole is 200 kilolines. Find the average voltage generated by the armature if it is wound:

(a) Simplex lap

(b) Simplex wave

Solution

The problem is solved by direct substitution into Eq. 3-1a. However, the number of parallel paths must first be determined.

(a) For the simplex lap the number of parallel paths is equal to the number of poles. Thus **a** = 6.

$$E_g = \frac{300\,(200\text{k})\,(6)\,(1800)\times 10^{-8}}{60\,(6)}$$

$$= 18 \text{ V}$$

(b) For the simplex wave the number of parallel paths is equal to two. Thus **a** = 2.

$$E_g = \frac{300\,(200\text{k})\,(6)\,(1800)\times 10^{-8}}{60\,(2)}$$

$$= 54 \text{ V}$$

Example 3-4 (SI)

An eight-pole generator has a total of 300 conductors. It is driven at 180 rad/s and the flux per pole is 2×10^{-3} Wb. Find the average voltage generated by the armature if it is wound:

(a) Simplex lap

(b) Simplex wave

Solution

The problem is solved by direct substitution into Eq. 3-1b. However, the number of parallel paths must first be determined.

(a) For the simplex lap the number of parallel paths is equal to the number of poles. Thus **a** = 8.

$$E_g = \frac{300\,(2 \times 10^{-3})\,(8)\,(180)}{2\,(\pi)\,(8)}$$

$$= 17.2 \text{ V}$$

(b) For the simplex wave the number of parallel paths is equal to two. Thus **a** = 2.

$$E_g = \frac{300\,(2 \times 10^{-3})\,(8)\,(180)}{2\,(\pi)\,(2)}$$

$$= 68.75 \text{ V}$$

Example 3-5 (English)

A 32-pole generator has a wave-wound armature (**a** = 2) with 800 conductors. It is

driven at 60 rev/min. Find the flux per pole necessary to generate an ave
voltage of 120 V.

Solution

Equation 3-1a will be used; however, it will first be rearranged to solve for th

$$E_g = \frac{z\phi PS \times 10^{-8}}{60a}$$

Solving for ϕ, we get

$$\phi = \frac{60aE_g}{zPS \times 10^{-8}}$$

$$= \frac{60(2)(120)}{800(32)(60) \times 10^{-8}}$$

$$= 937.5 \text{ kilolines}$$

Example 3-6 (SI)

A 32-pole generator has a wave-wound armature ($\mathbf{a} = 2$) with 800 conductors. It is driven at 6 rad/s. Find the flux per pole necessary to generate an average armature voltage of 120 V.

Solution

Equation 3-1b will be used; however, it will first be rearranged to solve for the flux, ϕ.

$$E_g = \frac{z\phi P\omega}{2\pi a}$$

Solving for ϕ, we get

$$\phi = \frac{2\pi aE_g}{zP\omega}$$

$$= \frac{2(\pi)(2)(120)}{800(32)(6)}$$

$$= 0.0098 = 9.8 \times 10^{-3} \text{ Wb}$$

SYMBOLS INTRODUCED IN CHAPTER 3

Symbol	Definition	Units: English and SI
a	Number of parallel paths	—
z	Number of armature conductors	—
V_a	Armature voltage	volts
R_a	Armature resistance	ohms

Symbol	Definition	Units: English and SI
I_a	Full-load armature current	amperes
E_g	Average (dc) generated voltage of a rotating armature (*Note:* Although neglected in this chapter, this is greater than V_a by an amount equal to the internal voltage drop across R_a)	volts

QUESTIONS

1. Define the following terms: armature; commutation; commutator segment; field poles.
2. What is the purpose of constructing the armature with laminated disks?
3. What is armature reaction?
4. Of what are brushes made?
5. What is the purpose of an interpole?
6. What are compensating windings?

PROBLEMS

(English and SI)

1. A four-pole generator has a simplex lap wound armature. It has a total of 480 conductors each rated 0.03 Ω/cond. The average generated voltage per conductor is 0.5 V. The wire used is rated 20 A. Find:
   (a) The armature resistance
   (b) The armature voltage
   (c) The rated armature current
   (d) The rated power of the armature
2. Repeat Problem 1 if the armature is simplex wave wound.
3. An eight-pole armature is wound simplex lap. It has a total of 1200 conductors, each having a resistance of 0.035 Ω. The average generated voltage per conductor is 0.56 V. If the conductors are rated 15 A, find:
   (a) The armature resistance
   (b) The armature voltage
   (c) The rated armature current
   (d) The rated power of the armature
4. Repeat Problem 3 if the armature is simplex wave wound.
5. A 12-pole armature wound simplex lap has 1620 conductors. Each conductor has a resistance of 0.1 Ω and generates an average voltage of 0.66 V. The conductors are rated 8 A. Find:
   (a) The armature resistance
   (b) The armature voltage
   (c) The rated armature current
   (d) The rated power of the armature

6. Repeat Problem 5 if the armature is simplex wave wound.

7. The armature in an eight-pole generator is wound simplex lap. It is rated 230 V, 5 kW. If it has a total of 1200 conductors and the total armature resistance is 0.8 Ω, find:
(a) The rated current per conductor
(b) The generated voltage per conductor
(c) The resistance per conductor

8. Repeat Problem 7 if the armature is simplex wave wound.

9. A 120-V 1-kW four-pole lap wound generator has a total of 720 conductors. If the total armature resistance is 1.0 Ω, find:
(a) The rated current per conductor
(b) The generated voltage per conductor
(c) The resistance per conductor

10. Repeat Problem 9 if the armature is simplex wave wound.

(English)

11. A 12 pole generator has a total of 600 conductors. It is driven at 1200 rev/min and the flux per pole is 350 kilolines. Find the generated armature voltage if it is wound:
(a) Simplex lap
(b) Simplex wave

12. A four-pole generator has a total of 800 conductors. When it is driven at 1800 rev/min, its armature generates 440 V. Find the required flux per pole if the armature is wound:
(a) Simplex lap
(b) Simplex wave

13. A six-pole generator has a total of 420 conductors and a flux per pole of 400 kilolines. Find the speed at which it must be driven to generate 120 V if the armature is:
(a) Simplex lap
(b) Simplex wave

14. A four-pole generator has a total of 1000 conductors and a flux per pole of 500 kilolines. When it is driven at 1200 rev/min the generated voltage is 200 V. Is the armature wound simplex lap, simplex wave, or neither?

15. A 12-pole generator has a total of 1800 armature conductors, flux per pole of 500 kilolines, and is driven at 1000 rev/min. Find the generated voltage if the armature is wound:
(a) Simplex lap
(b) Simplex wave

(SI)

16. A 16-pole generator has a total of 800 conductors. It is driven at 120 rad/s and the flux per pole is 0.004 Wb. Find the generated armature voltage if it is wound:
(a) Simplex lap
(b) Simplex wave

17. A six-pole generator has a total of 900 conductors. When it is driven at 60π rad/s, its armature generates 230 V. Find the required flux per pole if the armature is wound:
(a) Simplex lap
(b) Simplex wave

18. An eight-pole generator has a total of 640 conductors and a flux per pole of 5.5×10^{-3} Wb. Find the speed it must be driven at to generate 460 V if the armature is wound:

(a) Simplex lap

(b) Simplex wave

19. A four-pole generator has a total of 1440 conductors and a flux per pole of 4.6×10^{-3} Wb. When it is driven at 40π rad/s the generated voltage is 265 V. Is the armature wound simplex lap, simplex wave, or neither?

20. A 16-pole generator has a total of 2240 armature conductors, flux per pole of 0.005 Wb, and is driven at 100 rad/s. Find the generated voltage if the armature is wound:

(a) Simplex lap

(b) Simplex wave

Chapter 4

DC Generator Characteristics

In this chapter we analyze the dc generator using an equivalent circuit and some simple equations. The equivalent circuit will resemble the circuits analyzed in any basic dc circuit analysis course. The equations used will be simplifications of equations already presented in Chapters 1, 2, and 3, or will be derived by applying Kirchhoff's voltage and current laws to the equivalent circuit.

4-1   BASIC GENERATOR EQUATION

Upon examination of Eqs. 3-1 we see that once a generator has been built, there are only two variables that can affect the generated voltage: the flux and the speed. If the number of conductors (z), poles ($\mathbf{P}$), and parallel paths ($\mathbf{a}$) are lumped together with the numerical constants, Eqs. 3-1 can be rewritten in a very simple form.

(English)

$$E_g = K\phi S \tag{4-1a}$$

(SI)

$$E_g = K'\phi\omega \tag{4-1b}$$

where $K = \dfrac{z\mathbf{P} \times 10^{-8}}{60\mathbf{a}}$

$K' = \dfrac{z\mathbf{P}}{2\pi\mathbf{a}}$

As before, S and ω represent the speed in rev/min and rad/s, respectively, and ϕ is the flux in lines (Eq. 4-1a) and webers (Eq. 4-1b).

Equations 4-1 show that the generated voltage varies directly (proportionately) with the flux and the speed at which the armature is turning. Thus if the flux *or* speed doubles, the voltage will double as well. If both the flux and speed double, the voltage will increase by a factor of 4.

If the flux is some value ϕ and the speed is S_1, the generated voltage will be

$$E_{g1} = K\phi S_1$$

If the flux is held constant and the speed is changed to S_2, the generated voltage will become

$$E_{g2} = K\phi S_2$$

Dividing the first equation by the second, we get

$$\frac{E_{g1}}{E_{g2}} = \frac{K \phi S_1}{K \phi S_2}$$

(English)

$$\frac{E_{g1}}{E_{g2}} = \frac{S_1}{S_2} \tag{4-2a}$$

For SI this equation becomes

(SI)

$$\frac{E_{g1}}{E_{g2}} = \frac{\omega_1}{\omega_2} \tag{4-2b}$$

Similarly, if the speed were held constant and the flux varied, the following equation could be derived in the same way:

$$\frac{E_{g1}}{E_{g2}} = \frac{\phi_1}{\phi_2} \tag{4-3}$$

If both flux and speed are allowed to change, the following equations are obtained:

(English)

$$\frac{E_{g1}}{E_{g2}} = \frac{\phi_1 S_1}{\phi_2 S_2} \tag{4-4a}$$

(SI)

$$\frac{E_{g1}}{E_{g2}} = \frac{\phi_1 \omega_1}{\phi_2 \omega_2} \tag{4-4b}$$

Example 4-1 (English)

A generator rotating at 1800 rev/min develops 140 V. If the flux remains constant and the speed drops to 1650 rev/min, what will be the new generated voltage? By how much must the flux be changed to restore the generated voltage to 140 V at the new, slower speed?

Solution

For the first part of the problem, the flux is constant; therefore, Eq. 4-2a is used.

$$\frac{E_{g1}}{E_{g2}} = \frac{S_1}{S_2}$$

$$\frac{140 \text{ V}}{E_{g2}} = \frac{1800 \text{ rev/min}}{1650 \text{ rev/min}}$$

$$E_{g2} = \frac{1650}{1800}(140) = 128.33 \text{ V}$$

For the second part of the problem, both flux and speed can change; therefore, Eq. 4-4a is used.

$$\frac{E_{g1}}{E_{g2}} = \frac{\phi_1 S_1}{\phi_2 S_2}$$

Since E_{g2} is to be restored to 140 V, the left side of the equation above becomes unity.

$$\frac{140 \text{ V}}{140 \text{ V}} = 1 = \frac{\phi_1(1800)}{\phi_2(1650)}$$

Solving for ϕ_2 gives us

$$\phi_2 = \frac{1800}{1650}\phi_1 = 1.091\phi_1$$

Thus the flux must be increased by 9.1%.

Example 4-2 (SI)

A generator rotating at 200 rad/s develops 220 V. If the flux remains constant and the speed drops to 185 rad/s, what will be the new generated voltage? By how much must the flux be changed to restore the generated voltage to 200 V at the new, slower speed?

Solution

For the first part of the problem, the flux is constant; therefore, Eq. 4-2b is used.

$$\frac{E_{g1}}{E_{g2}} = \frac{\omega_1}{\omega_2}$$

$$\frac{220 \text{ V}}{E_{g2}} = \frac{200 \text{ rad/s}}{185 \text{ rad/s}}$$

$$E_{g2} = \frac{185}{200}(220) = 203.5 \text{ V}$$

For the second part of the problem, both flux and speed can change; therefore, Eq. 4-4b is used.

$$\frac{E_{g1}}{E_{g2}} = \frac{\phi_1 \omega_1}{\phi_2 \omega_2}$$

Since E_{g2} is restored to 220 V, the left side of the equation above becomes unity.

$$\frac{220 \text{ V}}{220 \text{ V}} = 1 = \frac{\phi_1(200)}{\phi_2(185)}$$

Solving for ϕ_2 yields

$$\phi_2 = \frac{200}{185}(\phi_1) = 1.081\phi_1$$

Thus the flux must be increased by 8.1%.

4-1.1 Graphical Analysis of the Generator Equation

If Eq. 4-1 (a or b) was plotted under conditions of constant flux, we would obtain the graph shown in Figure 4-1. The straight line obtained was expected since we already established that the voltage varied directly with speed.

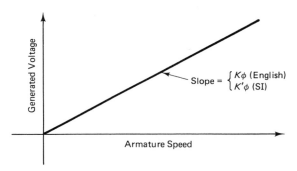

Figure 4-1 Generated voltage versus speed with flux constant.

Example 4-3 (English)

The data in Table 4-1 was obtained in the laboratory for a generator under test. The flux was kept constant.

Table 4-1

E_g (V)	115	120	124	126	130	137
S (rev/min)	1500	1550	1600	1650	1700	1750

(a) Plot the voltage versus speed.

(b) Calculate the slope ($K\phi$) of the line.

Solution

(a) The data is plotted and shown in Figure 4-2. Note that when graphing the data, the points are not all connected by the line. The line is drawn so that it passes as close as possible to the most points. It is drawn as a straight line because we know from theory that it should be.

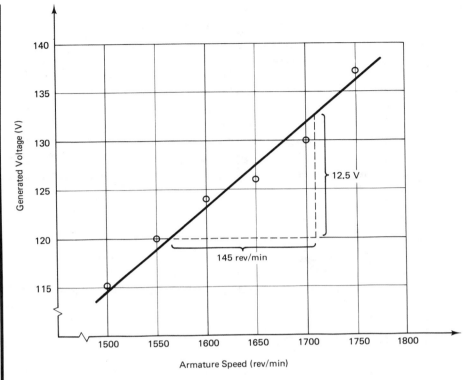

Figure 4-2 Plot of data for Example 4-3.

The fact that the data do not all lie on a straight line can be attributed to the following:

1. The accuracy of the voltmeter used to measure the voltage.
2. The accuracy of the tachometer used to measure the speed.
3. Misreading the data (human error); that is, 128 V was misread as 126 V.

(b) The slope is obtained by constructing a right triangle between any two points on the line (they do not have to be data points).

$$K\phi = \text{slope} = \frac{\text{change in voltage}}{\text{change in speed}}$$

$$= \frac{12.5 \text{ V}}{145 \text{ rev/min}} = 0.086 \text{ V/(rev/min)}$$

Example 4-4 (SI)

The data in Table 4-2 was obtained in the laboratory for a generator under test. The flux was kept constant.

Table 4-2

E_g (V)	202	210	214	220	227	237
ω (rad/s)	140	145	150	155	160	165

(a) Plot the voltage versus speed.

(b) Calculate the slope $(K'\phi)$ of the line.

Solution

(a) The data is plotted and shown in Figure 4-3. Note that when graphing the data, the points are not all connected by the line. The line is drawn so that it passes as close as possible to the most points. It is drawn as a straight line because we know from theory that it should be.

The fact that the data do not all lie on a straight line can be attributed to the following:

1. The accuracy of the voltmeter used to measure the voltage.
2. The accuracy of the tachometer used to measure the speed.
3. Misreading the data (human error); that is, 229 V was misread as 227 V.

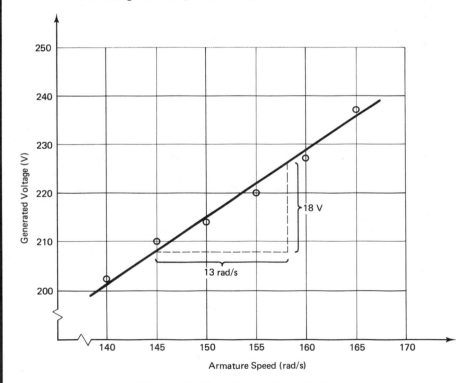

Figure 4-3 Plot of data for Example 4-4.

(b) The slope is obtained by constructing a right triangle between any two points on the line (they do not have to be data points).

$$K'\phi = \text{slope} = \frac{\text{change in voltage}}{\text{change in speed}}$$

$$= \frac{18 \text{ V}}{13 \text{ rad/s}} = 1.38 \text{ V/(rad/s)}$$

4-2 *EQUIVALENT CIRCUIT OF A DC GENERATOR*

In any field of science, the analysis of a device or process becomes much simpler when an accurate diagram can be used to represent it. Such is the case with the generator, and as will be seen shortly, its equivalent circuit will be nothing more than a simple series or series–parallel circuit.

Figure 4-4 is a very simple circuit representing a dc generator. As the various forms of the dc generator are encountered, Figure 4-4 will be modified somewhat to represent the type of generator being discussed. The field circuit represents the electromagnet used to generate the flux (ϕ) in the magnetic core, much like those discussed in Chapter 1. The total resistance in the field circuit is denoted by R_f and the current in the field is I_f.

The armature circuit has two basic parts. The first part is a symbol for the dc voltage generated by the rotating armature, E_g. Note that this voltage (E_g) is the one calculated using Eqs. 4-1. The second part is the total resistance of the armature winding, R_a. Note that this resistance is the same as the one calculated in part (a) of Examples 3-1 and 3-2. I_a represents the armature current.

The quantity I_L represents the current delivered to the load and V_t is the output (or terminal) voltage of the generator. It is through these terminals that **electrical**

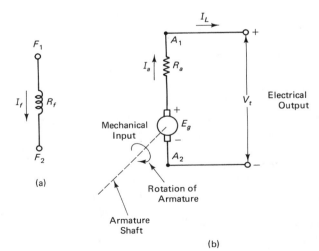

(a)

(b)

Figure 4-4 General circuit representation of a dc generator: (a) field circuit; (b) armature circuit.

power is fed **out** of the generator to a load. In this case the load current is equal to the armature current.

The dashed line in Figure 4-4 represents the armature shaft. It is through this shaft that **mechanical power** is fed **into** the generator. Rotation of this shaft is provided by a **prime mover,** which is any device (motor, waterwheel, etc.) that supplies the turning force.

4-3 SEPARATELY EXCITED GENERATOR

A simple but unpopular type of generator is the separately excited generator shown in Figure 4-5. It is unpopular because it requires a separate dc voltage source to supply current to the field winding. Note that the field circuit contains a rheostat which is used to vary the field current. By varying the field current, the flux (ϕ) in Eqs. 4-1 is varied. In this way the generated voltage is adjusted.

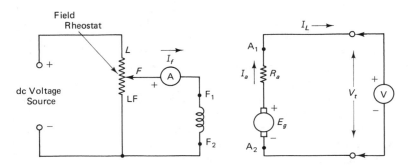

Figure 4-5 Separately excited generator.

The ammeter (A) and voltmeter (V) can be used to measure the field current and terminal voltage whenever testing is done. The rheostat connection shown in Figure 4-5 is used here because it allows us to reduce the field current to zero (i.e., when the arm of the rheostat is set to the end labeled LF).

4-3.1 No-Load Magnetization Curve

When electronic circuits are designed, a transistor's characteristics are often used as an aid. In the same way that a transistor's characteristic curves are used to identify the internal behavior of a particular device, the magnetization curve is used to identify the internal characteristics of a particular generator. The magnetization curve is a plot of generated voltage (E_g) versus field current (I_f) at a given speed. If the field current is increased, the flux will increase, and the generated voltage will increase as well. Equations 4-1 imply that our plot would be linear. If, however, we recall from Section 1-2 the nonlinear behavior of ferromagnetic materials, we can expect a characteristic similar to Figure 1-5. The curve obtained is shown in Figure 4-6.

The circuit shown in Figure 4-5 could be used to obtain the data for the curve.

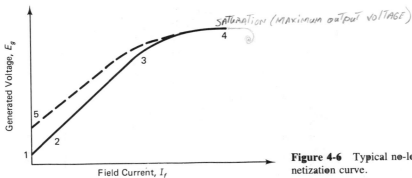

Figure 4-6 Typical no-load magnetization curve.

A **no-load** condition is one in which the current drawn from the generator is zero ($I_L = 0$). In this particular case I_a equals I_L, which is zero; therefore, there will be no internal voltage drop across the armature resistance. Kirchhoff's voltage law tells us that V_t must then be the same as E_g. Although the voltmeter is connected to measure V_t, it is in effect reading E_g, which is the quantity desired.

As the testing is begun, although the field current is zero, the flux starts out with some initial value (**residual magnetism**) from a previous use. At zero field current a small voltage is obtained (point 1 on Figure 4-6). As I_f increases, the flux, hence the voltage, increases linearly as well (point 2 to 3 on Figure 4-6). Beyond point 3 on the curve, the core is beginning to saturate and large increases in I_f will produce only small increases in flux (or voltage). A generator is normally operated in this saturation region. If the current is decreased to zero, the plot returns along the dashed line (point 4 to point 5). This is due to hysteresis, described in Section 1-2.1. The fact that hysteresis is present means that we should always increase the field current to its maximum value without decreasing it in between. If while taking data I_f was decreased occasionally, the result would be data points that would not lie on a smooth curve.

Example 4-5 (English)

The data in Table 4-3 was obtained for the test setup in Figure 4-5. The speed of the generator during the test was 1700 rev/min.

(a) Plot the magnetization curve at 1700 rev/min.

(b) Calculate a new set of data for a speed of 1400 rev/min and plot it on the same graph as that of part (a).

Table 4-3

I_f (A)	0	0.1	0.16	0.22	0.3	0.35	0.45
E_{g1} (V)	5	22	40	50	80	88	91

Solution

(a) The data in Table 4-3 is plotted and shown in Figure 4-7 as the solid curve.

(b) At any particular value of flux (or field current) the generated voltage is proportional

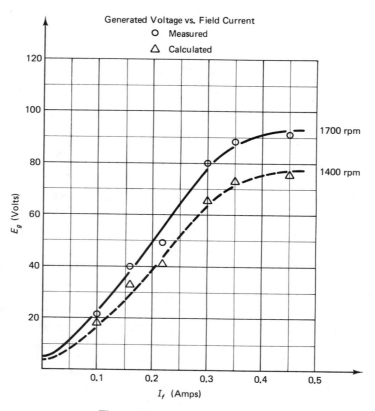

Figure 4-7 Graph for Example 4-5.

to speed. Therefore, at each data point in Table 4-3 a new voltage E_{g2} can be calculated at the reduced speed using Eq. 4-2a.

$$\frac{E_{g1}}{E_{g2}} = \frac{S_1}{S_2}$$

$$E_{g2} = \frac{S_2}{S_1}(E_{g1})$$

Some sample calculations are shown below and entered in Table 4-4. The result is plotted in Figure 4-7 as the dashed curve.

$$\text{At } I_f = 0: \qquad E_{g2} = \frac{1400}{1700}(5) = 4.1 \text{ V}$$

$$\text{At } I_f = 0.1: \qquad E_{g2} = \frac{1400}{1700}(22) = 18.1 \text{ V}$$

etc.

Table 4-4

I_f (A)	0	0.1	0.16	0.22	0.3	0.35	0.45
E_{g2} (V)	4.1	18.1	32.9	41.2	65.9	72.5	74.9

Example 4-6 (SI)

The data in Table 4-5 was obtained for the test setup in Figure 4-5. The speed of the generator during the test was 200 rad/s.

(a) Plot the magnetization curve at 200 rad/s.

(b) Calculate a new set of data for a speed of 120 rad/s and plot it on the same graph as that of part (a).

Table 4-5

I_f (A)	0	0.2	0.36	0.5	0.65	0.82	1.0
E_{g1} (V)	8	30	65	102	118	132	137

Solution

(a) The data in Table 4-5 is plotted and shown in Figure 4-8 as the solid curve.

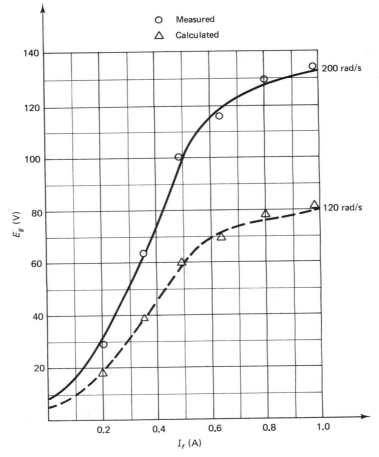

Figure 4-8 Graph for Example 4-6.

(b) At any particular value of flux (or field current) the generated voltage is proportional to speed. Therefore, at each data point in Table 4-5 a new voltage E_{g2} can be calculated at the reduced speed using Eq. 4-2b.

$$\frac{E_{g1}}{E_{g2}} = \frac{\omega_1}{\omega_2}$$

$$E_{g2} = \frac{\omega_2}{\omega_1}(E_{g1})$$

Some sample calculations are shown below and entered in Table 4-6. The result is plotted in Figure 4-8 as the dashed curve.

At $I_f = 0$: $\quad E_{g2} = \dfrac{120}{200}(8) = 4.8 \text{ V}$

At $I_f = 0.2$: $\quad E_{g2} = \dfrac{120}{200}(30) = 18 \text{ V}$

etc.

Table 4-6

I_f (A)	0	0.2	0.36	0.5	0.65	0.82	1.0
E_{g2} (V)	4.8	18	39	61.2	70.8	79.2	82.2

4-4 VOLTAGE REGULATION

Under typical conditions a generator is operated **under load.** In other words, its terminals are connected to a device that draws current (electrical power) from it. The load can be anything (resistors, motors, lights, etc.) that can operate from a dc voltage equal in magnitude to that supplied at the generator's terminals. A typical connection is shown in Figure 4-9 for a separately excited generator. In this case the field rheostat has been connected in a way different from Figure 4-5; however, it can still control the flux. The quantity R_f represents the total field circuit resistance (field coil plus rheostat setting). Since a load is connected to the generator, a load current (I_L) will flow. The armature is in series with the load; therefore, I_a will equal I_L. Because of the armature current that is flowing, there will be a voltage drop within the armature equal in magnitude to $I_a R_a$. As a result of this internal voltage drop, the terminal voltage (V_t) will be less than the generated voltage (E_g) by an amount equal to $I_a R_a$. It is given by

$$V_t = E_g - I_a R_a \tag{4-5}$$

Voltage regulation is a measure of the change in terminal voltage as the load (I_L) increases. Although it can be calculated at any load condition, it is usually calculated at **rated** conditions and called the **full-load voltage regulation.** The rated conditions are listed on a generator's nameplate. The ratings include the rated (maximum) power that the generator can safely supply, in kilowatts; the terminal voltage when supplying rated power; the speed at which the generator is to be driven; and

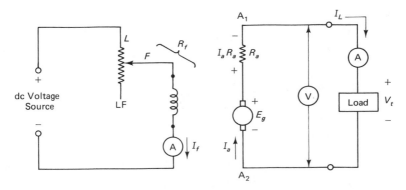

Figure 4-9 Separately excited generator under load.

sometimes the mechanical power input necessary to supply the rated power out. The power input is specified in horsepower (1 hp = 746 W) or kilowatts.

The voltage regulation is given by Eq. 4-6 and is expressed as a percentage:

$$\frac{\%\ \text{voltage}}{\text{regulation}} = \frac{V_{NL} - V_{FL}}{V_{FL}} \times 100 \tag{4-6}$$

In Eq. 4-6 V_{NL} is the no-load terminal voltage and V_{FL} is the full-load or rated terminal voltage.

Example 4-7

A 5-kW 120-V generator is under test. When the load is removed, the terminal voltage is found by measurement to be 138 V. Calculate the voltage regulation.

Solution

The full-load voltage is given by the rating (V_{FL} = 120 V). The no-load voltage was found by measurement (V_{NL} = 138 V). Using Eq. 4-6 yields

$$\%\ \text{V.R.} = \frac{138 - 120}{120} \times 100$$

$$= 15\%$$

The magnetization curve is a picture of a generator's internal characteristics. A curve useful in analyzing a generator that is externally connected to a load is its **load characteristic.** The load characteristic is a plot of terminal voltage versus load current. A typical one is shown in Figure 4-10 for a separately excited generator rated 1 kW, 125 V. A dashed line is usually placed at the rated current point. If the difference between no-load and full-load voltage were truly the $I_a R_a$ drop as given by Eq. 4-5, the characteristic would be a falling straight line. However, due to the effects of armature reaction (see Section 3-1.3), the flux does not remain constant but decreases slightly with the increasing armature current. This causes the downward curving of the characteristic.

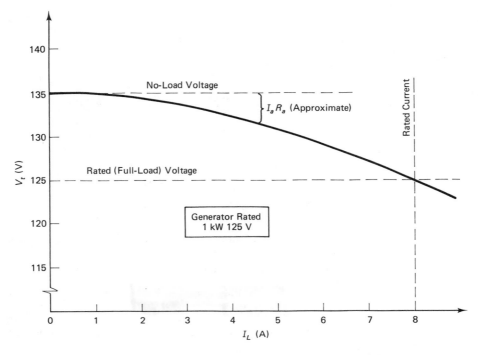

Figure 4-10 Separately excited generator load characteristic.

Example 4-8

For the generator whose characteristic is shown in Figure 4-10:

(a) Calculate the percent voltage regulation.

(b) Verify that the rated current is 8 A, as shown by the dashed line in Figure 4-10.

Solution

(a) The no-load voltage is the terminal voltage at $I_L = 0$; therefore, $V_{NL} = 135$ V. The full-load voltage is the terminal voltage at rated current ($I_L = 8$ A); therefore, $V_{FL} = 125$ V. (*Note:* The full-load voltage is also indicated by the nameplate rating.) Using Eq. 4-6, we get

$$\% \text{ V.R.} = \frac{135 - 125}{125} \times 100$$

$$= 8\%$$

(b) The rated current can be obtained from the nameplate rating:

$$\text{rated power output} = \text{rated voltage} \times \text{rated current}$$

$$1 \text{ kW} = 125 \text{ V} \times I_{\text{rated}}$$

$$I_{\text{rated}} = \frac{1000 \text{ W}}{125 \text{ V}} = 8 \text{ A}$$

4-5 GENERATOR EFFICIENCY

The efficiency of any device is a measure of how much useful power can be gotten out of something for a given amount of input power supplied. In its most basic form, efficiency can be solved for using Eq. 4-7, where the symbol η used is the Greek letter eta:

$$\text{efficiency} = \eta(\%) = \frac{P_o}{P_i} \times 100 \qquad (4\text{-}7)$$

In Eq. 4-7, P_o and P_i represent the output and input power, respectively. They can be expressed in any form (watts, kW, hp, etc.); however, both must have the same units. As is the case with any nonideal (real) device, there will always be internal losses; therefore, we can never get out as much power as we put in. Thus P_o will always be less than P_i and the efficiency must always be less than 100%.

Example 4-9

A 1-kW 125 V generator requires 2 hp to supply rated output. Calculate the generator efficiency.

Solution

The output power is given as 1 kW. The input power is given as 2 hp. Both will be converted to watts.

$$P_o = 1 \text{ kW} = 1000 \text{ W}$$

$$P_i = 2 \text{ hp} = 2 \text{ hp}(746 \text{ W/hp}) = 1492 \text{ W}$$

$$\eta = \frac{1000}{1492} \times 100 = 67\%$$

4-5.1 Stray Power Losses

Before any conversion to electrical power takes place in the generator, some of the mechanical power supplied by the prime mover is lost to what are called **stray power** (or **rotational**) losses. The following are the different losses which are included in the general term "stray power loss":

1. *Friction loss.* Any time there is contact between moving parts, friction which produces an energy loss in the form of heat will be present. In dc machines there is **brush friction.** This is due to the brushes rubbing against the commutator as the armature rotates. It is fairly constant over all speeds and is the same as coulomb friction studied in any elementary physics course.

Also present in any rotating machine is **viscous friction.** Any rotating machine will have its shaft set in well-lubricated bearings. As smoothly as the bearings appear to turn, there is friction present. This friction is called viscous friction and is proportional to speed. Although negligible on starting and at slow speeds, it becomes significant at high speeds.

2. *Windage.* This loss refers to the energy expended in overcoming wind re-

sistance. Not only is there wind resistance between the rotating armature and the air inside the machine, but many larger machines have cooling fans attached to the rotating shaft. The purpose of the fan is to prevent overheating; however, in return a windage loss is suffered.

3. *Hysteresis loss.* When an armature is in any given position, it is magnetized. If we try to turn the armature, the flux within it must change position as well. To do this, energy must be expended. This effort required is very similar to the energy required to separate (or turn) a north pole from a south pole when they are attracted together. This cyclic changing of the flux within the armature causes a hysteresis loss. It is **directly proportional to speed and flux.**

4. *Eddy current loss.* In Chapter 2 we learned that if a conductor moved through a magnetic field cutting the lines of flux, a voltage would be induced (Faraday's law). When an armature rotates, not only are the windings cutting flux but the armature core itself cuts flux as well. Since the core is an electrical conductor, voltage is induced in it, causing small currents (**eddy currents**) to flow within. Also, since the core has resistance, there is an I^2R power loss that causes the armature to heat up.

By making the armature with laminated disks, instead of being solid, the resistance of the core is greatly increased. This reduces the eddy currents, which in turn decreases the loss. It should be noted that although the laminated disks increase resistance, they do not significantly change reluctance; therefore, the flux will not be decreased. The eddy current loss is **directly proportional to the square of the speed and flux.**

The stray power losses above are subdivided into two groups. The friction and windage losses are called **mechanical losses,** while the hysteresis and eddy current losses are called **core losses.**

4-5.2 Copper Losses

Whenever current flows through resistance, there will be an I^2R power loss in the form of heat. This power loss is called a **copper loss.** If, for example, we examine Figure 4-9, we see that there are copper losses present in two locations. First, there is a copper loss in the field circuit equal to $R_f I_f^2$. Second, there is a copper loss in the armature equal to $R_a I_a^2$.

As the generator circuits become more involved, to enhance their characteristics and control them, more copper losses will appear and will readily be seen and calculated from the equivalent circuit drawn.

4-5.3 Stray Load Loss

Stray load losses are due to the distortion of the flux as it passes through the machine. The distortion is mainly due to armature reaction and leakage around the armature slots and field poles. It is difficult to calculate or measure; however, since it is very small, it is neglected. In large generators rated 150 kW or more, the stray load loss is assumed to be about 1% of rated power.

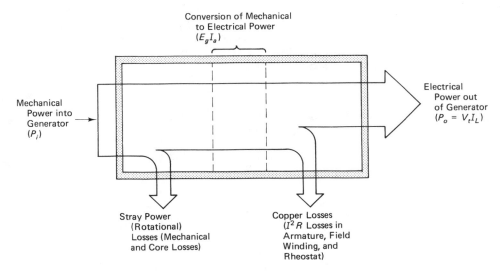

Figure 4-11 Power flow in a generator.

4-5.4 Power Flow Diagram

To understand the power flow more clearly, the diagram in Figure 4-11 can be analyzed. Through its shaft, mechanical power is fed into the generator. Before any conversion from mechanical power to electrical power takes place, the generator suffers rotational losses. After these rotational losses are subtracted from the input, we are left with the mechanical power, which is converted to electrical power ($E_g I_a$) by the armature. Of the electrical power that has been generated ($E_g I_a$), some is lost internally. This loss is the copper loss. Upon subtracting the copper losses from the generated power, we are finally left with the net power out of the machine.

The following equations can now be written. It should be fairly clear that if we add up the power out with all the losses, it will total the power into the machine.

$$P_i = P_o + \text{losses} \tag{4-8}$$

If Eq. 4-8 is substituted into Eq. 4-7, we get

$$\text{efficiency} = \eta(\%) = \frac{P_o}{P_o + \text{losses}} \times 100 \tag{4-9}$$

Equation 4-9 is fairly valuable. With it we can calculate efficiency without having the input power, which is a mechanical quantity and therefore difficult to measure. The power output and copper losses can be easily calculated and/or measured. The stray power loss can be found easily by running the generator as a motor under no-load conditions. This test will be discussed in detail in Section 5-5.2.

It should be mentioned at this point that any generator can be run as a motor, and vice versa. If its shaft is mechanically driven, it will supply electrical power out (generator). If, on the other hand, a dc voltage is supplied to the terminals, it will develop mechanical power and the shaft will turn (motor).

Example 4-10

A 2-kW 220-V generator has constant stray power losses of 150 W. The copper losses at rated load are calculated to be 280 W. If it is driven by a motor, find the power required from the motor and the efficiency of the generator, both at rated load.

Solution

The power input of the generator is the total power required from the motor as shown in Figure 4-12. Using Eq. 4-8, we have

$$P_i = P_o + \text{losses}$$

where P_o is the rated power of the generator.

$$P_i = 2000 \text{ W} + \underbrace{150 \text{ W} + 280 \text{ W}}_{\text{total losses}}$$

$$= 2430 \text{ W}$$

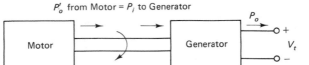

Figure 4-12 Diagram for Example 4-10.

This is the power required from the motor; however, the motor output since it is a mechanical quantity is specified in horsepower. Therefore,

$$\text{(from motor)} \quad P_o' = \frac{2430 \text{ W}}{746 \text{ W/hp}} = 3.26 \text{ hp}$$

The efficiency of the generator can be calculated using Eq. 4-7 since P_o and P_i are now both known.

$$\eta = \frac{2000 \text{ W}}{2430 \text{ W}} \times 100 = 82.3\%$$

4-6   SHUNT GENERATOR

The most popular generators fall into a category called **self-excited.** That is, they do not need an external source of voltage (as in Figure 4-9) to supply current to the field winding. Instead, these generators derive their field excitation from the voltage they are generating. You might be asking the question: "Which comes first, the chicken or the egg?" In other words, does the voltage come first which causes a magnetic field which further builds up the voltage; or does the field come first, causing a voltage to build up which when applied to the field winding creates a larger magnetic field? In this case the riddle does have an answer. When a self-excited generator is driven, there will always be some flux in the core of the generator from a previous use. This flux is the residual magnetism described in Section 1-2.1 under the topic of hysteresis. The

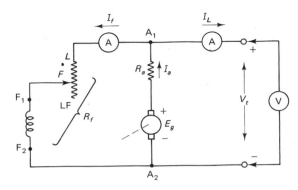

Figure 4-13 Shunt generator.

most common self-excited generator is the **shunt generator** shown in Figure 4-13. It is called a shunt generator because the field winding shunts (is in parallel with) the armature. As before, a rheostat (now called the **shunt-field rheostat**) is inserted in series with the field winding. Its purpose is to vary the field current, thus controlling the generated voltage.

When the generator is driven, the residual magnetism in the core is sufficient to cause a small initial generated voltage. This voltage will cause a current to flow out of the armature and through the field circuit, causing the flux to increase. The larger flux will make the voltage increase, causing more current in the field circuit. This build-up process continues until a stable situation is reached when the current I_f causes a flux, which produces a voltage just sufficient to cause the current I_f. The stable point is determined by the amount of resistance in the field circuit, R_f. It can be found easily by plotting the **field resistance line** on the same graph with the magnetization curve. The intersection of the two determines the stable point. Figure 4-14 shows a typical graph.

When a particular voltage buildup is desired, the resistance required to produce that voltage is called the **critical resistance.** It can be found by dividing the desired voltage by the corresponding current on the magnetization curve. It should be noted that a **larger resistance** will lower the field current, thus decreasing the voltage.

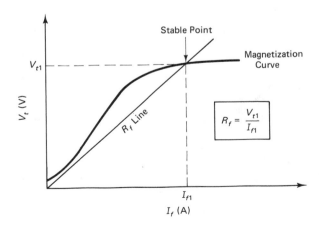

Figure 4-14 Graph of field resistance line and magnetization curve.

Example 4-11

The magnetization curve for a shunt generator is shown in Figure 4-15. Plot the field resistance line on the graph and calculate the resistance necessary to get a build up of 120 V.

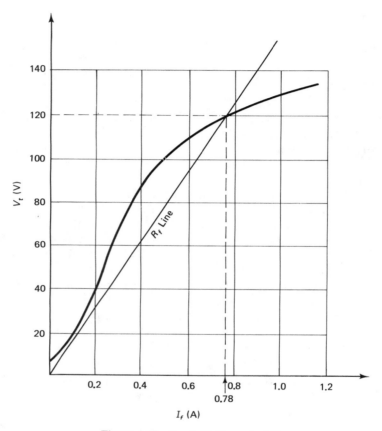

Figure 4-15 Graph for Example 4-11.

Solution

First find the point on the magnetization curve where it crosses the desired voltage (120 V). The field resistance line is obtained by drawing a straight line between this point and the origin. The value of field resistance needed is calculated from the slope of the R_f line.

$$R_f = \text{slope} = \frac{120 \text{ V}}{0.78 \text{ A}} = 153.85 \; \Omega$$

$$R_f \approx 154 \; \Omega$$

4-6.1 Failure of a Generator to Build Up

There are times when a generator, in the lab or field, will fail to build up a voltage when driven at some speed. This problem can be due to any one of several possibilities:

1. The most common problem in the lab occurs when the initial voltage due to residual magnetism creates a flux that opposes the residual flux in the core. The voltage will end up at one-half its initial value, and variation of the field rheostat will have no effect. There are two ways to solve the problem. First, if the direction of rotation of the generator is reversed, the voltage will be of opposite polarity (recalling Fleming's right-hand rule). The flux produced will now aid the residual flux and a buildup will occur if the rheostat is varied. If the direction of rotation is desired or deemed correct, the alternative solution is to reverse the shunt field connections. That is, F_1 and F_2 in Figure 4-13 are interchanged.

2. The problem can also be due to an open circuit somewhere in the generator. In the lab this is usually due to a burned-out fuse in the ammeter, which is being used to measure the field current. In the field the open circuit could be due to an open solder connection, a dirty commutator, worn-out brushes, field rheostat set too high, or more seriously a burned-out armature. To prevent the last of these from occurring, generators are fuse protected so that no more than rated current will be drawn from them.

3. Finally, if there is little or no residual magnetism in the core, a buildup will not occur. This can be caused from contact with something that can demagnetize the core or simply from long periods of inactivity. This is corrected by **flashing the field,** a process whereby an external source is connected to the field winding, giving it some magnetization.

4-6.2 Shunt Generator Load Characteristic

As a shunt generator is loaded down, that is, more and more current is drawn from it, it will exhibit the same characteristic as the separately excited generator shown in Figure 4-10. This time, however, the decrease in voltage as load current increases will be more severe and poorer voltage regulation will result. Not only will the voltage decrease because of the increasing $I_a R_a$ drop and armature reaction, but there is now an additional factor. As the voltage decreases for the reasons given above, less field current is produced (remember, it is self-excited), causing a lower flux. The lower flux will now generate a slightly lower voltage.

As the voltage continues to drop, the core of the generator comes out of saturation. At this point, the **breakdown point,** changes in voltage will produce a proportional change in flux and the voltage will quickly decay toward zero volts.

Figure 4-16 shows the load characteristics of a generator both self- and separately excited.

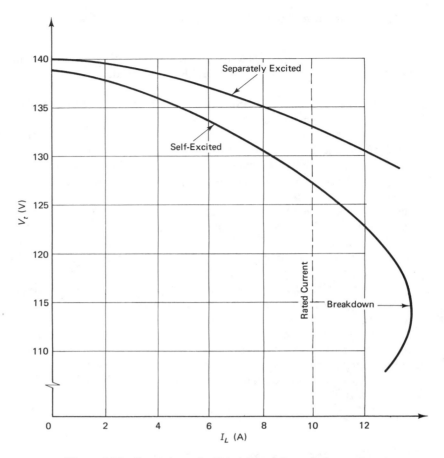

Figure 4-16 Comparison of self and separately excited generators.

Example 4-12

Find the percent voltage regulation for the two generator characteristics in Figure 4-16.

Solution

For the separately excited generator,

$$V_{NL} = 140 \text{ V} \quad \text{and} \quad V_{FL} \approx 133 \text{ V}$$

$$\% \text{ V.R.} = \frac{140 - 133}{133} \times 100$$

$$= 5.26\%$$

For the self-excited generator,

$$V_{NL} = 139 \text{ V} \quad \text{and} \quad V_{FL} \approx 128 \text{ V}$$

$$\% \text{ V.R.} = \frac{139 - 128}{128} \times 100$$

$$= 8.59\%$$

It should be noted that the smaller number is more desirable.

Example 4-13

A 5.5-kW 220-V shunt generator has a field resistance of 140 Ω, armature resistance of 0.5 Ω, and stray power losses of 95 W at rated conditions. Find at rated load:

(a) The field current

(b) The load current

(c) The armature current

(d) The generated voltage

(e) The total copper loss

(f) The efficiency

(g) The voltage regulation (for this, assume that the flux, hence E_g, remains constant)*

Solution

The problem can be solved easily by drawing the diagram shown in Figure 4-17 and including all of the known quantities. For simplicity the shunt-field rheostat has been left out of the picture.

(a) The full-load terminal voltage is given by the rating (220 V). Since this is the voltage across the field winding,

$$I_f = \frac{V_t}{R_f} = \frac{220 \text{ V}}{140}$$

$$= 1.57 \text{ A}$$

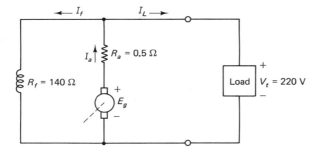

Figure 4-17 Diagram for Example 4-13.

*This assumption is a reasonable engineering approximation. A generator is normally operated in a saturated state. Therefore, the small change in V_t (hence I_f) due to the change from full load to no load would produce an insignificant change in the flux.

(b) The load current at rated conditions is obtained from the ratings as well.

$$\text{Current} = \frac{\text{power}}{\text{voltage}}$$

$$I_L = \frac{5.5 \text{ kW}}{220 \text{ V}} = 25 \text{ A}$$

(c) The armature current can be found by applying Kirchhoff's current law.

$$I_a = I_L + I_f$$

$$= 25 \text{ A} + 1.57 \text{ A} = 26.57 \text{ A}$$

(d) The generated voltage is found by applying Kirchhoff's voltage law.

$$E_g = I_a R_a + V_t$$

$$= 26.57 \text{ A} \times 0.5 \text{ } \Omega + 220 \text{ V}$$

$$= 13.3 + 220 = 233.3 \text{ V}$$

(e) In this case there are two copper losses to be calculated: in the armature and in the shunt field.

$$P_{\text{arm}} = R_a I_a^2 = 0.5 \text{ } \Omega \times (26.57 \text{ A})^2$$

$$P_{\text{arm}} \approx 353 \text{ W}$$

$$P_f = R_f I_f^2 = 140 \text{ } \Omega \times (1.57 \text{ A})^2$$

$$P_f \approx 345 \text{ W}$$

The total copper loss equals the sum of the two:

$$\text{total } P_{\text{Cu}} = 353 \text{ W} + 345 \text{ W} = 698 \text{ W}$$

(f) $P_i = P_o + P_{\text{Cu}} + \text{stray power loss}$

$$= 5.5 \text{ kW} + 698 \text{ W} + 95 \text{ W}$$

$$= 6293 \text{ W}$$

$$\text{efficiency} = \eta = \frac{P_o}{P_i} \times 100$$

$$= \frac{5500 \text{ W}}{6293 \text{ W}} \times 100$$

$$= 87.4\%$$

(g) Assuming that the flux remains constant, E_g will be the same value at no load as it is at full load. At no load Figure 4-17 reduces to the series circuit shown in Figure 4-18.

$$I_a = I_f = \frac{E_g}{R_a + R_f}$$

$$= \frac{233.3 \text{ V}}{0.5 \text{ } \Omega + 140 \text{ } \Omega}$$

$$= 1.66 \text{ A}$$

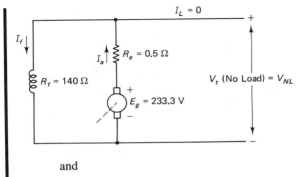

Figure 4-18 No-load condition for Example 4-13.

and

$$V_{NL} = I_f R_f = 1.66 \text{ A} \times 140 \text{ } \Omega$$

$$= 232.4 \text{ V}$$

$$\% \text{ V.R.} = \frac{V_{NL} - V_{FL}}{V_{FL}} \times 100$$

$$= \frac{232.4 - 220}{220} \times 100$$

$$= 5.64\%$$

4-7 SERIES GENERATOR

If the field winding is connected in series (rather than parallel) with the armature, we will have the **series generator** shown in Figure 4-19. Aside from its characteristics being different from the shunt generator, there is an important difference in construction. The shunt-field winding is made up of many turns of thin wire. In this way sufficient field intensity (H) is produced with a small current. In a series motor, since the armature current is large under load conditions, very few turns are needed. The wire, however, must be thick so that it can handle the large current.

Referring to Figure 4-19, we see that the three currents are identical

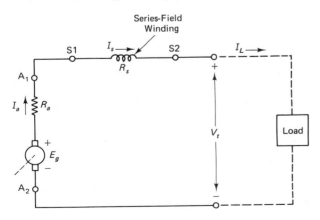

Figure 4-19 Schematic of a series generator.

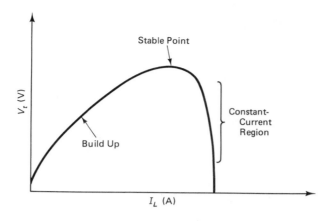

Figure 4-20 Load characteristic of a series generator.

$(I_a = I_s = I_L)$. Under no-load conditions, the current will be zero and V_t will equal E_g. The generated voltage will be small, due only to residual magnetism. If we connect the terminals to a load and start to draw current, the flux will begin to increase. The larger flux will increase E_g, causing more current to flow, increasing the field even further. This buildup continues until the core is saturated and a stable situation is reached. If the load is increased beyond this stable point, the voltage begins to drop rapidly since the flux can no longer increase (i.e., saturation). The load characteristic is shown in Figure 4-20.

In the event that the generator does not build up a voltage when an attempt is made to increase the load, the problem is most likely the field polarity. If the flux produced by the current flow in the winding opposes the residual magnetism, a buildup will not occur. As with the shunt generator, the field winding (S1 and S2 in Figure 4-19) should be reversed. Alternatively, the direction of rotation could be reversed if it were found to be incorrect. This would reverse the polarity of E_g, causing the current to flow in the opposite direction through the field winding. The field flux would now aid the residual flux and a buildup would occur.

As can be seen from the load characteristic, the variation of terminal voltage with load current is quite severe (poor voltage regulation). For this reason the series generator does not find use as a constant-voltage source. It does, however, find use as a voltage booster, where a dc voltage must be boosted in response to large currents. It has also found use as a constant-current generator. In the region indicated in Figure 4-20, the load current remains relatively constant with wide variations in load voltage.

As a means of adjusting the voltage, a rheostat, called a **series-field diverter,** is connected in parallel with the series winding. The control obtained, however, is minor compared with the variations produced by load changes.

4-8   COMPOUND GENERATOR

In an effort to improve the falling load characteristic of the shunt generator, a series winding was added to it, forming a **compound generator.** In this way the voltage boost capability of the series generator would counteract the falling characteristic of

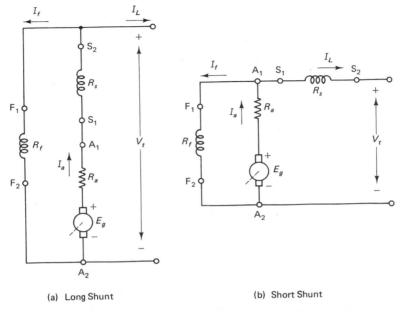

(a) Long Shunt   (b) Short Shunt

Figure 4-21 Compound generators: (a) long shunt; (b) short shunt.

the shunt generator with increasing load. There are two forms of the compound generator and they are shown in Figure 4-21. The only difference between the two is the current that flows through the series winding. In the **short shunt** it is I_L and in the **long shunt** it is I_a. Thus under no load conditions, the series field in the long shunt will be excited, whereas that in the short shunt will not. Their characteristics are essentially similar.

Under laboratory conditions the series-field winding is initially shorted out with a switch, a diverter set to zero ohms, or both. Under these conditions we essentially have a shunt generator. Once the proper polarity on the shunt field has been established and the rheostat has been adjusted, giving a buildup of voltage, the series-field winding is energized. At this point the correct polarity for the series field is not known. If the series field opposes the shunt field, the generator is said to be **differentially compounded.** If it aids the shunt field, the generator is **cumulatively compounded.**

4-8.1 Differentially Compounded Generator

When a generator is differentially compounded and current begins to flow through the series field, the flux (series plus shunt) will decrease since they are opposing. As a result, the generated voltage will drop. As more and more current is drawn, the voltage will drop sharply, due to the decreasing flux. The characteristic is shown in Figure 4-22. Upon comparison with Figure 4-20, we see that it resembles somewhat the series generator characteristic. The differentially compounded generator thus finds use in constant-current applications as well. It does not find practical use as a source of dc voltage.

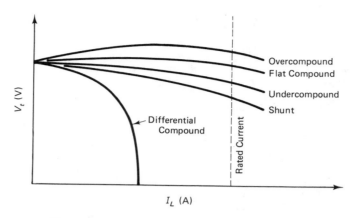

Figure 4-22 Comparison of generator characteristics.

4-8.2 Cumulatively Compounded Generator

When the series field has the proper polarity, the two fields will add up. The generator, now cumulatively compounded, will have a load characteristic which does not fall as sharply as the shunt generator. In fact, if the series field is strong enough, the characteristic may even rise. This condition, when the full-load voltage is greater than the no-load voltage, is called **overcompounding.** When the full-load voltage is about equal to the no-load voltage, we have **flat compounding.** Finally, when the full-load voltage is less than the no-load voltage, we have **undercompounding.** The three characteristics are shown in Figure 4-22.

The differing characteristics are arrived at by changing the magnetomotive force (MMF) of the series-field winding. Once a generator has been built, the number of turns in the winding cannot be changed. In practice, therefore, the different degrees of compounding are obtained by adjusting the current in the winding with a diverter, as shown in Figure 4-23. The shunt-field rheostat has been left out of the picture, even though it would most likely be used.

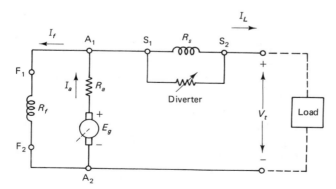

Figure 4-23 Short shunt generator with series-field diverter.

Example 4-14

A short shunt compound generator rated 3 kW, 200 V has stray power losses of 120 W at full load. If $R_f = 100\ \Omega$, $R_a = 0.9\ \Omega$, and $R_s = 0.2\ \Omega$ (a diverter is not used), find at full load:

(a) The load current $-\boxed{I_L} = \dfrac{P}{V} = \dfrac{3000\ W}{200\ V} = 15A$

(b) The shunt field current $-\boxed{V_f} =$

(c) The armature current

(d) The generated voltage

(e) The mechanical power converted to electrical power

(f) The copper losses

(g) The efficiency

Solution

(a) The load current at full load is calculated from the rating

$$I_L = \frac{3\ kW}{200\ V} = \frac{3000\ W}{200\ V}$$

$$= 15\ A$$

To simplify the solution of the remaining parts, Figure 4-24 is drawn.

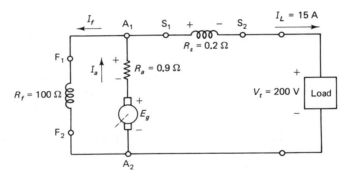

Figure 4-24 Circuit diagram for Example 4-14.

(b) To get I_f, we must first find the voltage across the shunt field (V_f). Using Kirchhoff's voltage law, it must equal 200 V plus the drop across the series field.

$$V_f = 200\ V + R_s(I_L)$$

$$= 200\ V + (0.2\ \Omega \times 15\ A)$$

$$= 200 + 3 = 203\ V$$

$$I_f = \frac{V_f}{R_f} = \frac{203\ V}{100\ \Omega} = 2.03\ A$$

(c)
$$I_a = I_L + I_f = 15 \text{ A} + 2.03 \text{ A}$$
$$= 17.03 \text{ A}$$

(d) Again using Kirchhoff's voltage law,
$$E_g = (17.03 \text{ A} \times 0.9 \text{ } \Omega) + (15 \text{ A} \times 0.2 \text{ } \Omega) + 200 \text{ V}$$
$$= 15.33 \text{ V} + 3 \text{ V} + 200 \text{ V}$$
$$= 218.33 \text{ V}$$

(e) Mechanical power converted to electrical power is given by $E_g I_a$:
$$E_g I_a = 218.33 \text{ V} \times 17.03 \text{ A}$$
$$= 3718.2 \text{ W}$$

(f) Shunt-field copper loss, P_f:
$$P_f = R_f I_f^2 = 100 \text{ } \Omega \times (2.03 \text{ A})^2 = 412 \text{ W}$$
Armature copper loss, P_a:
$$P_a = R_a I_a^2 = 0.9 \text{ } \Omega \times (17.03 \text{ A})^2 = 261 \text{ W}$$
Series-field copper loss, P_s:
$$P_s = R_s I_L^2 = 0.2 \text{ } \Omega \times (15)^2 = 45 \text{ W}$$
total copper losses $= P_f + P_a + P_s = 718 \text{ W}$

(g) $P_i = P_o +$ copper losses $+$ stray power loss
$$= 3 \text{ kW} + 718 \text{ W} + 120 \text{ W}$$
$$= 3000 \text{ W} + 718 \text{ W} + 120 \text{ W}$$
$$= 3838 \text{ W}$$
$$\eta = \frac{P_o}{P_i} \times 100 = \frac{3000}{3838} \times 100$$
$$= 78.2\%$$

It should be noted that if the stray power loss is subtracted from the input power we will be left with the mechanical power converted to electrical power. This is an alternative way to solve part (e)

$$3838 \text{ W} - 120 \text{ W} = 3718 \text{ W}$$

Example 4-15

A long shunt compound generator rated 5 kW, 125 V has an efficiency of 80% when supplying rated load. If $R_f = 125 \text{ } \Omega$, $R_a = 0.2 \text{ } \Omega$, and $R_s = 0.05 \text{ } \Omega$, find at rated load:

(a) The load current

(b) The field current

(c) The armature current

(d) The copper losses

(e) The stray power loss

Solution

(a) From the rating

$$I_L = \frac{5 \text{ kW}}{125 \text{ V}} = \frac{5000 \text{ W}}{125 \text{ V}}$$

$$= 40 \text{ A}$$

To simplify the solution of the remaining parts, Figure 4-25 is drawn.

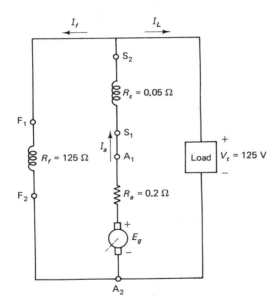

Figure 4-25 Circuit diagram for Example 4-15.

(b) The voltage across the shunt field in this case is V_t.

$$I_f = \frac{V_t}{R_f} = \frac{125 \text{ V}}{125 \text{ }\Omega} = 1 \text{ A}$$

(c)

$$I_a = I_L + I_f$$
$$= 40 \text{ A} + 1 \text{ A} = 41 \text{ A}$$

(d) Shunt-field loss, P_f:

$$P_f = V_t I_f = 125 \text{ V} \times 1 \text{ A} = 125 \text{ W}$$

Armature loss, P_a:

$$P_a = R_a I_a^2 = 0.2 \text{ }\Omega \times (41 \text{ A})^2 = 336.2 \text{ W}$$

Series-field loss, P_s:

$$P_s = R_s I_a^2 = 0.05 \ \Omega \times (41 \ \text{A})^2 = 84.05 \ \text{W}$$

total copper loss $= P_f + P_a + P_s = 545.25 \ \text{W}$

(e) To get the stray power loss, we must first find the input power. The efficiency that is given is used to do this.

$$\eta(\%) = \frac{P_o}{P_i} \times 100$$

$$80 = \frac{5 \ \text{kW}}{P_i} \times 100$$

Solving for P_i, we have

$$P_i = 5000 \times \frac{100}{80}$$

$$= 6250 \ \text{W}$$

Using Eq. 4-8, we can solve for the total losses:

$$\text{total losses} = P_i - P_o = 6250 \ \text{W} - 5000 \ \text{W}$$

$$= 1250 \ \text{W}$$

But

$$\text{total losses} = \text{copper losses} + \text{stray power loss}$$

Therefore,

$$\text{stray power loss} = 1250 \ \text{W} - 545.25 \ \text{W}$$

$$= 704.75 \ \text{W}$$

4-9 PARALLEL OPERATION

In many areas of industry it is desirable to have a **fail-operative** system. A "fail-op" system (as it is called) is one in which the system continues to operate normally even after a component has failed. In an effort to obtain this fail-op capability in dc power systems, generators with similar or identical characteristics are placed in parallel. In this way they can share load requirements. If one of the generators should fail, it could be switched out automatically and the remaining generators would supply all or most of the failed unit's load.

In addition to fail-op systems, generators are paralleled for the following reasons:

1. When routine maintenance must be performed on a particular unit, it can be disconnected. The other generators would pick up the slack and there would be no loss of power.

2. The most efficient operation of a generator occurs at or near its full-load rating. This efficient operation can be maintained as load requirements vary. At times of great power demand, many generators will be connected in parallel. As demand decreases, generators can be disconnected and those that remain would continue to operate efficiently at their rated output.

3. As demand increases over long periods of time, small units can be added at a minimum cost.

4. If a large amount of power is needed, a single large generator with that capability may not be readily available. Two or more smaller units may be available, and when connected in parallel could supply the large load requirement.

Some general rules must be obeyed when generators are connected in parallel. First, the generators should be of the same type: shunt or cumulative compound. Their voltage rating should be about the same to ensure that they all tend to contribute power. A **bus** is a conducting line along which all points are at the same voltage with respect to ground. The line is usually a solid rectangular bar called a **bus bar.** An **infinite bus** is one that can theoretically supply infinite current without its voltage dropping. In other words, it would have 0% voltage regulation. Therefore, a generator whose voltage is lower than the **bus voltage** will receive power rather than deliver it. It will be operating as a motor. When the no-load voltage, which is in effect the generated voltage, is greater than the bus voltage, generators will contribute power to the bus.

In addition, like polarities of the generators must all be connected together; that is, all positive terminals to the bus and all negative terminals to ground for a positive bus voltage.

4-9.1 Shunt Generators in Parallel

The parallel connection of shunt generators is a very stable one and easy to analyze, whether to a bus or a resistive load. In either case, as long as the terminal voltage of the generator being connected is equal to or slightly above that of the generator already supplying power, a stable situation is reached with the new unit picking up some of the load. If two generators with identical characteristics are connected in parallel, they will each supply half of the load. If their characteristics are different, they will each supply a different portion of the load. It is important to remember that when the stable situation is reached, the terminal voltages of all the generators will be equal. In addition, a generator whose terminal voltage is less than that of the existing system should never be connected. If it were, it would draw power, putting an additional load on the system and thus end up being driven as a motor.

Example 4-16

A shunt generator whose no load voltage is 230 V is connected to a 210-V bus. If its armature resistance is 0.5 Ω and the field resistance is 115 Ω, find the currents in the generator and the power it will deliver to the bus after it is connected.

Solution

The first step is to draw a picture of the system and calculate E_g. It is shown in Figure 4-26. The generated voltage is calculated before the switch is closed. Since $V_t = 230$ V,

$$I_f = \frac{V_t}{R_f} = \frac{230 \text{ V}}{115 \text{ }\Omega} = 2 \text{ A}$$

$$I_a = I_f = 2 \text{ A}$$

and

$$E_g = V_t + I_a R_a$$

$$E_g = 230 \text{ V} + 2 \text{ A} \times 0.5 \text{ }\Omega = 231 \text{ V}$$

After the switch is closed, the terminal voltage will equal the bus voltage.

$$I_f = \frac{V_t}{R_f} = \frac{210 \text{ V}}{115 \text{ }\Omega} = 1.83 \text{ A}$$

$$I_a = \frac{E_g - V_t}{R_a} = \frac{(231 - 210) \text{ V}}{0.5 \text{ }\Omega}$$

$$= \frac{21 \text{ V}}{0.5 \text{ }\Omega} = 42 \text{ A}$$

$$I_L = I_a - I_f = 42 \text{ A} - 1.83 \text{ A} = 40.17 \text{ A}$$

$$P = V_t I_L = 210 \text{ V} \times 40.17 \text{ A}$$

$$= 8435.7 \text{ W}$$

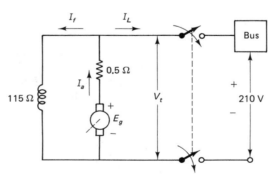

Figure 4-26 Diagram for Example 4-16.

Example 4-17

A shunt generator (number 1) is supplying power to a variable load. When the load increases to 10 A at 120 V, a second shunt generator is connected in parallel to pick up some of the load and prevent overheating. Find the load supplied by both generators after the second one is connected. Table 4-7 contains information about the two generators.

Table 4-7

Generator 1	Generator 2
$R_f = 100 \ \Omega$	$R_f = 100 \ \Omega$
$R_a = 0.98 \ \Omega$	$R_a = 1 \ \Omega$
$E_g = 131 \ \text{V}$	$E_g = 128 \ \text{V}$

Solution

Since the generators are connected to a load rather than a bus, the terminal voltage will change. To solve the problem, the resistive load at the time the generators were connected must be calculated. From the information given, we know the voltage across the load and the current through it. Therefore, referring to Figure 4-27, we have

$$R_L = \frac{V_t}{I_L} = \frac{120 \ \text{V}}{10 \ \text{A}} = 12 \ \Omega$$

Now that the resistance is known, an equivalent circuit with all known quantities inserted can be drawn. It is shown in Figure 4-28. Upon examination of the circuit, we see that it is nothing more than a two-source series–parallel circuit. There are several techniques that can be used to solve for V_t and I_L. The one used here will be to convert the voltage

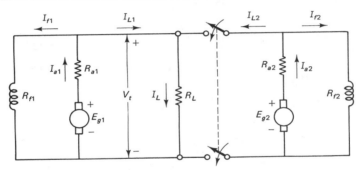

Figure 4-27 Diagram for Example 4-17.

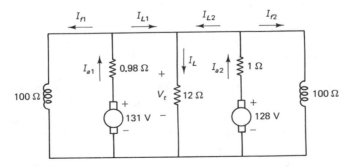

Figure 4-28 Equivalent circuit for Figure 4-27.

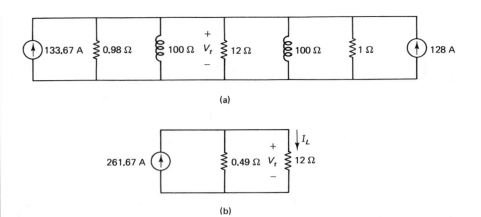

(a)

(b)

Figure 4-29 (a) Converted circuit; (b) simplification of converted circuit.

generators with series resistors to current generators with parallel resistors using Norton's theorem. The converted circuit is shown in Figure 4-29a with its simplification in Figure 4-29b. The circuit was simplified by adding the current generators and combining all resistors in parallel with the exception of the load. The load current is found using current division.

$$I_L = \frac{261.67 \text{ A} \times 0.49 \text{ }\Omega}{12.49 \text{ }\Omega} = 10.266 \text{ A}$$

$$V_t = R_L I_L = 12 \text{ }\Omega \times 10.266 \text{ A} = 123.2 \text{ V}$$

Going back to Figure 4-28, we can now calculate the remaining currents.

$$I_{f1} = I_{f2} = \frac{V_t}{100 \text{ }\Omega} = 1.232 \text{ A}$$

$$I_{a1} = \frac{131 \text{ V} - V_t}{0.98 \text{ }\Omega} = 7.96 \text{ A}$$

$$I_{L1} = I_{a1} - I_{f1} = 7.96 \text{ A} - 1.232 \text{ A} = 6.73 \text{ A}$$

$$I_{a2} = \frac{128 \text{ V} - V_t}{1 \text{ }\Omega} = 4.8 \text{ A}$$

$$I_{L2} = I_{a2} - I_{f2} = 4.8 \text{ A} - 1.232 \text{ A} = 3.57 \text{ A}$$

4-9.2 Compound Generators in Parallel

As long as the compound generators have a falling characteristic like the shunt generator, the connection is a stable one. The generators will share the load, preventing any one unit from overloading. However, if their characteristics are rising (overcompound), a very unstable situation arises. Suppose that two generators are supplying power to a load and a disturbance occurs, causing the generated voltage of unit 1 to increase slightly. As a result of the increase in its generated voltage, unit 1 will tend to supply more load. The increase in load causes an increase in series-field flux, which in turn increases the generated voltage even more and the cycle will repeat itself. While unit 1 is taking on more and more load, unit 2 has less of a demand on it and its load will tend to decrease. As unit 2 loses load, its series-field flux will decrease, causing its generated voltage to decrease as well. The decrease in generated voltage will cause unit 2 to supply even less current, which decreases the flux even further. Once these two cycles have started, they continue until unit 1 has assumed all of the load and unit 2 supplies none. In extreme cases where overload protection (a circuit breaker) is not provided, unit 2 may start to operate as a motor and draw current or before this happens unit 1 may saturate its core, causing the cycle to halt.

To stabilize the foregoing condition, provision must be made so that an increase in either unit's generated voltage will cause an equal increase in the series flux of both units. This is accomplished by shorting the armatures of the generators with an **equalizer,** as shown in Figure 4-30. The equalizer is nothing more than a large cable having almost zero resistance.

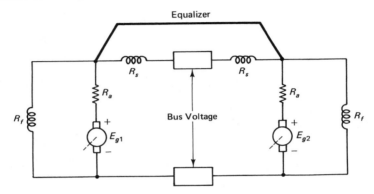

Figure 4-30   Compound generators connected to a bus.

Since the armatures are at the same voltage at all times, the series fields will always have the same voltage across each other. Under normal conditions the current in the equalizer is zero. If a disturbance now occurs, causing E_{g1} to increase, its armature voltage will increase. Since the armature voltage of unit 1 is bigger than that of unit 2, current will now flow through the equalizer into unit 2's series field. This will cause E_{g2} to increase until the two armature voltages are equal.

Example 4-18

Two overcompound short shunt generators are connected to a 210-V bus as shown in Figure 4-30. Their characteristics are slightly different; however, due to the equalizer the armature voltages are equal. Table 4-8 contains data on the two units. Find the load they each supply.

Table 4-8

Unit 1	Unit 2
$R_f = 100\ \Omega$	$R_f = 100\ \Omega$
$R_a = 1\ \Omega$	$R_a = 1\ \Omega$
$R_s = 0.5\ \Omega$	$R_s = 0.4\ \Omega$
$E_g = 234\ V$	$E_g = 234\ V$

Solution

Figure 4-30 is redrawn inserting all the known and unknown quantities. It is shown in Figure 4-31. The following equations can be written for generator 1.

$$I_{f1} = \frac{V_a}{100 \ \Omega} \tag{4-10}$$

$$I_{L1} = \frac{V_a - 210 \ \text{V}}{0.5 \ \Omega} \tag{4-11}$$

$$I_{a1} = \frac{234 \ \text{V} - V_a}{1 \ \Omega} \tag{4-12}$$

Since the equalizer current is zero, we can write

$$I_{a1} = I_{L1} + I_{f1} \tag{4-13}$$

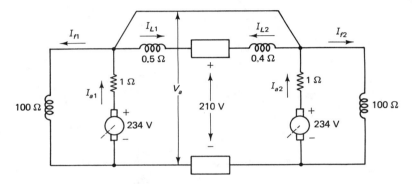

Figure 4-31 Circuit diagram for Example 4-18.

Substituting Eqs. 4-10, 4-11, and 4-12 into Eq. 4-13, we get

$$\frac{234 - V_a}{1} = \frac{V_a - 210}{0.5} + \frac{V_a}{100}$$

Solving for V_a yields

$$234 - V_a = 2V_a - 420 + \frac{V_a}{100}$$

$$654 = 3V_a + \frac{V_a}{100}$$

$$65,400 = 300V_a + V_a$$

$$65,400 = 301V_a$$

and

$$V_a = 217.28 \ \text{V}$$

We can now solve for the respective loads.

$$I_{L1} = \frac{V_a - 210 \text{ V}}{0.5}$$

$$= \frac{217.28 - 210}{0.5}$$

$$= 14.56 \text{ A}$$

and

$$I_{L2} = \frac{V_a - 210 \text{ V}}{0.4}$$

$$= \frac{217.28 - 210}{0.4}$$

$$= 18.2 \text{ A}$$

Although in practice identical units may be chosen, there will always be small differences present which will cause the imbalance in load distribution seen in Example 4-18. To eliminate the imbalance, series diverters are placed in parallel with the series-field windings. By adjusting (fine tuning) the diverters, the units can be balanced so that they both supply the same load.

SYMBOLS INTRODUCED IN CHAPTER 4

Symbol	Definition	Units: English and SI
E_g	Generated voltage	volts
V_t	Terminal voltage	volts
V_{NL}	Terminal voltage at no load	volts
V_{FL}	Terminal voltage at full (rated) load	volts
V_a	Armature voltage	volts
I_f	Shunt-field current	amperes
I_a	Armature current	amperes
I_s	Series-field current	amperes
I_L	Load current	amperes
R_f	Shunt-field resistance	ohms
R_a	Armature resistance	ohms
R_s	Series-field resistance	ohms
P_f	Shunt-field power loss	watts
P_{arm}	Armature power loss	watts
P_{Cu}	Total copper losses	watts
P_s	Series-field power loss	watts
η	Efficiency	percent

Symbol	Definition	Units: English and SI
V.R.	Voltage regulation	percent
P_i	Generator input power	hp or watts
P_o	Generator output power	watts

QUESTIONS

1. Why should a graph of generated voltage versus speed be a straight line for a separately excited generator?

2. Define the following terms: prime mover; self-excited; separately excited; no load; full load; voltage regulation; load characteristic.

3. Why can efficiency never be more than 100%?

4. Define the following terms: stray power; viscous friction; windage; hysteresis; eddy current; core loss; copper loss; stray load loss.

5. What does the term "critical resistance" mean when referring to a self-excited generator?

6. Give three reasons why a generator will fail to build up a voltage.

7. Define the following terms: series-field diverter; compound generator; short shunt; long shunt; differential compounding; cumulative compounding.

8. If a compound generator has negative voltage regulation, how is it compounded (i.e., over-, under-, or flat)?

9. Define the following terms: fail operative; bus; infinite bus; equalizer.

10. Why are generators paralleled?

11. Why can overcompounded generators never be paralleled?

12. Using Figures 4-11 and 4-13 as an aid, derive the fact that the mechanical power converted to electrical power is equal to $E_g I_a$.

PROBLEMS

(English)

1. A generator develops 135 V when rotating at 1650 rev/min. If the flux is held constant, what will be the generated voltage at each of the following speeds?
(a) 1450 rev/min
(b) 1000 rev/min
(c) 1800 rev/min
(d) 2000 rev/min

2. A generator turning at 1800 rev/min develops 128 V. As a result of a short in the field winding the flux drops by 30%. This causes a drop in voltage. How fast should it turn to restore the voltage to its original value?

3. A generator turning at rated speed develops rated voltage. If the speed drops by 10%, the voltage will also drop by 10%. What percent increase in flux will restore the voltage to its rated value?

4. The data in Table 4-9 was obtained in the laboratory for a generator under test. The flux was kept constant.
(a) Plot the voltage versus speed.
(b) Calculate the slope of the graph.

Table 4-9

E_g (V)	315	375	450	525	600	675
S (rev/min)	900	1100	1400	1600	1800	2000

5. While the speed of a generator was held constant at 1200 rev/min, its field current was increased, giving a buildup of voltage. The data is given in Table 4-10.
(a) Plot the magnetization curve at 1200 rev/min.
(b) Calculate and plot the magnetization curve for a speed of 900 rev/min.

Table 4-10

I_f (A)	0	0.5	1.0	1.5	2.0	2.5
E_g (V)	20	160	350	440	460	510

(SI)

6. A generator develops 248 V when rotating at 170 rad/s. If the flux is held constant, what will be the generated voltage at each of the following speeds?
(a) 140 rad/s
(b) 100 rad/s
(c) 185 rad/s
(d) 200 rad/s

7. A generator turning at 200 rad/s develops 252 V. As a result of a short in the field winding, the flux drops by 30%. This causes a drop in voltage. How fast should it turn to restore the voltage to its original value?

8. A generator turning at rated speed develops rated voltage. If the speed drops by 12%, the voltage will also drop by 12%. What percent increase in flux will restore the voltage to its rated value?

9. The data in Table 4-11 was obtained in the laboratory for a generator under test. The flux was kept constant.
(a) Plot the voltage versus speed.
(b) Calculate the slope of the graph.

Table 4-11

E_g (V)	120	180	210	240	290
ω (rad/s)	80	110	130	150	180

10. While the speed of a generator was held constant at 40π rad/s, its field current was increased, giving a buildup of voltage. The data is given in Table 4-12.
(a) Plot the magnetization curve at 40π rad/s.
(b) Calculate and plot the magnetization curve for a speed of 30π rad/s.

Table 4-12

I_f (A)	0	0.4	0.7	1.0	1.3	1.9
E_g (V)	10	72	140	195	225	250

(English and SI)

11. A 220-V 10-kW generator has 255 V measured at its terminals under a no-load condition. What is the percent voltage regulation?

12. A 1-kW 120-V generator has a full-load regulation of 18%. What voltage should be measured at its terminals at no load?

13. For the generator characteristics shown in Figure 4-32, calculate the percent voltage regulation. For each one, the full-load current is 12 A.

14. A 5-kW 230-V generator requires 9 hp when supplying rated load. What is its full-load efficiency?

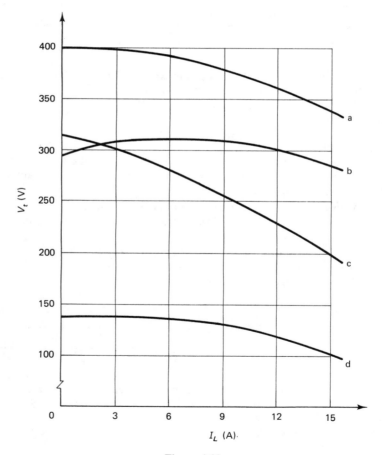

Figure 4-32

15. A 1-kW 120-V generator has constant stray power losses of 75 W. At full load the copper losses are 175 W. Find:

(a) Its full-load input in watts and horsepower

(b) Its full-load efficiency

16. A 10-kW 440-V generator has 80% efficiency at full load. Its rated copper losses are calculated to be 1550 W. What are its stray power losses?

> **Problems 17 to 21 refer to Figure 4-33, which shows magnetization curves for four different shunt generators.**

17. What value of field resistance will allow the generators to build up to the following voltages?

(a) Generator 1, 400 V

(b) Generator 2, 330 V

(c) Generator 3, 230 V

(d) Generator 4, 120 V

18. If the field resistance for generator 1 is 200 Ω, to what voltage will it build up?

19. If the field resistance for generator 2 is 140 Ω, to what voltage will it build up?

20. If the field resistance for generator 3 is 92 Ω, to what voltage will it build up?

21. What value of field resistance will allow generator 4 to build up to 160 V?

22. A 5-kW 230-V shunt generator has a field resistance of 130 Ω, an armature resistance of 0.8 Ω, and stray power losses of 150 W at rated load. Find at full load:

(a) The field current

(b) The load current

(c) The armature current

(d) The generated voltage

(e) The shunt-field loss

(f) The armature loss

(g) The efficiency

23. A 15-kW 230-V shunt generator has a field resistance of 115 Ω, an armature resistance of 1.0 Ω, and stray power losses of 1300 W. Find the rated:

(a) Field current

(b) Load current

(c) Armature current

(d) Generated voltage

(e) Shunt-field loss

(f) Armature loss

(g) Efficiency

24. A 6-kW 120-V shunt generator has a field resistance of 100 Ω, an armature resistance of 1.2 Ω, and stray power losses of 220 W. Find the rated:

(a) Field current

(b) Load current

(c) Armature current

(d) Generated voltage

(e) Shunt-field loss

(f) Armature loss

(g) Efficiency

(h) Voltage regulation (assume that the generated voltage remains constant)

25. A 20-kW 230-V shunt generator generates 250 V at rated load while drawing a field current of 2.8 A. What is its armature resistance?

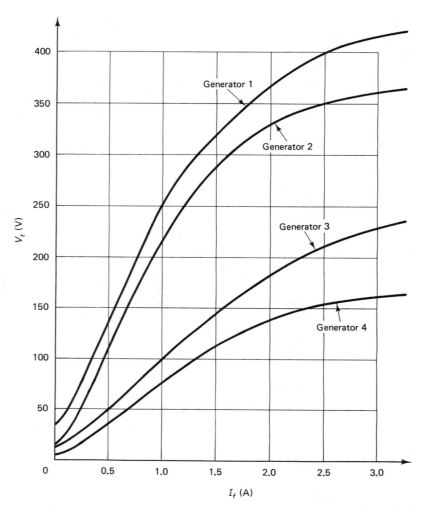

Figure 4-33

26. A 10-kW 440-V shunt generator has a rated efficiency of 82%. If its field resistance is 200 Ω and the armature resistance is 0.9 Ω, find the stray power loss.

27. What value load resistance should be connected to a 1-kW 100-V series generator so that it supplies rated output?

28. A 2-kW 230-V series generator has a field resistance of 0.1 Ω, an armature resistance of 0.6 Ω, and stray power losses of 110 W. Find the rated:
(a) Field current
(b) Load current
(c) Armature current
(d) Series-field loss
(e) Armature loss

(f) Generated voltage

(g) Efficiency

29. A 4-kW 200-V short shunt compound generator has constant stray power losses of 150 W. If $R_f = 110\ \Omega$, $R_a = 1.2\ \Omega$, and $R_s = 0.1\ \Omega$, find at full load:

(a) The load current

(b) The shunt-field current

(c) The armature current

(d) The generated voltage

(e) The shunt-field loss

(f) The series-field loss

(g) The armature loss

(h) The efficiency

30. Repeat Problem 29 at half-rated load. Assume that the generated voltage is the same as that calculated in part (d).

31. A 4.8-kW 120-V long shunt compound generator has $R_f = 90\ \Omega$, $R_a = 0.4\ \Omega$, $R_s = 0.08\ \Omega$, and constant stray power losses of 200 W. Find at full load:

(a) The load current

(b) The shunt-field current

(c) The armature current

(d) The generated voltage

(e) The shunt-field loss

(f) The series-field loss

(g) The armature loss

(h) The efficiency

32. In Problem 31, how much mechanical power is converted to electrical power?

33. A 100-kW 600-V short shunt compound generator has $R_f = 300\ \Omega$, $R_a = 0.1\ \Omega$, $R_s = 0.05\ \Omega$, and constant stray power losses of 2400 W. Find at full load:

(a) The load current

(b) The shunt-field current

(c) The armature current

(d) The generated voltage

(e) The shunt-field loss

(f) The series-field loss

(g) The armature loss

(h) The efficiency

34. In Problem 33, find the percent voltage regulation. Assume that the generated voltage remains constant from no load to full load at the value calculated in part (d).

35. Repeat Problem 33, but this time the generator is wired long shunt.

36. A long shunt compound generator is rated 12 kW, 120 V. When supplying rated load, its efficiency is 75%. If $R_f = 100\ \Omega$, $R_a = 0.16\ \Omega$, $R_s = 0.04\ \Omega$, find at rated load:

(a) The load current

(b) The field current

(c) The armature current

(d) The copper loss

(e) The stray power loss

37. Repeat Problem 36, but this time the generator is wired short shunt.

38. A shunt generator whose no-load voltage is 140 V is connected to a 120-V bus. If

$R_a = 0.8 \ \Omega$ and $R_f = 60 \ \Omega$, find:

(a) Its generated voltage

(b) Its field current

(c) Its armature current

(d) Its load current

(e) Its power output

39. A shunt generator whose no-load voltage is 480 V is connected to a 440-V bus. If $R_a = 0.6 \ \Omega$ and $R_f = 165 \ \Omega$, find the power it will supply to the bus.

40. A 230-V 4.6-kW shunt generator (unit 1) is supplying power to a load. As soon as its output reaches full load, a second shunt generator (unit 2) with the same rating is connected in parallel to help share the load. Find the load current supplied by each unit after they are connected. Table 4-13 contains information about the two generators.

Table 4-13

	Unit 1	Unit 2
R_f (Ω)	120	118
R_a (Ω)	1.15	1.1
E_g (V)	248	252

41. Two overcompound short shunt generators are connected to a 330-V bus as shown in Figure 4-30. An equalizer is used to connect the two armatures. Table 4-14 contains information about the two generators. Find the load they each supply to the bus.

Table 4-14

	Unit 1	Unit 2
R_f (Ω)	110	110
R_a (Ω)	1.0	1.0
R_s (Ω)	0.12	0.15
E_g (V)	350	350

Chapter 5

DC Motor Characteristics

In analyzing the dc motor we will use the same basic techniques used with the generator. Some simple equations and an equivalent circuit will be introduced. However, with the motor there are some basic differences. The power flow will be opposite to that of the generator; that is, the input power to the motor is electrical and the output power is mechanical. Since the ratings are in terms of useful output, motors are rated in horsepower, a mechanical unit, rather than watts (kilowatts are used in SI). In addition, the direction of current will be different since the electrical power flow is into the machine rather than out as it was with the generator.

5-1 BASIC MOTOR EQUATION

In Chapter 2 some basic concepts on the generation of torque were introduced. Equations 2-9, which solved for the maximum torque developed by a rotating conductor, were derived. These equations could be further modified by averaging the torque over a full revolution. In addition, they could be simplified by grouping all the constant parameters (length and number of conductors, radius of armature, area of core, etc.) into one constant. If this were done, they would appear as Eq. 5-1, expressing the average torque developed by a rotating armature.

$$T = K\phi I_a \tag{5-1}$$

In Eq. 5-1, K is the constant dependent on the physical parameters of the motor. Naturally, its value would be different depending on the system of units used (English or SI). However, since this equation will be used qualitatively rather than quantitatively, its numerical value is not important here. The same is true for the flux (ϕ), which would be in kilolines or webers, depending on the system of units. As before, I_a represents the armature current.

Equation 5-1 is extremely important in understanding the behavior of a motor and will be referred to frequently in this chapter. It simply says that the torque developed by a motor is directly proportional to the armature current and the flux in the core. Another important relationship from rotational mechanics is expressed by

$$\alpha = \frac{T}{J} \tag{5-2}$$

This equation states that the angular acceleration (α) of a rotating body (such as an armature) is directly proportional to the applied torque (T) and inversely proportional to the total moment of inertia (J) of the rotating body. Here again, since Eq. 5-2 will be used qualitatively, the units are not important and have been left out to avoid confusion.

5-1.1 Relationship between Torque and Power

When dealing with motors the output quantity specified can be either torque or power, depending on the particular application. Often, knowing one, we must be able to calculate the other. Although the two terms are sometimes used interchangably, they are very different. Power is a measure of the time it takes to do a specific amount of work. Torque, on the other hand, is a measure of how rapidly a rotating body can be turned, as shown by Eq. 5-2. The two are related by the speed at which a rotating body is turning. The exact relationship is given by

(English)

$$T = \frac{7.04\,P}{S} \tag{5-3a}$$

(SI)

$$T = \frac{1000\,P}{\omega} \tag{5-3b}$$

The units in Table 5-1 should be used with Eqs. 5-3. It should be noted that if P is the output power, T will be the output torque. Similarly, if P is the developed power, T will be the developed torque. It will be seen in Section 5-5 that the two quantities (output power and developed power) are different.

Table 5-1

English (Eq. 5-3a)		SI (Eq. 5-3b)	
P	watts	P	kilowatts
S	rev/min	ω	rad/s
T	ft-lb	T	N-m

Example 5-1 (English)

What must be the power rating in hp of a motor which at full load supplies 50 ft-lb of torque at

(a) 1800 rev/min

(b) 600 rev/min

Solution

Equation 5-3a is used; however, it is first rearranged to solve for P.

$$P = T \times \frac{S}{7.04}$$

(a) $P = 50 \text{ ft-lb} \times \dfrac{1800 \text{ rev/min}}{7.04}$

$= 12{,}784 \text{ W}$

$P = \dfrac{12{,}784 \text{ W}}{746 \text{ W/hp}}$

$= 17.14 \text{ hp}$

(b) $P = 50 \text{ ft-lb} \times \dfrac{600 \text{ rev/min}}{7.04}$

$= 4261.4 \text{ W}$

$P = \dfrac{4261.4 \text{ W}}{746 \text{ W/hp}}$

$= 5.71 \text{ hp}$

Example 5-2 (SI)

What must be the power rating in kW of a motor which at full load supplies 70 N-m of torque at

(a) 210 rad/s

(b) 70 rad/s

Solution

Equation 5-3b is used; however, it is first rearranged to solve for P.

$$P = T \times \frac{\omega}{1000}$$

(a) $P = 70 \text{ N-m} \times \dfrac{210 \text{ rad/s}}{1000}$

$= 14.7 \text{ kW}$

(b) $\quad P = 70 \text{ N-m} \times \dfrac{70 \text{ rad/s}}{1000}$

$\qquad\quad = 4.9 \text{ kW}$

5-1.2 Measurement of Torque

Whenever a motor is tested, of primary concern is its power output. This is a difficult quantity to measure. However, if the torque output at a given speed is known, the power output can be calculated using Eqs. 5-3. It is therefore desirable at times to measure the output torque of a motor. The simplest and most common way to measure torque is with a **prony brake.** A typical one is shown in Figure 5-1. A motor is positioned behind the brake and its shaft is connected to the drum. For the connection shown, the motor must rotate clockwise. As the motor turns the drum, it is loaded down with a friction belt. The knob is turned tightening the belt. To keep turning, the motor must put out a torque to overcome the opposing torque due to friction between the belt and drum. This torque ($f \times r$) is transmitted to the arm by the cable. The arm is kept from rotating by the force (F) applied by the scale. Since the arm is in equilibrium, the moment due to the cable ($f \times r$) must equal that due to the scale ($F \times d$). But $f \times r$ is equal to the output torque of the motor; therefore, $F \times d$ is also equal to the output torque. The quantity d is a fixed distance on the prony brake and F is measured at any particular load setting by the scale. Thus, by knowing the speed, scale force, and distance d, the output torque, hence power, can be calculated.

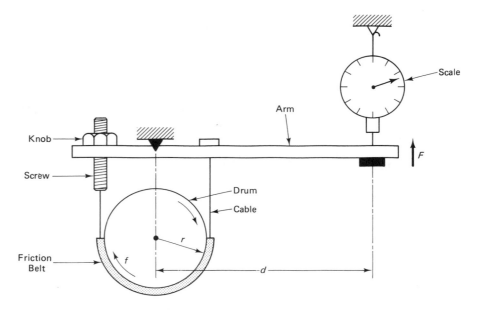

Figure 5-1 Typical prony brake arrangement.

Example 5-3 (English)

The prony brake arrangement shown in Figure 5-1 is used to measure the output power of a motor. At a particular load setting, the speed is 1200 rev/min, the force F is 16 lb, and d is 0.5 ft. Find the output power in hp.

Solution

The output torque is first calculated.

$$T_o = F \times d = 16 \text{ lb} \times 0.5 \text{ ft}$$

$$= 8 \text{ ft-lb}$$

Now using Eq. 5-3a, we have

$$P_o = T_o \times \frac{S}{7.04}$$

$$= 8 \text{ ft-lb} \times \frac{1200 \text{ rev/min}}{7.04}$$

$$= 1363.64 \text{ W}$$

$$= \frac{1363.64 \text{ W}}{746 \text{ W/hp}}$$

$$= 1.83 \text{ hp}$$

Example 5-4 (SI)

The prony brake arrangement shown in Figure 5-1 is used to measure the output power of a motor. At a particular load setting, the speed is 100 rad/s, the force F is 50 N and d is 0.2 m. Find the output power in kW.

Solution

The output torque is first calculated.

$$T_o = F \times d = 50 \text{ N} \times 0.2 \text{ m}$$

$$= 10 \text{ N-m}$$

Now using Eq. 5-3b, we have

$$P_o = T_o \times \frac{\omega}{1000}$$

$$= 10 \text{ N-m} \times \frac{100 \text{ rad/s}}{1000}$$

$$= 1 \text{ kW}$$

Another technique used to measure the power output of a motor makes use of a **cradle dynamometer.** This cumbersome method has the shaft of the motor under test connected to a calibrated generator (see Figure 4-12). As the generator is loaded down electrically, it draws power from the motor. In other words, the output power of the motor is equal to the input power of the generator. Since the generator is calibrated, its losses at any load setting are accurately known. The generator output being electrical in nature is easily measured. Therefore, by adding the two (see Eq. 4-8), the generator input power (hence motor output power) can be obtained.

5-2   *BACK ELECTROMOTIVE FORCE (BACK EMF)*

The theory we have learned so far tells us that if a voltage is applied to an armature, current will flow. Furthermore, if the armature is in a magnetic field, a torque will be developed. This is more or less a verbal description of Eq. 5-1. In addition, Eq. 5-2 tells us that if a torque is applied to something that can rotate, it will begin to accelerate and build up speed. It appears that we have satisfactorily described the motor. That is, a voltage applied to the motor terminals causes current flow. The current produces a torque, and the torque in turn accelerates the motor shaft. There is, however, a very important concept that has been left out of this description. The need for further explanation arises when the question is asked: "Does the motor accelerate forever as long as the voltage is applied?" If the answer is yes, further explanation is not needed. This conclusion is absurd, however, since it implies that infinite speed can be obtained with the smallest amount of voltage. How, then, can we explain how the motor accelerates up to a certain speed and then continues to run at that speed without any further acceleration? We can explain this behavior by reasoning in reverse. If the speed is constant, the acceleration must be zero, which says (Eq. 5-2) that the torque must be zero. If we assume for the moment that the flux is present, we must conclude (Eq. 5-1) that the current in the armature has gone to zero.* Since the voltage is still applied to the motor, we must assume that another voltage equal and opposite to the applied voltage has entered the picture. This opposing voltage (EMF) is called the **counter** or **back electromotive force.** It is due to the generator action taking place in the armature. Our theory on generators tells us that if conductors are rotating in a magnetic field, a voltage will be generated. As they cut the flux faster and faster, the generated voltage will continue to increase. Furthermore, application of Fleming's right-hand rule would tell us that this voltage opposes the current flow that made the armature turn to begin with.

Theoretically, we have answered our question. A voltage applied to a motor causes armature current, which in turn produces torque. The torque accelerates the motor. As the motor builds up speed, the back EMF increases until it is equal and opposite to the applied voltage. At this point the current (hence torque) has gone to zero, the acceleration is zero, and the motor runs at a constant speed. In reality, the current does not go to zero but rather some small amount producing just enough torque necessary to overcome the motor's internal friction.

*Ideally, this would be the case; however, we shall see in the following pages of this chapter that a current does indeed flow even though the acceleration has gone to zero.

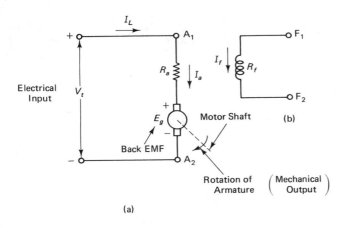

(a)

(b)

Figure 5-2 General circuit representation of a dc motor: (a) armature circuit; (b) field circuit.

5-3   EQUIVALENT CIRCUIT OF A DC MOTOR

With an understanding of the back EMF concept and the motor's behavior, we can construct a general equivalent circuit for the motor. It is shown in Figure 5-2. If we carefully examine this circuit with the generator's equivalent circuit (Figure 4-4), we see that they are identical with two exceptions. First, in order to be consistent with the power flow (i.e., left to right), the terminal voltage, which for the motor is the input, is on the left side of the diagram. The shaft, which now represents the mechanical power output, is shown on the right side. Second, and most important, is the fact that the current is flowing into the armature and against the generated voltage. This is an indication that V_t is greater than E_g. For the motor I_L will be called the **line current**. The quantities R_f, R_a, I_f, I_a, and E_g will have the same meaning as with the generator (see Section 4-2); however, for the motor E_g will be called the **back EMF**.

5-3.1 Behavior of the Motor under Load

Let us assume that the motor in Figure 5-2 has nothing connected to its shaft (no-load condition) and has built up to its steady-state speed. At this point E_g will be slightly less than V_t and the armature current will be

$$I_a = \frac{V_t - E_g}{R_a} \tag{5-4}$$

producing just enough torque (Eq. 5-1) to overcome the internal friction of the motor. In addition, we will assume that the field flux will remain constant.* If a load is now placed on the motor, it will slow down. Since the back EMF is proportional to speed (Eqs. 4-1), E_g will decrease. Equation 5-4 tells us that as E_g decreases, the armature current (I_a) will increase. As the current increases, so does the torque. This process continues until the motor reaches a speed such that the increase in torque is equal to

---

*This assumption is accurate for permanent-magnet and shunt motors.

the load placed on the motor's shaft. If the load is now removed, the motor, which is now developing too much torque, will accelerate until it reaches the original (no-load) speed.

5-3.2 Torque–Speed Curve

In order to obtain a graphical display of the preceding discussion a plot of load torque versus motor speed can be drawn. First we must derive a new equation. If the flux is constant, Eq. 5-1 can be written as Eq. 5-5.

$$T = K'I_a \tag{5-5}$$

where

$$K' = K\phi$$

When the motor is running at constant speed (zero acceleration), the developed torque will equal the load torque (T_L). Therefore, Eq. 5-5 can be rewritten as

$$T_L = K'I_a \tag{5-6}$$

Substituting Eq. 5-4 into Eq. 5-6, we get

$$T_L = \frac{K'(V_t - E_g)}{R_a} \tag{5-7}$$

Now Eq. 4-1a (or 4-1b) is substituted for E_g, remembering that the flux is constant. Upon rearranging, we get

$$T_L = \frac{K'V_t}{R_a} - \frac{K'K''S}{R_a}$$

If we simplify this equation by forming two new constants, we obtain Eqs. 5-8, which is what we are looking for.

(English)

$$T_L = K_mV_t - K_BS \tag{5-8a}$$

(SI)

$$T_L = K_mV_t = K_B\omega \tag{5-8b}$$

Naturally, all quantities in Eq. 5-8a would have different values and units from the corresponding ones in Eq. 5-8b with the exception of V_t. For a given value of V_t, Eq. 5-8 can be plotted. This is called the **torque–speed curve** and is shown in Figure 5-3. Upon examining the torque–speed curve, we can note the following. When the load torque is zero, the motor will run at its no load speed. As a load is applied and it increases to a value T_{L1}, the speed decreases to a speed S_1 (ω_1 in SI). As the load increases further, the speed continues to decrease until the motor finally stalls. The minimum load torque that will stall the motor is called the **stall torque.**

Thus we can see that the torque–speed curve is a powerful tool. With it, we can predict a motor's speed for any load.

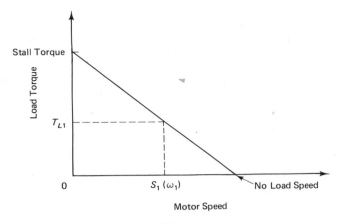

Figure 5-3 Typical motor torque–speed curve.

Example 5-5 (English)

A permanent-magnet dc motor has a no-load speed of 1200 rev/min. Its stall torque is 220 ft-lb. At what speed will it run if a 50-ft-lb load is applied? What is its output power under this load condition?

Solution

To find the speed under load, construct a torque–speed curve. This is done by locating the stall torque on the ordinate and the no-load speed on the abscissa, then connecting the two points with a straight line. It is shown in Figure 5-4. Next, at the load torque of

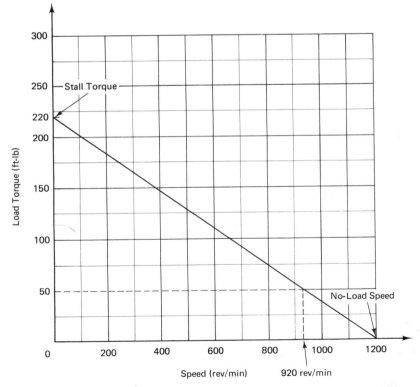

Figure 5-4 Torque–speed curve for Example 5-5.

129

50 ft-lb a horizontal line is drawn until it touches the torque–speed curve. Then a vertical line is drawn down to the speed axis and the load speed is found (920 rev/min). To find the output power under this load condition, use Eq. 5-3a.

$$P_o = \frac{TS}{7.04}$$

$$= \frac{50 \text{ ft-lb} \times 920 \text{ rev/min}}{7.04}$$

$$= 6534 \ W$$

or

$$P_o = \frac{6534 \ W}{746 \ W/hp} = 8.76 \text{ hp}$$

Example 5-6 (SI)

A permanent-magnet dc motor has a no-load speed of 120 rad/s. Its stall torque is

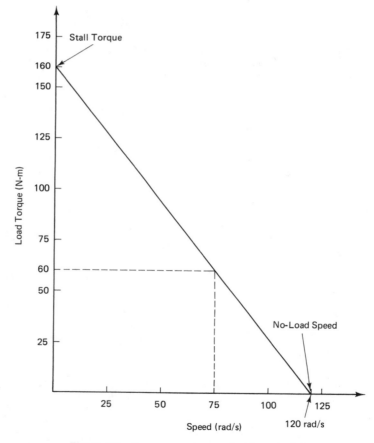

Figure 5-5   Torque–speed curve for Example 5-6.

160 N-m. At what speed will it run if a 60-N-m load is applied? What is its output power under this load condition?

Solution

To find the speed under load, construct a torque–speed curve. This is done by locating the stall torque on the ordinate and the no-load speed on the absicssa, then connecting the two points with a straight line. It is shown in Figure 5-5. Next, at the load torque of 60 N-m, a horizontal line is drawn until it touches the torque–speed curve. Then a vertical line is drawn down to the speed axis and the load speed is found (75 rad/s). To find the output power under this load condition, use Eq. 5-3b.

$$P_o = \frac{T\omega}{1000}$$

$$= \frac{60 \text{ N-m} \times 75 \text{ rad/s}}{1000}$$

$$= 4.5 \text{ kW}$$

5-4   *SPEED REGULATION*

When dealing with generators the voltage regulation is an indication of how much the output (terminal voltage) changes as we go from a no-load to a full-load condition. Similarly, with motors we can obtain an indication of how much the output (shaft speed) changes as we go from a no-load to a full-load condition. This is called the **speed regulation** and is calculated in exactly the same way as shown by Eqs. 5-9.

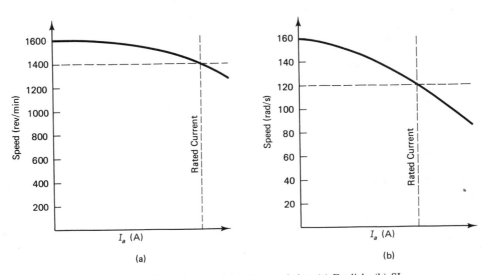

Figure 5-6   Typical motor load characteristics: (a) English; (b) SI.

(English)

$$\frac{\% \text{ speed}}{\text{regulation}} = \frac{S_{NL} - S_{FL}}{S_{FL}} \times 100 \tag{5-9a}$$

(SI)

$$\frac{\% \text{ speed}}{\text{regulation}} = \frac{\omega_{NL} - \omega_{FL}}{\omega_{FL}} \times 100 \tag{5-9b}$$

As before, the smaller the percent regulation (0% would be perfect) the better the motor's characteristic. A typical load characteristic of a motor is shown in Figure 5-6. It is a plot of motor speed versus armature current. Armature current is used because it is an indication of load (see Section 5-3.1) and is easily obtained.

Example 5-7 (English)

For the motor whose characteristic is shown in Figure 5-6a, find the percent speed regulation.

Solution

Equation 5-9a is used. The no-load and full-load speeds are obtained from the characteristic at the zero and rated current points, respectively.

$$\% \text{ S.R.} = \frac{1600 - 1400}{1400} \times 100$$

$$= 14.3\%$$

Example 5-8 (SI)

For the motor whose characteristic is shown in Figure 5-6b, find the percent speed regulation.

Solution

Equation 5-9b is used. The no-load and full-load speeds are obtained from the characteristic at the zero and rated current points, respectively.

$$\% \text{ S.R.} = \frac{160 - 120}{120} \times 100$$

$$= 33.3\%$$

5-5   MOTOR EFFICIENCY

Having already discussed the topic of efficiency in Chapter 4, the coverage here will be brief. The treatment for the motor will be identical to that of the generator with a few exceptions.

With the motor, input power is electrical and output power is mechanical. Therefore, the energy conversion taking place in the armature is electrical to mechanical (opposite to that of the generator). Equation 4-7 will be rewritten here as Eq. 5-10 so that it can be referenced easily.

$$\text{efficiency} = \eta(\%) = \frac{P_o}{P_i} \times 100 \tag{5-10}$$

In the case of the generator, power output is used in the efficiency calculations since it is electrical in form and easily measured. The opposite is true for the motor; power input is used in efficiency calculations since now it is the electrical quantity. Equations 5-11 and 5-12 can be written with little explanation. Note that

$$P_o = P_i - \text{losses} \tag{5-11}$$

is really the same as Eq. 4-8; however, we are now solving for P_o. Equation 5-12 is obtained by substituting Eq. 5-11 into Eq. 5-10. Note the difference between

$$\text{efficiency} = \eta(\%) = \frac{P_i - \text{losses}}{P_i} \times 100 \tag{5-12}$$

and Eq. 4-9. Very often a motor's nameplate will have written on it all the information necessary to solve for efficiency. This quantity is called the **nameplate efficiency** and is very close to that actually obtained when the motor is run at rated load.

Example 5-9 (English)

Find the nameplate efficiency of a dc motor rated 2 hp, 120 V, 15A.

Solution

The 120 V refers to the terminal voltage that the motor has been designed for. The 15 A represents the line current that the motor will draw when it puts out 2 hp. Therefore,

$$P_i = V_t I_L = 120 \text{ V} \times 15 \text{ A} = 1800 \text{ W}$$

$$P_o = 2 \text{ hp} \times 746 \text{ W/hp} = 1492 \text{ W}$$

Using Eq. 5-10, we have

$$\eta = \frac{1492 \text{ W}}{1800 \text{ W}} \times 100 = 82.9\%$$

Example 5-10 (SI)

Find the nameplate efficiency of a dc motor rated 2 kW, 220 V, 12 A.

Solution

The 220 V refers to the terminal voltage for which the motor has been designed. The 12 A represents the line current that the motor will draw when it puts out 2 kW. Therefore,

$$P_i = V_t I_L = 220 \text{ V} \times 12 \text{ A} = 2640 \text{ W}$$

$$P_o = 2 \text{ kW} = 2000 \text{ W}$$

Using Eq. 5-10, we have

$$\eta = \frac{2000 \text{ W}}{2640 \text{ W}} \times 100 = 75.8\%$$

5-5.1 Power Flow Diagram

Figure 5-7 shows the power flow diagram for a dc motor. Upon comparison with Figure 4-11, we see that it is the same diagram with the power terms interchanged. Through its terminals, electrical power is fed into a motor. If we subtract (Eq. 5-13) all the electrical (copper) losses from the input, we are left with the electrical power, which is converted to mechanical power by the motor's armature.

$$P_i - \text{copper losses} = E_g I_a \tag{5-13}$$

The converted power ($E_g I_a$) is also called the **developed power.** Although this is the mechanical power developed by the motor, it is not equal to what we can get out of the motor. From the developed power we must subtract the internal mechanical and core losses (stray power) before we can get anything out.

$$P_o = P_i - \text{copper losses} - \text{stray power losses} \tag{5-14}$$

5-5.2 Measurement of Stray Power Losses

If Eq. 5-14 is rearranged so that we are solving for the stray power loss, and in addition we specify that the motor is running at no load (i.e., $P_o = 0$), Eq. 5-15 is obtained.

$$\text{stray power loss} = P_i - \text{copper losses} \tag{5-15}$$

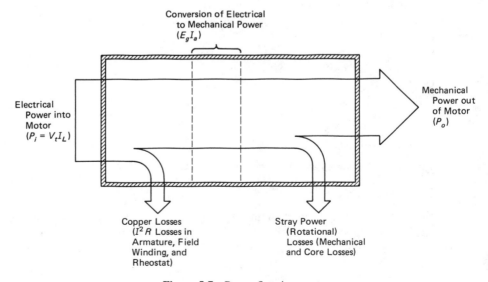

Figure 5-7 Power flow in a motor.

The input power and copper losses are electrical quantities which can be easily calculated from some simple measurements. Although the stray power loss calculated with Eq. 5-15 is at a no-load condition, it is representative of the stray power loss at rated load for the following reasons:

1. The mechanical losses are proportional to speed; therefore, if the motor is run at its rated speed for the no-load test, the mechanical losses will be the same as at rated load.
2. The core losses are proportional to speed and flux. Therefore, by running the motor at rated speed and keeping the flux at its rated load value, the no-load core losses would be the same as at rated load. It will be seen (Section 5-6) that for the shunt motor the flux is independent of load and therefore remains unchanged as the load is varied.

Example 5-11

The motor shown in Figure 5-2 is run at rated speed with nothing connected to its shaft (no-load condition). The copper losses, calculated from measured currents and resistances, are found to be 210 W. The terminal voltage and line current are measured as 120 V and 2.5 A, respectively. Find the stray power losses.

Solution

$$P_i = V_t I_L = 120 \text{ V} \times 2.5 \text{ A} = 300 \text{ W}$$

Using Eq. 5-15, we have

$$\text{stray power} = P_i - \text{copper losses}$$
$$= 300 \text{ W} - 210 \text{ W}$$
$$\text{stray power loss} = 90 \text{ W}$$

5-5.3 Maximum Efficiency

A typical motor has an efficiency curve as shown in Figure 5-8. It is a plot of efficiency versus power output. A motor is designed so that its peak efficiency occurs when its

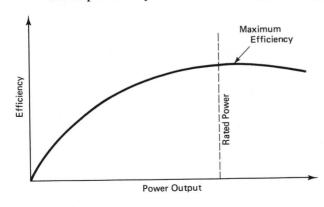

Figure 5-8 Efficiency versus power output for a motor.

output is close to rated output. The actual point at which maximum efficiency occurs can be derived using calculus. The derivation will not be shown here; however, the results will.

If a motor is running at constant speed, the stray power losses will be constant as power output changes. In addition, if the motor is a shunt motor, the field loss will be constant. The only loss that will vary is the armature copper loss ($I_a^2 R_a$), which, since it depends on I_a, will increase as the load increases. Maximum efficiency will occur when the fixed losses (stray power + shunt field) equal the variable loss (armature copper):

$$\overbrace{\underbrace{\text{stray power} \atop \text{loss}} + \underbrace{\text{shunt field} \atop \text{loss}}}^{\text{fixed losses}} = \overbrace{\underbrace{\text{armature} \atop \text{loss}}}^{\substack{\text{variable} \\ \text{loss}}} \qquad (5\text{-}16)$$

5-6 SHUNT MOTOR

As its name indicates, a shunt motor is a motor whose field winding shunts (is in parallel with) the armature. A diagram of a shunt motor is shown in Figure 5-9. Comparison of this motor with the shunt generator (Figure 4-13) shows that they are identical. The only differences are the current directions.

As can be seen from Figure 5-9, the line voltage supplies the field excitation as well as the current to the armature. Since the line (terminal) voltage, which is constant, is directly across the field circuit, the field current will be constant. If the field current is constant, the flux will also be constant and independent of load variations. Equation 5-1 can thus be written as Eq. 5-5, which says that the torque developed by a shunt motor is directly proportional to the armature current. This is shown graphically in Figure 5-10.

5-6.1 Direction of Rotation

Once a motor has been energized, current will flow through the field circuit. A magnetic field will be set up with a certain polarity within the motor. At the same time,

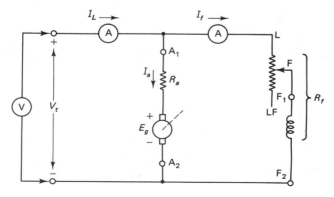

Figure 5-9 Shunt motor.

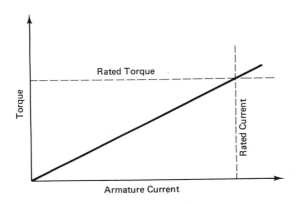

Figure 5-10 Graph of torque versus armature current for a shunt motor.

current will flow through the armature in a given direction. Based on the directions of the current in the armature and the flux, and application of Fleming's left-hand rule (see Section 2-4.2), the motor will rotate in a specified direction.

Let us take as an example the motor in Figure 5-9, with the voltage polarity and current directions as shown. Furthermore, we will assume that it is rotating in a clockwise direction. The following tests will be performed:

1. With the motor initially turning clockwise, power is turned off and the shunt field connection is reversed (i.e., F_1 and F_2 are interchanged). Since the field current, flux as well, has reversed its direction, the motor will turn counterclockwise when power is turned on.

2. With the motor now turning counterclockwise, power is turned off and the armature connection is reversed (i.e., A_1 and A_2 are interchanged). When power is turned on, the motor will have reversed and now turn clockwise again since the direction of current through the armature has reversed.

3. With the motor now turning clockwise, power is turned off and the polarity of the terminal voltage is reversed. When power is turned on, the motor will not reverse its direction and will still rotate in a clockwise direction. The reason for this behavior is clear when we look at Figure 5-9. By reversing the terminal voltage polarity, both field current and armature current reverse. Reversal of one current tends to make the motor reverse and turn counterclockwise. When the second current is reversed, the motor tends to change again and the original clockwise direction is maintained.

5-6.2 Effect of Load on the Shunt Motor

It has already been shown that the motor's torque varies with the load. That is, as the load increases, the motor develops more torque until its output equals the new load. We have also seen that the armature current is proportional to motor torque (Figure 5-10); therefore, it is proportional to load torque as well. From Figure 5-9 we can write the following equation by applying Kirchhoff's voltage law.

$$V_t - I_a R_a = E_g \qquad (5\text{-}17)$$

If Eqs. 4-1 are substituted into Eq. 5-17 for E_g, we can solve for the speed and get (English)

$$S = \frac{V_t - I_a R_a}{K\phi} \tag{5-18a}$$

(SI)

$$\omega = \frac{V_t - I_a R_a}{K'\phi} \tag{5-18b}$$

Since V_t, R_a, and ϕ in Eqs. 5-18 are constant, the only variables are I_a and the speed. If the load then increases, causing a corresponding increase in I_a, the numerator of Eqs. 5-18 will decrease. This means that the speed will also decrease. This is in accordance with the theory presented in Section 5-3.1. A typical speed characteristic for the shunt motor is shown in Figure 5-11.

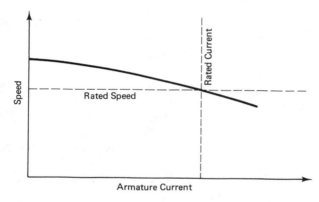

Figure 5-11 Graph of speed versus armature current for a shunt motor.

5-6.3 Effect of Flux on the Shunt Motor

If we assume in Eqs. 5-18 that the $I_a R_a$ drop is small (negligible) compared to V_t, we can write the approximate equation*

$$\text{speed} \approx \frac{V_t}{K\phi} \tag{5-19}$$

Equation 5-19 states that the motor speed is inversely proportional to the flux. By increasing the flux (or I_f) the speed will be lowered, and by decreasing the flux the speed will increase. It is for this reason that the rheostat is included in the field circuit of Figure 5-9. By **increasing** the field resistance, I_f and ϕ will decrease, causing the speed to **increase.** Conversely, by **decreasing** the field resistance, I_f and ϕ will increase, making the motor **slow down.**

The effect of flux on speed presents a potentially dangerous situation. If the field circuit is ever opened while the motor is running, the field current will go to zero. The

*This is a fair approximation, especially for light loads. Under these conditions, the back EMF (E_g) builds up to the point where it is almost equal to V_t and the $I_a R_a$ drop would thus be very small.

flux in turn will drop to the very small residual magnetism left in the core. In accordance with Eq. 5-19, a very small ϕ will cause a very high speed. In fact, the speed will be so high that if it continues for more than a moment, it might damage the surrounding area. At the least the motor would be damaged. In the laboratory the most common cause of a runaway motor is the ammeter fuse. That is, while adjusting the field rheostat to obtain a slow speed, too much current is drawn through the ammeter. The meter's protective fuse burns out, causing the open field circuit.

Example 5-12 (English)

The motor in Figure 5-9 is tested by varying the load and the data in Table 5-2 is obtained. It is rated 1 hp. Its armature resistance is 1.5 Ω and the stray power losses are constant at 50 W. Plot the line current, speed, efficiency, and output torque versus power output all on the same graph. Indicate the rated power point by a dashed line on the graph. The line voltage is 120 V. Also calculate the maximum efficiency point and compare it to the graph.

Table 5-2

I_L (A)	2.8	3.6	4.7	5.8	7.0	8.5
I_f (A)	0.5	0.5	0.5	0.5	0.5	0.5
S (rev/min)	1800	1780	1750	1720	1680	1650

Solution

From the data in Table 5-2 and the information given, Table 5-3 is constructed. The following sample calculations (done for the maximum load point) are used to generate the data for Table 5-3.

$$I_a = I_L - I_f = 8.5 \text{ A} - 0.5 \text{ A} = 8 \text{ A}$$

$$\text{field loss} = V_t I_f = 120 \text{ V} \times 0.5 \text{ A} = 60 \text{ W}$$

(constant at each point)

$$\text{armature loss} = R_a I_a^2 = 1.5 \; \Omega \times (8 \text{ A})^2 = 96 \text{ W}$$

$$\text{total losses} = \text{field loss} + \text{armature loss} + \text{stray power}$$

$$= 60 \text{ W} + 96 \text{ W} + 50 \text{ W} = 206 \text{ W}$$

$$P_i = V_t I_L = 120 \text{ V} \times 8.5 \text{ A} = 1020 \text{ W}$$

$$P_o = P_i - \text{total losses} = 1020 \text{ W} - 206 \text{ W} = 814 \text{ W}$$

In horsepower,

$$P_o = \frac{814 \text{ W}}{746 \text{ W/hp}} = 1.09 \text{ hp}$$

$$\eta = \frac{P_o}{P_i} \times 100 = \frac{814}{1020} \times 100 = 80\%$$

$$T_o = \frac{7.04 P_o}{S} = \frac{7.04 \times 814 \text{ W}}{1650 \text{ rev/min}} = 3.47 \text{ ft-lb}$$

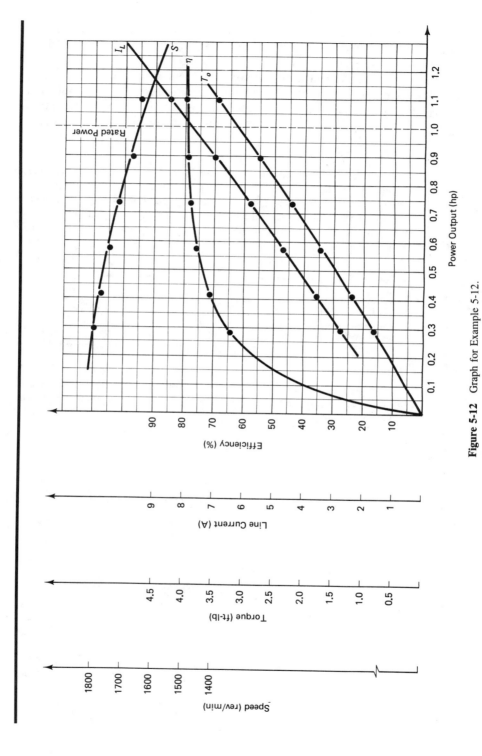

Figure 5-12 Graph for Example 5-12.

140

Table 5-3

I_L (A)	2.8	3.6	4.7	5.8	7.0	8.5
I_a (A)	2.3	3.1	4.2	5.3	6.5	8.0
Armature loss (W)	7.9	14.4	26.5	42.1	63.4	96
Total losses (W)	117.9	124.4	136.5	152.1	173.4	206
P_i (W)	336	432	564	696	840	1020
P_o (W)	218.1	307.6	427.5	543.9	666.6	814
P_o (hp)	0.29	0.41	0.57	0.73	0.89	1.09
η (%)	65	71	76	78	79	80
T_o (ft-lb)	0.85	1.22	1.72	2.23	2.79	3.47
S (rev/min)	1800	1780	1750	1720	1680	1650

The tabulated data is plotted and shown in Figure 5-12. The theoretical point at which maximum efficiency occurs can be calculated using Eq. 5-16.

$$\text{S.P. loss} + \text{field loss} = \text{armature loss}$$

$$50\text{ W} + 60\text{ W} = R_a I_a^2$$

$$110\text{ W} = 1.5\ \Omega \times I_a^2$$

$$I_a^2 = \frac{110\text{ W}}{1.5\ \Omega} = 73.33$$

$$I_a = 8.5\text{ A}$$

Although the calculation above is valid only for constant-speed tests, the change in speed encountered is minor. The data was not taken to a high enough point to verify the calculation. The curve for efficiency is peaking around 1.1 hp. This corresponds to 8.5 A of armature current, which is the same as the calculated value.

Example 5-13 (SI)

The motor in Figure 5-9 is tested by varying the load and the data in Table 5-4 is obtained. It is rated 1 kW. Its armature resistance is 1.8 Ω and the stray power losses are constant at 30 W. Plot the line current, speed, efficiency, and output torque versus power output all on the same graph. Indicate with a dashed line on the graph the rated power point. The line voltage is 230 V. Also compute the maximum efficiency point and compare it to the graph.

Table 5-4

I_L (A)	1.1	1.9	2.8	4.0	5.1	6.2	7.2
I_f (A)	0.2	0.2	0.2	0.2	0.2	0.2	0.2
ω (rad/s)	200	197	194	190	187	183	180

Solution

From the data in Table 5-4 and the information given, Table 5-5 is constructed. The following sample calculations (done for the maximum load point) are used to generate the data for Table 5-5.

$$I_a = I_L - I_f = 7.2 \text{ A} - 0.2 \text{ A} = 7 \text{ A}$$

field loss $= V_t I_f = 230 \text{ V} \times 0.2 \text{ A} = 46 \text{ W}$ (constant at each point)

armature loss $= R_a I_a^2 = 1.8 \ \Omega \times (7 \text{ A})^2 = 88.2 \text{ W}$

total losses $=$ field loss $+$ armature loss $+$ stray power

$$= 46 + 88.2 + 30 = 164.2 \text{ W}$$

$$P_i = V_t I_L = 230 \text{ V} \times 7.2 \text{ A} = 1656 \text{ W}$$

$$P_o = P_i - \text{total losses} = 1656 \text{ W} - 164.2 \text{ W} = 1491.8 \text{ W}$$

$$= 1.49 \text{ kW}$$

$$\eta = \frac{P_o}{P_i} \times 100 = \frac{1492}{1656} \times 100$$

$$= 90\%$$

$$T_o = 1000 \frac{P}{\omega} = 1000 \times \frac{1.49}{180}$$

$$= 8.3 \text{ N-m}$$

Table 5-5

I_L (A)	1.1	1.9	2.8	4.0	5.1	6.2	7.2
I_a (A)	0.9	1.7	2.6	3.8	4.9	6.0	7.0
Armature loss (W)	1.5	5.2	12.2	26	43.2	64.8	88.2
Total losses (W)	77.5	81.2	88.2	102	119.2	140.8	164.2
P_i (W)	253	437	644	920	1173	1426	1656
P_o (W)	176	356	556	818	1054	1285	1492
P_o (kW)	0.176	0.356	0.556	0.818	1.054	1.285	1.492
η (%)	69.5	81.5	86.3	88.9	89.8	90.1	90
T_o (N-m)	0.9	1.8	2.9	4.3	5.6	7.0	8.3
ω (rad/s)	200	197	194	190	187	183	180

The tabulated data is plotted and shown in Figure 5-13. The theoretical point at which maximum efficiency occurs can be calculated using Eq. 5-16.

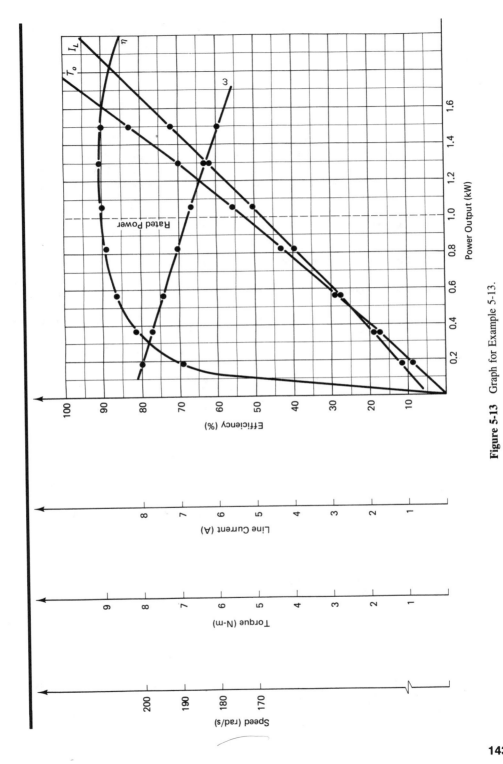

Figure 5-13 Graph for Example 5-13.

$$\text{S.P. loss} + \text{field loss} = \text{armature loss}$$

$$30 \text{ W} + 46 \text{ W} = R_a I_a^2$$

$$76 \text{ W} = 1.8 \ \Omega \times I_a^2$$

$$I_a^2 = \frac{76 \text{ W}}{1.8 \ \Omega} = 42.2$$

$$I_a = 6.5 \text{ A}$$

Although the calculation above is only valid for constant-speed tests, the change in speed encountered is minor. The curve for efficiency peaks around 1.3 kW. The line current for that output power is about 6.2 A, which is fairly close to the calculated value.

5-7   SERIES MOTOR

As its name indicates, the series motor is one whose field winding is in series with its armature. The armature current is then the field current. Therefore, the flux is proportional to the armature current. This leads to a distinct and different characteristic from the shunt motor. Since changes in load affect armature current, they will now also affect the flux. That is, doubling the current will double the flux, and vice versa. A typical diagram of a series motor is shown in Figure 5-14.

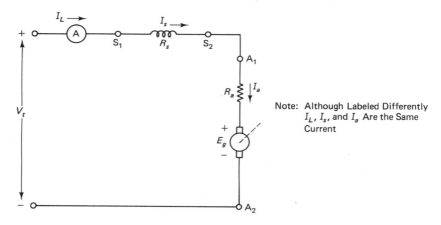

Figure 5-14   Series motor. (*Note:* Although labeled differently, I_L, I_S and I_a are the same current.)

If I_a is substituted for ϕ in the basic torque equation (Eq. 5-1), we obtain Eq. 5-20 for the series motor.

$$T = K' I_a \times I_a = K' I_a^2 \tag{5-20}$$

The torque characteristic for the series motor is thus parabolic, as shown in Figure 5-15. This is true as long as the field remains unsaturated.

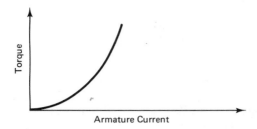

Figure 5-15 Torque versus armature current for a series motor.

5-7.1 Direction of Rotation

If a series motor like the one in Figure 5-14 is energized, it will rotate in a direction based on the flux polarity, armature current direction, and Fleming's left-hand rule.

If the field connection (S1–S2) is reversed, the direction of rotation will also change since the flux polarity will reverse. In addition, if the armature connection (A1–A2) is reversed, the speed will again change direction since the direction of current through the armature will reverse. If the polarity of the terminal voltage reverses, however, the motor will continue to turn in the same direction. This occurs because two changes are taking place. The flux polarity and armature current direction reverse simultaneously, thus nullifying their individual effects.

5-7.2 Effect of Load on a Series Motor

From Figure 5-14 and application of Kirchhoff's voltage law, we can write down

$$E_g = V_t - I_a R_s - I_a R_a \qquad (5\text{-}21)$$

Furthermore, by substituting Eqs. 4-1 into Eq. 5-21 and rearranging, we obtain the following speed equations:

(English)

$$S = \frac{V_t - I_a(R_a + R_s)}{K\phi_s} \qquad (5\text{-}22a)$$

(SI)

$$\omega = \frac{V_t - I_a(R_a + R_s)}{K'\phi_s} \qquad (5\text{-}22b)$$

The subscript s is used because the flux is due to a series field. Since ϕ_s is proportional to I_a for the series motor, we can rewrite the speed equations above as Eqs. 5-23. Note that the constants K and K' will be different now. For our discussion their actual value is not important.

(English)

$$S = \frac{V_t - I_a(R_a + R_s)}{KI_a} \qquad (5\text{-}23a)$$

(SI)

$$\omega = \frac{V_t - I_a(R_a + R_s)}{K'I_a} \qquad (5\text{-}23\text{b})$$

Examination of Eqs. 5-23 reveals that I_a is the only variable quantity affecting speed. At no load I_a is very small. If I_a is made small in Eqs. 5-23, a very large speed will result. If the load, hence I_a, increases, the speed will decrease. This is obvious from Eqs. 5-23 if we make the denominator larger and the numerator smaller. A typical speed characteristic is shown in Figure 5-16. As the armature current gets smaller and smaller (approaching a no-load condition), the speed gets so high that it is potentially dangerous. It is for this reason that series motors are never started or used without some sort of load directly connected to them. They find use in applications such as electric trains, where a heavy load is always connected.

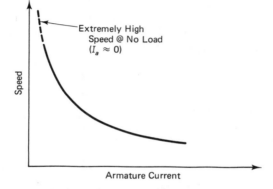

Figure 5-16 Speed versus armature current for a series motor.

The high torque developed under large load conditions produces good acceleration. In addition, the falling speed characteristic combined with the rising torque characteristic provides a fairly constant power output as load is varied. That is, if the torque–speed product is constant, the power will be constant (see Eqs. 5-3).

As was the case with the shunt motor, the speed of the series motor is inversely proportional to flux. That is, by decreasing the flux, the speed will increase, and vice versa. However, the potential danger present in the shunt motor does not exist here. If the series field were suddenly lost, due to an open circuit, the flux would become very small. The armature, being in series with the field, would lose its current and the motor would stop rather than run away.

5-8 COMPOUND MOTOR

The compound motor, like the compound generator, combines the effects of both a shunt and a series field. It, too, can be wired in either a long shunt or a short shunt configuration, as shown in Figure 5-17.

Under no-load or light-load conditions, the series field current will be very small; however, the shunt field will have its full excitation. As a result, at no load the motor

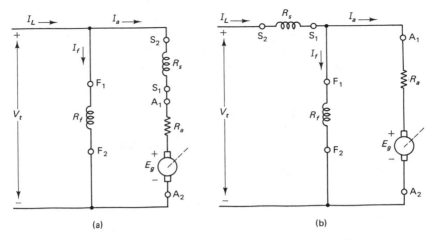

Figure 5-17 Compound motors: (a) long shunt; (b) short shunt.

behaves like a shunt motor and does not approach dangerously high speeds (see Figure 5-16) as the series motor does.

5-8.1 Differential Compound Motor

If the series- and shunt-field windings are connected in such a way that their fields oppose each other, the motor is said to be **differentially compounded.** The basic torque and speed equations (Eqs. 5-1 and 5-19) become Eqs. 5-24 and 5-25, respectively. In these equations, ϕ_f represents the shunt-field flux and ϕ_s represents the series-field flux.

$$T = K(\phi_f - \phi_s)I_a \qquad (5\text{-}24)$$

$$\text{speed} \approx \frac{V_t}{K(\phi_f - \phi_s)} \qquad (5\text{-}25)$$

As the load increases, both I_a and ϕ_s will increase, causing the term $(\phi_f - \phi_s)$ to decrease. These equations indicate that as the net flux $(\phi_f - \phi_s)$ decreases, the torque will decrease and the speed will increase. This behavior is generally undesirable and potentially dangerous. That is, a large enough load will cause the series-field current to increase to the point where the denominator of Eq. 5-25 approaches zero. This, in turn, will produce a dangerously high speed. The differential compound motor has very few practical applications.

When a differential compound motor is started under load in the laboratory, the series winding (S1–S2) should initially be shorted out using a switch. This is done to prevent the series field from overpowering the shunt field, thus causing the motor to turn in the wrong direction.

A series-field shorting switch is used on starting for two other reasons:

1. By shorting out the series field, proper shunt-field polarity can be established to give the desired direction of rotation.

2. A shorting switch is used in a test for proper compounding. With the series field shorted out and the motor turning in the proper direction, a load is applied. Upon opening the switch, if the speed increases, the motor is differentially compounded. This is caused by the sudden decrease in flux. If the speed decreases, the net flux has increased and the motor is cumulatively compounded.

5-8.2 Cumulative Compound Motor

When the shunt and series windings are connected so that their respective fields aid each other, the motor is said to be **cumulatively compounded.** The torque and speed equations can then be rewritten as Eqs. 5-26 and 5-27.

$$T = K(\phi_f + \phi_s)I_a \tag{5-26}$$

$$\text{speed} \approx \frac{V_t}{K(\phi_f + \phi_s)} \tag{5-27}$$

Examination of these equations reveals that they behave in a manner similar to the shunt motor. An increase in load causes I_a to increase, producing a larger torque. The increase in torque for the cumulative compound motor will be greater than the shunt, however, since the net flux will increase as well.

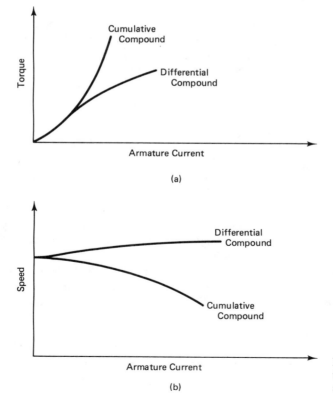

Figure 5-18 Compound motor characteristics: (a) torque characteristics; (b) speed characteristics.

Although compounding improves the torque characteristic, it degrades the speed characteristic. Since the denominator in Eq. 5-27 increases with load, the speed will fall off more sharply here than for the shunt motor. Nonetheless, the trade-off is a beneficial one when dealing with loads that can increase suddenly. The torque and speed characteristics for compound motors are shown in Figure 5-18.

Example 5-14 (English)

A long shunt cumulatively compound motor is run at full load. It is rated 240 V, 10 A, 1200 rev/min. If $R_f = 180\ \Omega$, $R_a = 2\ \Omega$, $R_s = 1.22\ \Omega$, and the stray power losses are 160 W, find:

(a) The back EMF

(b) The power rating

(c) The efficiency

(d) The torque rating

(e) The developed torque

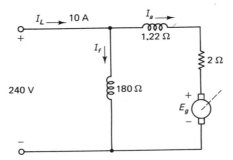

Figure 5-19 Diagram for Example 5-14.

Solution

The first step is to draw a diagram of a long shunt motor inserting all of the known quantities. This is shown in Figure 5-19. The next step is to solve for the unknown currents.

$$I_f = \frac{V_t}{R_f} = \frac{240\ \text{V}}{180\ \Omega} = 1.33\ \text{A}$$

$$I_a = I_L - I_f = 10\ \text{A} - 1.33\ \text{A} = 8.67\ \text{A}$$

(a) $E_g = V_t - I_a(R_a + R_s)$

$$= 240\ \text{V} - 8.67\ \text{A}(2\ \Omega + 1.22\ \Omega)$$

$$= 240 - 27.92 = 212.08\ \text{V}$$

(b) The power rating is the output power at full load. To obtain it, we will use Eq. 5-11. The losses must first be calculated.

$$\text{shunt loss} = V_t I_f = 240\ \text{V} \times 1.33\ \text{A} = 319.2\ \text{W}$$

$$\text{armature loss} = R_a I_a^2 = 2\ \Omega \times (8.67)^2 = 150.34\ \text{W}$$

$$\text{series loss} = R_s I_a^2 = 1.22 \; \Omega \times (8.67)^2 = 91.71 \text{ W}$$

$$\text{stray loss} = 160 \text{ W}$$

$$P_i = V_t I_L = 240 \text{ V} \times 10 \text{ A} = 2400 \text{ W}$$

$$P_o = P_i - \text{losses} = 2400 \text{ W} - 721.25 \text{ W}$$

$$= 1678.75 \text{ W}$$

or

$$P_o = \frac{1678.75 \text{ W}}{746 \text{ W/hp}} = 2.25 \text{ hp}$$

(c)

$$\eta = \frac{P_o}{P_i} \times 100$$

$$= \frac{1678.75}{2400} \times 100$$

$$= 70\%$$

(d) The torque rating is obtained by using Eq. 5-3a and the power rating.

$$T = 7.04 \times \frac{P}{S}$$

$$= 7.04 \times \frac{1678.75}{1200}$$

$$= 9.85 \text{ ft-lb}$$

(e) The developed torque is obtained by using Eq. 5-3a and the developed power. The developed power can be gotten easily by two techniques;

(1) $P_{\text{dev}} = P_o + \text{stray power loss}$

$$= 1678.75 + 160 = 1838.75 \text{ W}$$

(2) $P_{\text{dev}} = \text{converted power} = E_g I_a \quad (\text{see Eq. 5-13})$

$$= 212.08 \text{ V} \times 8.67 \text{ A}$$

$$= 1838.73 \text{ W}$$

$$T = 7.04 \times \frac{P}{S}$$

$$= 7.04 \times \frac{1838.75}{1200}$$

$$= 10.79 \text{ ft-lb}$$

Example 5-15 (SI)

A long shunt cumulatively compound motor is run at full load. It is rated 120 V, 15 A, 140 rad/s. If $R_f = 100 \; \Omega$, $R_a = 0.8 \; \Omega$, $R_s = 0.2 \; \Omega$, and the stray power losses are 65 W, find;

(a) The back EMF

(b) The power rating

(c) The efficiency

(d) The torque rating

(e) The developed torque

Solution

The first step is to draw a diagram of a long shunt motor inserting all of the known quantities. This is shown in Figure 5-20. the next step is to solve for the unknown currents.

$$I_f = \frac{V_t}{R_f} = \frac{120 \text{ V}}{100 \text{ }\Omega} = 1.2 \text{ A}$$

$$I_a = I_L - I_f = 15 \text{ A} - 1.2 \text{ A} = 13.8 \text{ A}$$

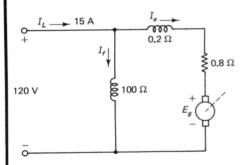

Figure 5-20 Diagram for Example 5-15.

(a) $E_g = V_t - I_a(R_a + R_s)$

$\qquad = 120 \text{ V} - 13.8 \text{ A}(0.8 \text{ }\Omega + 0.2 \text{ }\Omega)$

$\qquad = 120 - 13.8 = 106.2 \text{ V}$

(b) The power rating is the output power at full load. To obtain it, we will use Eq. 5-11. The losses must first be calculated.

$$\text{shunt loss} = V_t I_f = 120 \text{ V} \times 1.2 \text{ A} = 144 \text{ W}$$

$$\text{armature loss} = R_a I_a^2 = 0.8 \text{ }\Omega \times (13.8 \text{ A})^2 = 152.35 \text{ W}$$

$$\text{series loss} = R_s I_a^2 = 0.2 \text{ }\Omega \times (13.8 \text{ A})^2 = 38.1 \text{ W}$$

$$\text{stray loss} = 65 \text{ W}$$

$$P_i = V_t I_L = 120 \text{ V} \times 15 \text{ A} = 1800 \text{ W}$$

$$P_o = P_i - \text{losses} = 1800 \text{ W} - 399.45 \text{ W}$$

$$= 1400.55 \text{ W} = 1.4 \text{ kW}$$

(c)
$$\eta = \frac{P_o}{P_i} \times 100$$

$$= \frac{1400}{1800} \times 100$$

$$= 78\%$$

(d) The torque rating is obtained by using Eq. 5-3b and the power rating.

$$T = 1000 \times \frac{P}{\omega}$$

$$= 1000 \times \frac{1.4}{140}$$

$$= 10 \text{ N-m}$$

(e) The developed torque is obtained by using Eq. 5-3b and the developed power. Developed power can be gotten easily by two techniques:

(1) $P_{\text{dev}} = P_o + \text{stray power loss}$

$$= 1400.55 \text{ W} + 65 \text{ W} = 1465.55 \text{ W}$$

(2) $P_{\text{dev}} = \text{converted power} = E_g I_a$ (see Eq. (5-13)

$$= 106.2 \text{ V} \times 13.8 \text{ A}$$

$$= 1465.56 \text{ W}$$

$$T = 1000 \times \frac{P}{\omega}$$

$$= 1000 \times \frac{1.465}{140}$$

$$= 10.46 \text{ N-m}$$

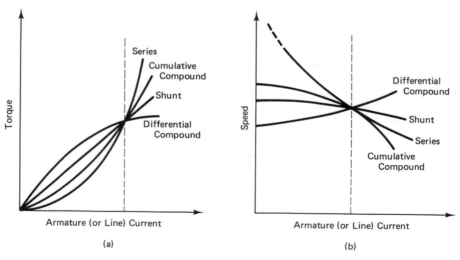

Figure 5-21 Comparison of motors: (a) torque characteristic; (b) speed characteristic.

5-8.3 Comparison of Motors

As an aid in determining which type of motor is best suited for a given application, the characteristics of similar motors can be compared. This can be done easily by examining the torque–speed characteristics shown in Figure 5.21. Since the motors are similar (equal rating), their characteristic curves intersect at the rated point. Depending on the range of load to be driven by the motor, proper selection can be made.

5-9   PERMANENT-MAGNET MOTORS

A permanent-magnet motor is one that does not require excitation to derive its field. Part of its core is made of an alloy which retains its magnetism almost indefinitely. A diagram of such a motor is shown in Figure 5-22. Since a field winding does not exist, this motor will be more efficient than a shunt or compound motor with similar rating. In addition, the speed and direction of rotation can be very easily controlled by the magnitude and polarity of the terminal voltage. That is, reversing the polarity of V_t will change only the direction of armature current, not the flux. It is for this reason that permanent-magnet motors find most of their use in toys and in electromechanical control systems called servomechanisms. Their characteristics are similar to those of a shunt motor with constant flux, with the exception of those mentioned above.

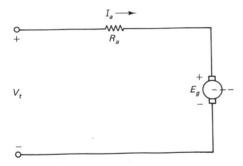

Figure 5-22   Schematic of a permanent-magnet motor.

5-10   STARTING DC MOTORS

Up until this point, no mention has been made of the procedure or special device needed to start a dc motor. If we examine the motor shown in Figure 5-23 and calculate the currents at starting, the need for a special device becomes apparent.

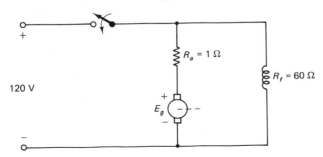

Figure 5-23   Shunt motor improperly started.

At starting the motor is not yet turning; therefore, the back EMF (E_g) is zero. An instant after the switch is closed, the currents will be*

$$I_a = \frac{120 \text{ V}}{1 \text{ }\Omega} = 120 \text{ A}$$

$$I_f = \frac{120 \text{ V}}{60 \text{ }\Omega} = 2 \text{ A}$$

$$I_L = 120 \text{ A} + 2 \text{ A} = 122 \text{ A}$$

If 120 A were to flow through the armature, the results would be catastrophic. Not only might the source of power be damaged, but the motor brushes would most likely burn up. In addition, the motor's armature winding, perhaps rated at 10 to 20 A, would burn. Protective devices such as fuses and circuit breakers would probably disconnect the power before damage could be done. In either event, the motor could not be used.

It can be seen that a device is needed which can gradually increase the voltage across the armature until the motor's speed and back EMF have built up.

5-10.1 Three-Point Starter

One type of manual starter is shown within the dashed line in Figure 5-24. It is called a **three-point starter** because there are three terminals to make connections to; L, A, and F. As the handle is pushed to the start position, the field winding is energized. Simultaneously, the armature circuit is energized with a starting resistance in series with the armature. This will limit the armature current.

As the motor builds up speed, the handle is gradually pushed toward the ON position. In this way, more and more voltage is applied to the armature until the handle reaches the ON position. At this point the full line voltage is across the armature. Since the motor has now built up to a high speed, the back EMF is big and the armature current will not be excessive. The handle is held in place by an electromagnet which is energized by the field current. As soon as the power is switched off, all currents will go to zero and the electromagnet will release the handle, which is pulled back to the OFF position by a spring.

It is important that the handle be moved slowly to the ON position. Moving it quickly will defeat the purpose of the starter by applying a large voltage to the armature before the back EMF has built up.

The fact that the electromagnet is energized by the field current presents a problem. In an effort to increase the motor's speed, the flux must be decreased by decreasing the field current. This is accomplished by increasing the field circuit resistance. If the decreasing field current weakens the electromagnet sufficiently, the

---

*In reality, the armature current will not change instantaneously. Since the armature is made up of wire wrapped around a magnetic core, it will have inductance. The armature current will therefore rise exponentially. However, the electrical time constant (L/R_a) will be much shorter than the mechanical time constant. Thus the current will reach its final value before the motor starts to move.

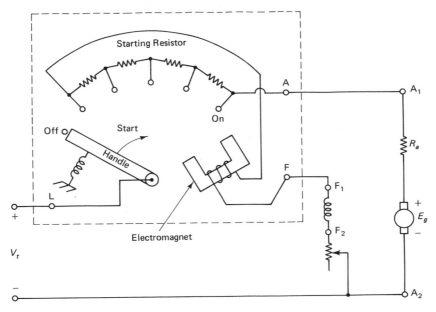

Figure 5-24 Three-point starter.

handle will be released and the motor will shut down. For this reason the three-point starter is undesirable.

5-10.2 Four-Point Starter

The **four-point starter** is shown in Figure 5-25. As its name indicates, there are four terminal connections to be made. Note that for this starter the electromagnet is not excited with the field current. Varying the flux will not affect the electromagnetic force, thus eliminating the problem described for the three-point starter.

5-10.3 Automatic Starters

An automatic starter is operated simply by pushing a button. There are two popular schemes used. A simplified diagram of the first of these is shown in Figure 5-26. It uses time-delay relays (B, C, and D) and is independent of the motor's speed.

When the start button is pressed, relay A is energized. Its contacts A_1, A_2, and A_3 will close. Contact A_1 will energize time delay relay B, A_2 will energize the field and armature circuits, and A_3 will latch the power to the relay logic. After a specified time delay, relay B will close its contacts. Contact B_1 will energize time-delay relay C, and B_2 will short out starting resistor R_1. After another time delay, relay C will close its contacts. Contact C_1 will energize time delay relay D, and C_2 will short out starting resistor R_2. Finally, when relay D closes its contact D_1, starting resistor R_3 is shorted out. The motor is now running will full terminal voltage applied to its armature. Relays A, B, C, and D must all fire at a voltage slightly less than V_t.

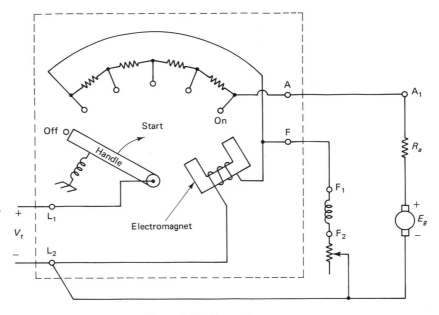

Figure 5-25 Four-point starter.

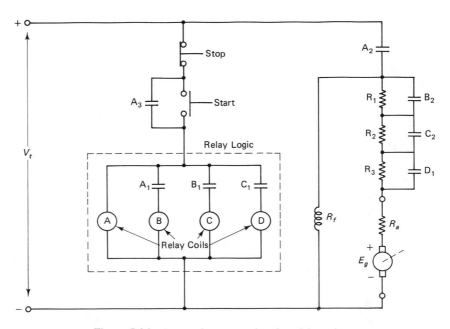

Figure 5-26 Automatic starter using time-delay relays.

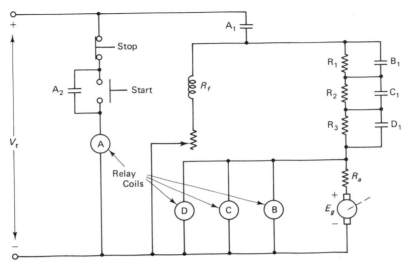

Figure 5-27 Automatic starter sensing back EMF.

The second scheme does not use time-delay relays but rather senses the back EMF and sequentially switches the relays as the speed (back EMF) builds up. Each relay is set to trip at a different voltage. A simplified diagram of one is shown in Figure 5-27. When the start button is pressed, relay A is energized. Contact A_1 will close, supplying power to the field and armature circuits. Contact A_2 will latch power to relay coil A. As the motor begins to build up speed, relays B, C, and D, which are connected directly across the armature, will sense an increasing voltage $(E_g + I_a R_a)$. When a predesigned armature voltage is reached (approximately 30% of V_t), relay B will be energized. Contact B_1 will short out resistor R_1 and the motor's speed will increase further. When a larger armature voltage is reached (approximately 60% of V_t), relay C will energize, closing contact C_1. This process continues, with relay D energizing at about 90% of V_t, closing contact D_1. At this point the full terminal voltage is applied to the armature and the motor will reach its final speed.

It should be pointed out that in the automatic schemes discussed, the relay time delays and firing voltages are based on the maximum permissible armature current and the minimum start up time required.

Both starters can be stopped by pressing the stop button. This deenergizes relay A, in both cases removing the power latch and disconnecting power from the motor. In addition, both schemes offer **undervoltage protection.** That is, if the line voltage (V_t) drops below a specified value, relay A will deenergize, causing the motor to shut down. To restart the motor, the start button would have to be pressed again.

A more elaborate scheme called **undervoltage release** could be designed using additional relays. In this case if the voltage drops below a specified value, the motor is shut down. However, when the voltage is restored, the motor will automatically start up again.

Example 5-16

A 120-V shunt motor is started with a four-point starter, as shown in Figure 5-25. The full-load armature current is 5 A, $R_a = 1$ Ω, and the starting resistance is 15 Ω. Calculate the starting armature current. What percent of rated current does this represent?

Solution

The instant power is applied, the total armature resistance will be $R_a + R_{starting}$.

$$R_{tot} = 1 \text{ Ω} + 15 \text{ Ω} = 16 \text{ Ω}$$

$$I_a = \frac{V_t}{R_{tot}} = \frac{120 \text{ V}}{16 \text{ Ω}}$$

$$= 7.5 \text{ A}$$

The starting armature current is 7.5 A. Dividing this by rated current and multiplying by 100 will give the starting current as a percent of rated current.

$$\frac{7.5 \text{ A}}{5 \text{ A}} \times 100 = 150\%$$

5-11 STOPPING A DC MOTOR

Stopping a motor can be done simply by pressing a button, as described in Section 5-10.3. However, big motors, because of their large inertia, may continue to rotate for quite a while. In cases where it could damage other equipment or injure people (emergency stops), this continued rotation cannot be allowed. A technique is needed in which the motor can be stopped abruptly.

One such technique, called **dynamic braking**, converts the mechanical energy of the motor to electrical energy, which is quickly dissipated. One way to do this is to switch a resistor in parallel with the armature on a command to stop. The back EMF of the motor will supply current to the resistor, which will dissipate the energy in the form of heat. A second method of dynamic braking has a generator mounted on the motor's shaft. Under normal running conditions, it represents a light load. On a command to stop, a resistor is connected to the generator's terminals. This draws electrical energy from the generator, quickly using up the motor's mechanical energy, which is the generator's input power.

Another technique used to stop a motor even faster is called **plugging.** This scheme is similar to putting a car into reverse when going forward. Although the car's transmission would be destroyed, the motor would not. Plugging is accomplished by reversing the polarity of the armature voltage in an emergency stop. By doing this the motor is commanded to turn in the opposite direction, causing it to slow down very quickly. Once it stops turning but before it actually reverses its motion, power is disconnected from the armature.

The only danger in plugging is the armature current that flows. Once the armature voltage is reversed, it will have the same polarity as the back EMF. The sum of the two voltages will cause an extremely large current to flow through the armature. To prevent this excessive current, a current-limiting resistor is placed in series with the armature when the voltage is reversed.

Example 5-17

A 120-V shunt motor has a rated armature current of 10 A and $R_a = 1\ \Omega$. It is running at rated speed and the back EMF is 110 V. If plugging is used to stop the motor, as shown in Figure 5-28, find the value of the current-limiting resistor R_x to limit the armature current to 15 A.

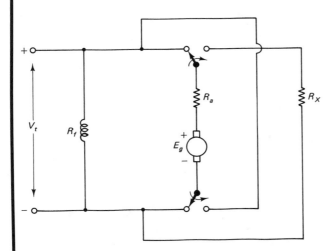

Figure 5-28 Diagram for Example 5-17.

Solution

On command to stop, both switches are thrown simultaneously. The armature current that flows will be

$$I_a = \frac{E_g + V_t}{R_a + R_x}$$

Substituting the known quantities and solving for R_x, we get

$$15\ A = \frac{110\ V + 120\ V}{1\ \Omega + R_x}$$

$$1 + R_x = \frac{330}{15} = 22$$

$$R_x = 21\ \Omega$$

SYMBOLS INTRODUCED IN CHAPTER 5

Symbol	Definition	Units	
		English	SI
T	Torque (developed or output)	lb-ft	N-m
α	Angular acceleration	rad/s^2	rad/s^2
J	Moment of inertia	$\dfrac{\text{ft-lb-s}^2}{\text{rad}}$	$\dfrac{\text{N-m-s}^2}{\text{rad}}$
P	Power (developed or output)	hp	watts
T_L	Load torque	lb-ft	N-m
K_m	Motor torque constant	ft-lb/V	N-m/V
K_B	Motor back EMF constant	$\dfrac{\text{ft-lb-min}}{\text{rev}}$	$\dfrac{\text{N-m-s}}{\text{rad}}$
S.R.	Speed regulation	percent	percent
ϕ_f	Shunt-field flux	lines	webers
ϕ_s	Series-field flux	lines	webers
R_x	External current-limiting resistor used in plugging	ohms	ohms

QUESTIONS

1. Under constant-power conditions, what is the relationship between torque and speed?

2. Define the following terms: prony brake; cradle dynamometer; back EMF; stall torque; no-load speed.

3. What is meant by "nameplate efficiency" and "speed regulation"?

4. What is the difference between output power and developed power?

5. Theoretically, when does maximum efficiency occur in a dc motor?

6. What is the effect on the direction of rotation of a shunt motor due to a reversal in field current; armature current; line voltage?

7. What is the effect on the direction of rotation of a series motor due to a reversal in field current; armature current; line voltage?

8. While a shunt motor is running, what will happen if the field circuit is broken?

9. While a series motor is running, what will happen if the field circuit is broken?

10. What is the difference between differential and cumulative compounding?

11. As the armature current increases beyond its full-load value, which motor has the greatest increase in torque?

12. What happens to the speed of a dc motor as the current it draws increases?

13. Why is a special starter needed for a dc motor?

14. What is the difference between undervoltage protection and undervoltage release?

15. Describe two techniques used for quickly stopping a motor.

16. Draw a picture of a four-point starter.

PROBLEMS

(English)

1. How much torque will a 5-hp motor supply while running at full load at:
(a) 900 rev/min?
(b) 1200 rev/min?
(c) 2000 rev/min?

2. A motor must supply 3500 ft-lb at 30 rev/min to operate a lift at full load. What should its rating be?

3. A 2-hp 1800-rev/min motor has 130 W of stray power losses. What is its full-load:
(a) Output torque?
(b) Developed torque?

4. If the motor in Problem 3 supplies half of its rated load at 2200 rev/min, how much torque is it putting out?

5. The arrangement shown in Figure 5-1 is used to measure the power output of a motor. While turning at 120 rev/min, the scale reads 27 lb. If the distance d is 15 in., find the output power.

6. A permanent-magnet dc motor has a no-load speed of 900 rev/min and its stall torque is 350 ft-lb. Draw the torque–speed curve and using it, find:
(a) The load torque for which it will run at 400 rev/min
(b) The output power if it is running at 750 rev/min

7. Figure 5-29 shows the torque–speed curves of four different motors. Answer the following questions by using the curves.
(a) What is the stall torque of motor 2?
(b) What is the no-load speed of motor 1?
(c) Which motor should you choose to drive a 100 ft-lb load at 90 rev/min?
(d) Which motor would have a 2-hp 60 rev/min rating?

8. A 1-hp 1800 rev/min motor has a no-load speed of 2100 rev/min. What is its percent speed regulation?

9. The motors whose speed characteristics are shown in Figure 5-30 all draw 10 A at full load. Find the percent speed regulation of:
(a) Motor 2
(b) Motor 4
(c) Motor 1
(d) Motor 3

10. A 120-V 2-hp shunt motor draws 17 A at full load. If $R_a = 0.7 \ \Omega$, $R_f = 90 \ \Omega$, and the rated speed is 1600 rev/min, find:
(a) The stray power losses
(b) The developed power
(c) The developed torque

11. A 240-V 1650-rev/min shunt motor draws 14 A at full load. If $R_a = 0.8 \ \Omega$, $R_f = 100 \ \Omega$, and the stray power losses are 438 W, find:
(a) The back EMF
(b) The power rating
(c) The torque rating
(d) The efficiency
(e) The developed power

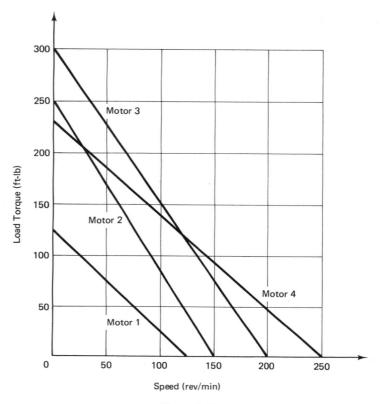

Figure 5-29

12. A 300-V 1450-rev/min shunt motor draws 15.5 A at full load. If $R_a = 0.5\ \Omega$, $R_f = 200\ \Omega$, and the stray power losses are 500 W, find:

(a) The back EMF

(b) The power rating

(c) The efficiency

13. A 1/2-hp 600 rev/min motor has a full-load speed regulation of 30%. What is its no-load speed?

14. What is the nameplate efficiency of a dc motor that is rated:

(a) 10 hp, 600 V, 15 A?

(b) 4 hp, 440 V, 8 A?

(c) 50 hp, 600 V, 75 A?

15. How much current will a 2-hp 120-V dc motor draw at full load if its rated efficiency is:

(a) 75%?

(b) 80%?

(c) 90%?

16. A 120-V dc motor draws 5 A at no load. From measurements and calculations the copper losses are found to be 340 W. What are its stray power losses?

17. A 440-V 10-hp shunt motor is tested as shown in Figure 5-9. Its armature resistance is

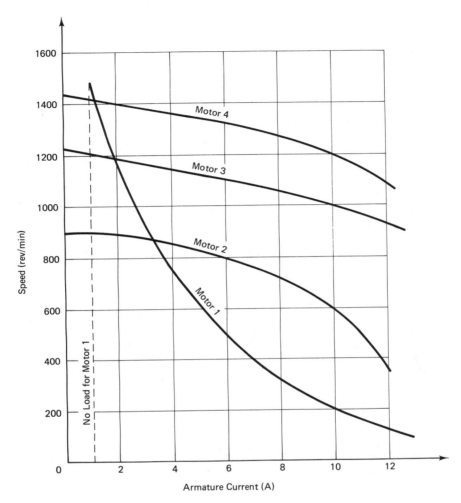

Figure 5-30

1.4 Ω and the stray power losses are constant at 220 W. The test results are given in Table 5-6.

(a) Plot the line current, speed, efficiency, and output torque versus power output, all on the same graph.

(b) At which point should maximum efficiency occur (i.e., at what line current and power output)? Compare it to the graph in part (a).

Table 5-6

I_L (A)	3	8	13	16	20	22
I_f (A)	0.8	0.8	0.8	0.8	0.8	0.8
S (rev/min)	1100	1040	980	940	900	870

18. A long shunt cumulatively compounded motor rated 115 V, 14 A, 600 rev/min is run at full load. If $R_f = 92\ \Omega$, $R_a = 0.9\ \Omega$, $R_s = 0.5\ \Omega$, and the stray power losses are 140 W, find:
(a) The back EMF
(b) The power rating
(c) The efficiency
(d) The torque rating
(e) The developed torque
(f) The electrical power converted to mechanical power (developed power)

19. A short shunt cumulatively compounded motor rated 300 V, 10 A, 900 rev/min is run at full load. If $R_f = 200\ \Omega$, $R_a = 0.8\ \Omega$, $R_s = 0.4\ \Omega$, and the stray power losses are 200 W, find:
(a) The back EMF
(b) The power rating
(c) The efficiency
(d) The torque rating
(e) The developed torque

20. A short shunt compound motor is rated 5 hp, 120 V, and draws a line current of 40 A. If $R_s = 0.04\ \Omega$, $R_f = 110\ \Omega$, and $R_a = 0.1\ \Omega$, find at rated load:
(a) The total copper losses
(b) The stray power losses
(c) The efficiency

21. Repeat Problem 20 if the motor is wired long shunt.

22. A 120-V 2000 rev/min series motor draws 60 A at full load. If $R_a = 0.1\ \Omega$, $R_s = 0.05\ \Omega$, and the stray power losses are 692 W, find at rated conditions:
(a) The total copper losses
(b) The electrical power converted to mechanical power (developed power)
(c) The rated power
(d) The rated torque
(e) The efficiency

23. A 140-V 12-A series motor has a full-load speed of 3000 rev/min. If $R_a = 0.8\ \Omega$, $R_s = 0.05\ \Omega$, and a light-load current is 3 A, find the speed at this light load. Assume that the flux is proportional to field current. (*Hint:* There are two changes taking place that affect speed; see Eq. 4-4a).

(SI)

24. How much torque will a 10-kW motor supply while running at full load at:
(a) 30π rad/s?
(b) 40π rad/s?
(c) 70π rad/s?

25. A motor must supply 5000 N-m at 3 rad/s to operate a lift at full load. What should its rating be?

26. A 1.5-kW 180 rad/s motor has 150 W of stray power losses. What is its full load:
(a) Output torque?
(b) Developed torque?

27. If the motor in Problem 26 supplies half of its rated load at 220 rad/s, how much torque is it putting out?

28. The arrangement shown in Figure 5-1 is used to measure the power output of a motor. While turning at 10 rad/s, the scale reads 100 N. If the distance d is 0.5 m, find the output power.

29. A permanent-magnet dc motor has a no-load speed of 90 rad/s and its stall torque is 500 N-m. Draw the torque–speed curve and using it find:

(a) The load torque for which it will run at 40 rad/s

(b) The output power if it is running at 75 rad/s

30. Figure 5-31 shows the torque–speed curves of four different motors. Answer the following questions by using the curves.

(a) What is the stall torque of motor 1?

(b) What is the no-load speed of motor 2?

(c) Which motor should you choose to drive a 150 N-m load at 12.5 rad/s?

(d) Which motor would have a 1.5-kW 5-rad/s rating?

31. A 1-kW 180 rad/s motor has a no-load speed of 200 rad/s. What is its percent speed regulation?

32. The motors whose speed characteristics are shown in Figure 5-32 all draw 6 A at full load. Find the percent speed regulation of:

(a) Motor 3

(b) Motor 1

(c) Motor 4

(d) Motor 2

33. A 500-W 60 rad/s motor has a full-load speed regulation of 25%. What is its no-load speed?

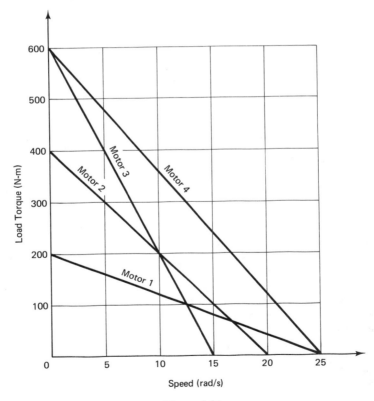

Figure 5-31

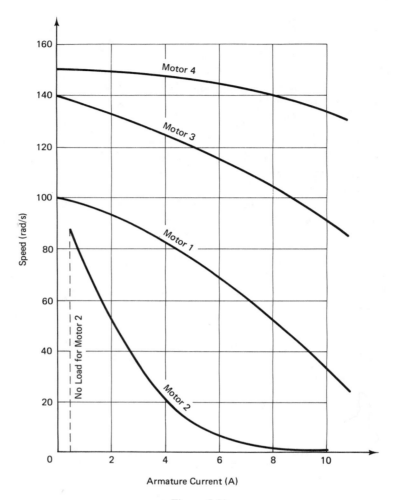

Figure 5-32

34. A 230-V 1.5-kW shunt motor draws 8.2 A at full load. If $R_a = 1.1\ \Omega$, $R_f = 200\ \Omega$, and the rated speed is 156 rad/s, find:
(a) The stray power losses
(b) The developed power
(c) The developed torque

35. A 460-V 83 rad/s shunt motor draws 7.5 A at full load. If $R_a = 1.2\ \Omega$, $R_f = 305\ \Omega$, and the stray power losses are 215 W, find:
(a) The back EMF
(b) The power rating
(c) The torque rating
(d) The efficiency
(e) The developed power

36. A 330-V 90 rad/s shunt motor draws 14.5 A at full load. If $R_a = 0.6\ \Omega$, $R_f = 220\ \Omega$, and the stray power losses are 690 W, find:
(a) The back EMF
(b) The power rating
(c) The efficiency
37. What is the nameplate efficiency of a dc motor that is rated:
(a) 3 kW, 230 V, 15 A?
(b) 40 kW, 600 V, 80 A?
(c) 5 kW, 330 V, 20 A?
38. How much current will a 1.5-kW 230-V dc motor draw at full load if its rated efficiency is:
(a) 92%?
(b) 85%?
(c) 77%?
39. A 230-V dc motor draws 3 A at no load. From measurements and calculations the copper losses are found to be 370 W. What are its stray power losses?
40. A 330-V 7.5-kW shunt motor is tested as shown in Figure 5-9. Its armature resistance is 1.2 Ω and the stray power losses are constant at 320 W. The test results are given in Table 5-7.
(a) Plot the line current, speed, efficiency, and output torque versus power output, all on the same graph.
(b) At which point should maximum efficiency occur (i.e., at what line current and power output)? Compare it to the graph in part (a).

Table 5-7

I_L (A)	4	10.5	17	21	26	30
I_f (A)	1.0	1.0	1.0	1.0	1.0	1.0
ω (rad/s)	120	112	104	99	93	85

41. A long shunt cumulatively compounded motor rated 220 V, 7.5 A, 60 rad/s is run at full load. If $R_f = 169\ \Omega$, $R_a = 1\ \Omega$, $R_s = 0.3\ \Omega$, and the stray power losses are 314 W, find:
(a) The back EMF
(b) The power rating
(c) The efficiency
(d) The torque rating
(e) The developed torque
(f) The electrical power converted to mechanical power (developed power)
42. A short shunt cumulatively compounded motor rated 330 V, 10 A, 90 rad/s, is run at full load. If $R_f = 164\ \Omega$, $R_a = 1.5\ \Omega$, $R_s = 0.2\ \Omega$, and the stray power losses are 528 W, find:
(a) The back EMF
(b) The power rating
(c) The efficiency
(d) The torque rating
(e) The developed torque
43. A short shunt compound motor is rated 5 kW, 240 V, and draws a line current of 24 A. If $R_s = 0.05\ \Omega$, $R_f = 200\ \Omega$, and $R_a = 0.4\ \Omega$, find at rated load:

(a) The total copper losses
(b) The stray power losses
(c) The efficiency

44. Repeat Problem 43 if the motor is wired long shunt.

45. A 230-V 200 rad/s series motor draws 30 A at full load. If $R_a = 0.2\ \Omega$, $R_s = 0.07\ \Omega$, and the stray power losses are 657 W, find at rated conditions:
(a) The total copper losses
(b) The electrical power converted to mechanical power (developed power)
(c) The rated power
(d) The rated torque
(e) The efficiency

46. A 250-V 8-A series motor has a full-load speed of 300 rad/s. If $R_a = 1.1\ \Omega$, $R_s = 0.06\ \Omega$, and a light-load current is 2 A, find the speed at this light load. Assume that the flux is proportional to field current. (*Hint:* There are two changes taking place that affect speed; see Eq. 4-4b.)

(English and SI)

47. A 230-V shunt motor is started with a four-point starter as shown in Figure 5-25. The full-load armature current is 8 A, $R_a = 0.8\ \Omega$, and the starting resistance is 20.5 Ω. Calculate:
(a) The starting armature current
(b) The percent of rated current that this represents

48. A 600-V 22-A shunt motor is started with a four-point starter as shown in Figure 5-25. Its field resistance is 300 Ω and $R_a = 0.7\ \Omega$. Find the starting resistance needed to limit the starting armature current to 145% of its rated value.

49. A 230-V shunt motor has a rated armature current of 8 A and $R_a = 2\ \Omega$. It is running at full load and plugging is used to stop the motor, as shown in Figure 5-28. Find the current-limiting resistor R_x needed to limit the armature current to 14 A.

50. With reference to Problem 49, how much armature current would flow on a command to stop if the external resistor were not used?

Chapter 6

Control of Direct-Current Motors

[Handwritten notes:]

3 PHASE INDUCTION MOTOR

P-300

ACTUAL SPEED OF MOTOR $S_{ACTUAL} = \frac{120 f}{f}(1-S)$

f REQ OF ROTOR = slip (S) × line freq (S line)

IMPEDANCE OF THE ROTOR = slip (S) × IMPEDANCE OF A blocked ROTOR

$X_R = S X_{BR}$ $S X_{BR} = 2\pi f L_R$

FULL IMPEDANCE $Z_R = \sqrt{R_R^2 + (S X_{BR})^2}$

INDUCED ER VOLTAGE (STANDSTILL) E_{BR}

$E_R = S E_{BR}$

In our industrial society the motor has become a workhorse. It has found its way into our homes, offices, and factories in almost every conceivable way. Motors are in toys, appliances, computer peripherals, typewriters, automobiles, aircraft, and are used to drive conveyor belts in the consumer product industry. The list can go on indefinitely; however, whatever the application we are always faced with the task of controlling their speed or position. This can be done manually or automatically. The speed control on a fan or blender in the home can be considered semiautomatic. The user must press a button or turn a knob to vary the motor speed. The speed of a motor that turns a generator is usually held constant. This is done automatically. The automatic controller can be an electronic circuit, or now that the microprocessor has become so inexpensive, a computer.

The electronic devices and techniques for controlling motors are so extensive that entire books on them have been written. The topic is only touched on in this chapter. However, computer control, in particular microprocessor control, is discussed more extensively since it will someday be the predominant technique if it is not already. It will be necessary to have a basic understanding of assembly language programming and some common electronic devices to benefit fully from the material in this chapter. Although any microprocessor can be used, the circuitry and software presented here will be tailored for the Motorola 6800 CPU, a very popular microprocessor.

6-1   TECHNIQUES OF SPEED CONTROL

There are three basic ways of controlling a motor's speed: field control, armature resistance control, and armature voltage control. No matter what the controlling

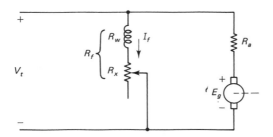

Figure 6-1 Field control of motor speed.

device is (computer, pushbutton, etc.), it will ultimately control the motor's speed by varying one of these three quantities.

6-1.1 Field Control

A simple circuit for manually controlling a shunt motor's speed by varying the flux is shown in Figure 6-1. We saw in Chapter 5 (see Eq. 5-19) that the speed is inversely proportional to flux. Since the flux is proportional to field current (I_f) and field current is inversely proportional to field resistance (R_f), we can say that the speed is directly proportional to R_f. That is, an increase in R_f will cause I_f to decrease, causing a corresponding decrease in flux. This will cause the speed to increase.

The converse is also true. As R_f decreases, the speed will also decrease. We are limited, however, when we wish to run the motor at slow speeds. Examination of Figure 6-1 reveals that the minimum R_f occurs when the rheostat (R_x) is at its zero setting. At this point R_f is equal to the field winding resistance R_w. Hence the minimum speed attainable with field control is limited by the winding resistance, which is constant. The minimum speed of the motor with field control is called its **basic speed.** Thus field control can be used only for speed variation above the basic speed of a motor.

Another drawback of field control presents itself at high speeds. To obtain high speeds the flux must be made very weak. Since the back EMF (E_g) is proportional to flux, it will decrease, causing the armature current to rise considerably. The low flux and high armature current can lead to a very unstable situation, due to armature reaction (see Section 3-1.3). In addition, brush arcing due to the high current will shorten the life of the commutator and brushes.

In spite of these disadvantages, field control is still widely used since it is inexpensive, easy to use, and provides for a continuous adjustment of speed (i.e., without steps).

6-1.2 Armature Resistance Control

A typical circuit used to adjust a motor's speed by armature resistance control is shown in Figure 6-2. By introducing a resistance (R_x) in series with the armature we will cause a decrease in armature current. The drop in I_a will cause a decrease in torque. Since the developed torque is now lower than that required to drive the present load, the motor will slow down. The speed reduction will cause the back EMF to drop, which in turn will increase I_a to the value it had before R_x was introduced. The motor

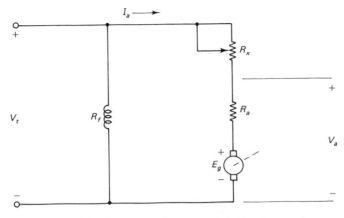

Figure 6-2 Armature resistance control of motor speed.

will now be developing sufficient torque to drive the load; however, it will be running at a slower speed.

We can therefore see that an increase in armature circuit resistance causes a decrease in motor speed. By making R_x very large, the motor could even be stopped. This is, in fact, a technique used to start and stop motors.

The maximum speed attainable occurs when the armature circuit resistance is a minimum. This minimum value is the armature resistance (R_a) and occurs when R_x is zero. Therefore, the maximum speed attainable with armature resistance control is the basic speed of the motor.

Thus we can see that armature resistance control is fairly easy to implement, permits starting without the need of a special starter, and allows speed control below the basic speed of a motor.

There are some disadvantages that should be mentioned. Since armature currents are normally high under load conditions, the rheostat used (R_x) would need a large power rating, making it quite expensive. In addition, the chain of events between the time when R_x is changed and when the speed stabilizes can take several seconds, especially in large machines. As a result, a few adjustments of R_x must be made before the selected speed is arrived at. In addition, the sensitivity to load will cause poor speed regulation.

6-1.3 Armature Voltage Control

We saw in Section 5-2 that a dc motor will run at a speed such that the back EMF is approximately equal to the armature voltage. The two voltages will never actually be equal, the difference between them being a function of the load. The fact remains, however, that as the armature voltage changes, the back EMF will similarly change. With constant flux, the speed of the motor must therefore change to produce the necessary back EMF. Since we want to vary the armature voltage without changing the flux, the motor must be either separately excited or a permanent-magnet type. A simplified diagram is shown in Figure 6-3.

As the armature voltage (V_a) is varied, the speed of the motor will vary proportionately. In addition, if the polarity of V_a is reversed, the direction of rotation of the

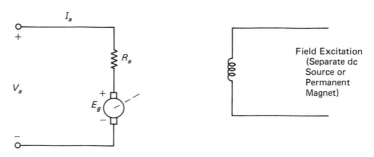

Figure 6-3 Armature voltage control of motor speed.

motor will also reverse, since E_g must eventually oppose the armature voltage. The source of V_a and the method used for varying it depend on the power rating of the motor. Very small motors such as those used in toys can be powered and controlled with simple flashlight batteries. As the power requirement rises, electronic power supplies are used and the variation in voltage is accomplished with operational amplifiers, power transistors, or power thyristors. For large power applications a dc generator is used to provide the armature voltage. In a technique called the **Ward–Leonard system** the generator and motor field windings are both excited with a second generator called an exciter. In addition, a second motor serves as the prime mover for the generator and exciter. A simplified diagram of the system is shown in Figure 6-4.

The generator as it is used in the Ward–Leonard system is called a **rotary amplifier.** Although this system is quite expensive, the cost is justified when motors in the 100-hp range are being controlled. The benefits are full range of speed control from standstill to higher than basic speed, easy reversal of speed (using armature voltage reversing switches not shown), and excellent speed regulation.

Figure 6-5 is a diagram comparing the speed control ranges of the three techniques described. There are many schemes used, however, they all ultimately use one of these three principles or a combination of them.

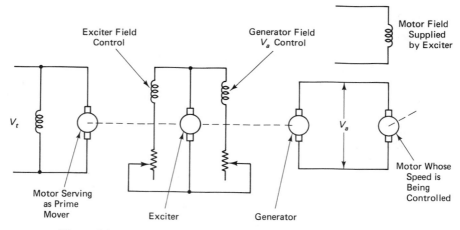

Figure 6-4 Simplified diagram of a Ward–Leonard speed control system.

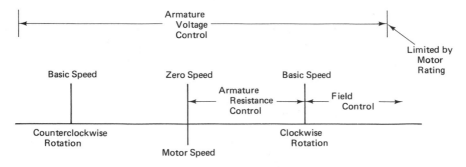

Figure 6-5 Comparison of motor speed control techniques.

6-2 *AUTOMATIC SPEED CONTROL*

An automatic speed control system will be defined here as one in which a motor will run at a preselected speed without the need for an operator to monitor it. Most automatic controllers use armature voltage control.* Speed control falls into two basic categories: open loop and closed loop.

An **open-loop** speed control system is one in which the motor's armature voltage is strictly a function of an input signal representing the commanded speed of the motor. A **closed-loop** speed control system is one in which the motors armature voltage is a function of both its input signal and a **feedback signal.** The feedback signal is a voltage representative of the motor's actual speed. In this way a closed-loop system can self-correct for changes in load, whereas an open-loop system cannot. Before we continue, some basic principles and devices must be reviewed.

6-2.1 DC Waveforms

Up to this point we have considered only dc voltages that are constant for all time, such as the one shown in Figure 6-6a. However, by definition, a dc voltage is one that produces a current in only one direction. Therefore, all of the waveforms shown in Figure 6-6 represent dc voltages and all could be used to drive a dc motor. The speed of the motor is proportional to the average (or dc) value of the armature voltage which is indicated for each of the waveforms.

6-2.2 Thyristors

This is a general term referring to a group of semiconductor devices which are in effect switches. Each of these devices is different in the way that it is constructed; thus each has different characteristics. Four of them will be discussed here: the silicon-controlled rectifier (SCR), bidirectional triode thyristor (TRIAC), the bidirectional diode thyristor (DIAC), and the optically coupled TRIAC driver. Their schematic symbols are shown in Figure 6-7.

---

*This is done because it is much easier and less expensive to control a voltage automatically rather than using a rheostat, which is necessary with field and armature resistance control.

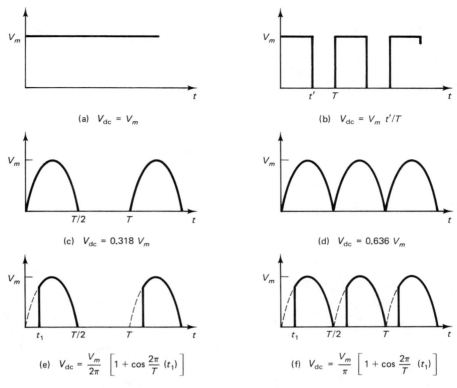

(a) $V_{dc} = V_m$

(b) $V_{dc} = V_m \, t'/T$

(c) $V_{dc} = 0.318 \, V_m$

(d) $V_{dc} = 0.636 \, V_m$

(e) $V_{dc} = \dfrac{V_m}{2\pi} \left[1 + \cos \dfrac{2\pi}{T} \, (t_1) \right]$

(f) $V_{dc} = \dfrac{V_m}{\pi} \left[1 + \cos \dfrac{2\pi}{T} \, (t_1) \right]$

Figure 6-6 Different dc waveforms: (a) constant; (b) pulse train; (c) half-wave rectified sine wave; (d) full-wave rectified sine wave; (e) portion of half-wave rectified sine wave; (f) portion of full-wave rectified sine wave.

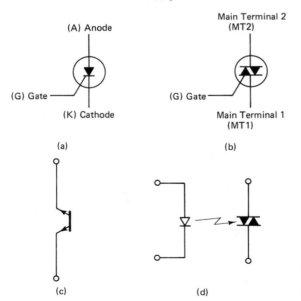

(A) Anode

(G) Gate

(K) Cathode

(a)

Main Terminal 2
(MT2)

(G) Gate

Main Terminal 1
(MT1)

(b)

(c)

(d)

Figure 6-7 Symbols for different thyristors: (a) SCR; (b) TRIAC; (c) DIAC; (d) optically coupled TRIAC.

6-2.2.1 SCR. The most popular thyristor is the SCR. It is a three-terminal device which can conduct in only one direction (like a diode). When the anode is positive with respect to the cathode, the SCR will start conducting when a specified amount of current flows into the gate. If the gate current is now removed, the SCR continues to conduct until the current from anode to cathode drops below some minimum value called the **holding current.** In ac applications the number of degrees of the ac cycle during which the SCR is on is called the **conduction angle.** The number of degrees of the ac cycle that pass before the SCR turns on is called the **firing delay angle.** The following example will help illustrate the SCR's behavior.

Example 6-1

An SCR is used to control the average voltage to a load. The device used is a 2N3871 with the following specifications: gate trigger current of 10 mA (based on gate pulse width of 1 ms), gate trigger voltage of 0.7 V. Analyze the circuit operation and design it. Figure 6-8 is a circuit diagram and Figure 6-9 shows the various waveforms. The supply V_s (120V rms, 60 Hz) and gate trigger voltage V_G are both given.

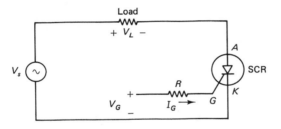

Figure 6-8 Circuit for Example 6-1.

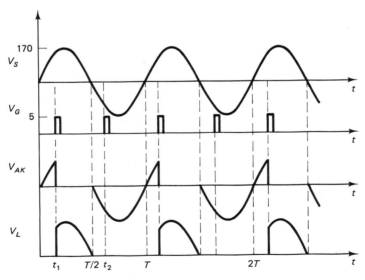

Figure 6-9 Waveforms for Example 6-1.

Solution

Before any calculations are made, we will describe the circuit operation, assuming that it will function properly. To simplify the problem, the pulse generating circuitry has been left out and we are given its output waveform (V_G), which is applied to the gate–cathode junction. The waveform V_G is a pulse of 0.1 ms width occurring 2 ms after every zero crossing of the supply voltage from which it is derived. Therefore, the time t_1 in Figure 6-9 is 2 ms.

Prior to the arrival of the first gate pulse the SCR is off and is therefore an open circuit. There is no current flow, hence V_L is zero and the supply voltage is entirely across the SCR (V_{AK}). When the first gate pulse arrives, the conditions to fire the SCR are met; sufficient current is injected into the gate and the anode is positive with respect to the cathode (V_{AK} is positive). The SCR is now on (a short circuit), V_{AK} is equal to zero, and the entire supply voltage will be across the load. When the first pulse is removed, the SCR remains on until the current through it drops below its holding current (about 5 mA for the 2N3871). This occurs just before time $T/2$, where V_S is approaching zero volts. When the second gate pulse arrives, the SCR remains off (does not fire). Even though the gate has been injected with sufficient current, the second condition is not being met. At this time (t_2) the anode is negative with respect to the cathode (V_{AK} is negative).

At time T the entire cycle will repeat itself; thus the load voltage V_L will be the repetitive waveform shown in Figure 6-9. Note the resemblance of this waveform to Figure 6-6e. It should also be noted that the SCR provides half-wave rectification (conversion to dc) automatically and by controlling the time t_1 the average value of the load voltage can be varied.

Calculations:

First the resistor R must be solved for from Figure 6-8 and the SCR specifications:

$$V_G = RI_G + V_{GK}$$

$$RI_G = V_G - V_{GK}$$

$$R = \frac{V_G - V_{GK}}{I_G}$$

$$= \frac{5\ \text{V} - 0.7\ \text{V}}{10\ \text{mA}}$$

$$= \frac{4.3\ \text{V}}{10\ \text{mA}}$$

$$= 430\ \Omega$$

This is a standard-size carbon composition resistor and could be used. However to ensure proper firing of the SCR at different temperatures, a smaller resistor could be used which would provide for a larger gate current. The limiting factor would be the peak forward gate current, which is 2 A for the 2N3871.

The maximum value of the supply voltage is obtained from its rms value.

$$V_m = \sqrt{2} \times V_{\text{rms}} = \sqrt{2} \times 120\ \text{V}$$

$$= 169.7 \approx 170\ \text{V}$$

Since the supply voltage is 60 Hz,

$$T = \frac{1}{60} = 16.67 \text{ ms}$$

$$\frac{T}{2} = 8.33 \text{ ms}$$

and t_1 is given as 2 ms. Using the equation given in Figure 6-6e yields

$$V_{dc} = \frac{170}{2\pi}\left[1 + \cos\frac{2}{16.67 \text{ ms}}(2 \text{ ms})\right]$$

$$= 27.1(1 + \cos 0.75)$$

$$= 27.1(1 + 0.73) = 46.9 \text{ V}$$

The firing delay angle (α_0) is that angle of the supply voltage corresponding to time t_1. For the supply

$$\omega = 2\pi f = 2\pi \times 60 \text{ Hz}$$

$$= 377 \text{ rad/s}$$

$$\alpha_0 = \omega t_1 = 377 \text{ rad/s} \times 2 \text{ ms}$$

$$= 0.75 \text{ rad}$$

or

$$\alpha_0 = 0.75 \text{ rad} \times 57.3 \text{ deg/rad}$$

$$= 43.2°$$

The conduction angle (α) is

$$\alpha = 180° - \alpha_0 = 180° - 43.2°$$

$$= 136.8°$$

6-2.2.2 DIAC. The DIAC is a two-terminal device whose characteristics make it particularly suitable as a triggering device for other thyristors. Its current–voltage characteristic is shown in Figure 6-10. As the voltage across its terminals increases, it is in a high-impedance state (open switch). In this state it actually passes a small current (10 to 100 μA). When the voltage reaches the device's switching voltage (V_d), the DIAC will fire. For 120-V ac applications, V_d is typically in the range 20 to 40 V. When the DIAC fires, its voltage will drop and the current through it will jump to a value dependent on the remainder of the circuit. The device has polar symmetry, that is, will behave the same way for negative voltages; however, the current flow will be opposite.

6-2.2.3 TRIAC. The TRIAC is quite similar to the SCR. It has three terminals and can be used to control a voltage. Unlike the SCR, however, the TRIAC will conduct in either direction. This makes it ideal for ac power control and the terms "anode" and "cathode" are not used.

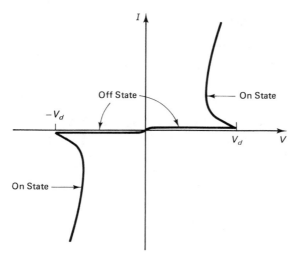

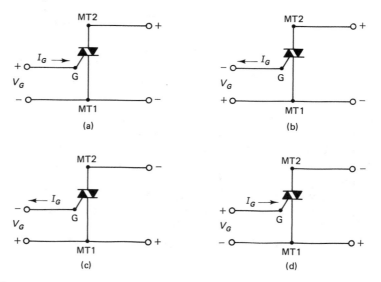

Figure 6-10 Typical volt–ampere characteristics of a DIAC.

When a voltage is applied to the main terminals of a TRIAC, it remains in a high-impedance state (open switch) until the appropriate voltage and current are applied to the gate. The amount of gate current needed to fire the TRIAC is a function of the polarity of the voltages applied to the gate and across the device's main terminals. It is also a function of temperature; however, we will not consider that variation here. There are four possible polarity conditions for firing a TRIAC. They are shown in Figure 6-11.*

Figure 6-11 The four possible voltage polarities which can be applied to a TRIAC: (a) MT2(+), G(+); (b) MT2(+), G(−); (c) MT2(−), V_G(−); (d) MT2(−), V_G(+).

*This is called four-quadrant triggering. Some TRIACS are guaranteed only for two-quadrant triggering: that is, MT2(+) with G(+) and MT2(−) with G(−) (Figure 6-11a and c).

To illustrate the different triggering conditions, a typical TRIAC's specifications are listed in Table 6-1. The TRIAC chosen is the 2N6346, which is well suited for low-power 120-V ac applications since it can block 200 V in the OFF state.

Table 6-1 2N6346 Specifications

Polarity conditions	Gate trigger current (mA)		Gate trigger voltage (V)	
	Typical	Max.	Typical	Max.
MT2(+), G(+)	12	50	0.9	2.0
MT2(+), G(−)	12	75	0.9	2.5
MT2(−), G(−)	20	50	1.1	2.0
MT2(−), G(+)	35	75	1.4	2.5

Note: The minimum gate pulse width is 2 μs.

6-2.2.4 *Optically Coupled TRIAC Driver.*

The TRIAC and its use in controlling ac power are discussed and applied in Chapter 13. It was introduced here so that we can understand the Optically Coupled TRIAC Driver which is used in dc control when isolation between circuits is necessary. When motors and other inductive loads are controlled by computers, care must be taken to isolate the computer. If this is not done, large voltage transients produced by the switching of currents in inductive loads can damage the computer and its input/output circuitry. The schematic symbol for the driver is shown in Figure 6-7d.

The device contains a light-emitting diode (LED) which emits infrared light when current flows through it. The infrared light strikes a bidirectional photodetector which fires (becomes a closed switch). The driver can be used in two ways: as a TRIAC controlling power to small loads with its triggering circuitry isolated, and as a means of triggering other high-power SCRs and TRIACs, which must be isolated from the triggering circuitry. An example of the first of these uses is shown in Figure 6-12.

When the input is high, the output of the TTL gate will go low, sinking the LED current (approximately 10 mA). This fires the driver and the lamp will turn on. Note

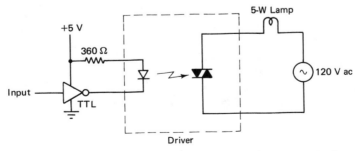

Figure 6-12 Optically coupled TRIAC driver used to interface TTL circuitry to a 120-V ac load.

that the driver's output current is about 42 mA (5 W/120 V), which is typical for a device like this.

6-2.3 Computer Speed Control Using an SCR

We saw in Section 6-2.2.1 that the SCR can be used to vary the average (dc) voltage to a load and simultaneously rectify an ac voltage. This makes it ideal for controlling the speed (using armature voltage control) of a dc motor when the source of power is ac. In this section we will see how a microprocessor (the Motorola 6800) could be used to do the job. The task will be broken up into three parts: first, we examine the circuitry needed to interface a motor with TTL circuitry (low-power integrated circuits); second, we examine the hardware necessary to connect the microprocessor via an output port to the TTL circuitry; and finally, we look at a computer program, written for the M6800 microprocessor in both assembly language and machine code, which could be used to control the motor's speed.

1. The circuitry used to interface the motor with the low-power TTL integrated circuitry will make use of an optically coupled TRIAC driver. In this way the computer circuitry will be optically isolated from the 120-V ac supply. An example of this arrangement is shown in Figure 6-13. In this scheme we are using half-wave control of the motor since it is easier to visualize. A scheme with two SCRs providing full-wave control could also be used.

When the input is low, the output of the inverter will be high and the driver LED will be off. This will keep the driver's main terminals open, preventing gate current flow into the SCR. Without gate current the SCR will be off (open switch). The motor will have zero volts across it and therefore will not run. The full 120 V will be across the SCR.

When the SCR turns on and off, it is switching an inductive load (the motor) on and off. When this happens, large voltage transients are induced, which could damage the SCR. Even if the SCR is not damaged, the transients could make the SCR operate improperly. That is, by placing a large voltage across the SCR, the required gate current to turn the device on and off will change. To eliminate the above problems, an *RC* snubber circuit is placed in parallel with the SCR. This is, in effect, a low-pass filter. It does not effect the line frequency; however, it will short out the high-

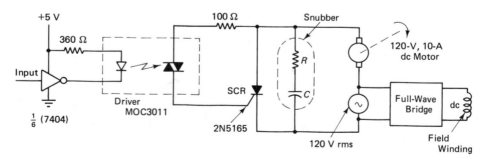

Figure 6-13 Circuit diagram showing the first part of the computer speed control task.

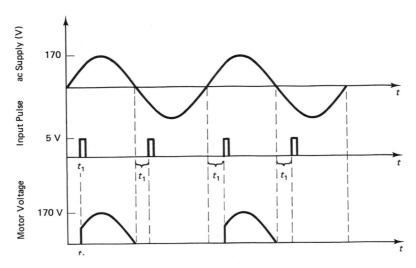

Figure 6-14 Typical waveforms for the speed control task.

frequency transients. Typical values for R and C are 100 Ω and 1 to 2 μF, respectively.

When the input is set high, the inverter output goes low, sinking current for the LED and turning it on. This fires the driver (closed switch), causing an inrush of current into the SCR's gate. Assuming the typical* on-state voltage of the driver (2.5 V) and gate trigger voltage of the SCR (1.5 V), there will be a maximum of 166 V (170 − 2.5 − 1.5) across the 100 Ω resistor. Remember that 120 V rms has a peak value of 170 V. The SCR gate current will be 1.66 A (166 V/100 Ω), neglecting the motor's armature resistance. This is below the maximum allowable forward gate current of the 2N5165 (2 A). Once the SCR fires, this current will drop to a very small value since the on-state voltage across the SCR will be about 1.5 V (taken from the data sheet).

An example of an input waveform and its effect on the average armature voltage is shown in Figure 6-14. The pulse width, which will be controlled by the computer program, will be 150 μs. This is an arbitrary choice and will work fine since it is greater than the minimum pulse width needed (2 μs) to fire the SCR. The time t_1, also controlled by the program, will vary the average armature voltage (see Figure 6-6e). The computer must be synchronized with the ac supply so that the pulse is generated at a time t_1 after every zero crossing. The synchronization technique will be covered in the next two parts.

2. This part of the task includes the computer I/O (input/output) circuitry. Both are necessary since the computer must sense when the ac supply is crossing zero (input port) and must generate the correct pulse width at the specified time t_1 to drive the TTL gate (output port). The circuitry for the I/O port is shown in Figure 6-15. Partial address decoding is used with three address bits (A15, A14, and A12) fed into the 7442 (a 4 line-to-10 line decoder). It is assumed that the memory map of our computer will permit this arrangement. The address decoder permits the input port and output

*The typical values for these devices were taken from their data sheets.

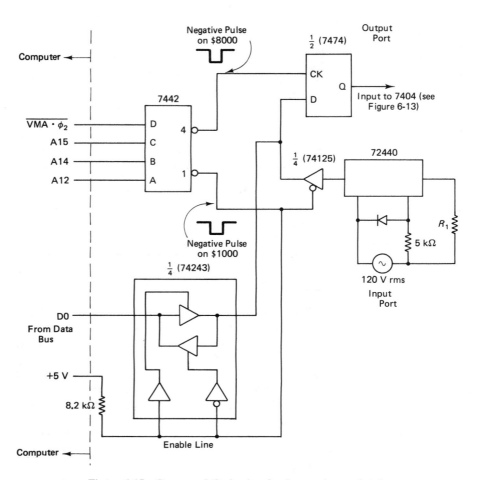

Figure 6-15 Computer I/O circuitry for the speed control task.

port to be addressed with $8000 and $1000, respectively. To detect the zero crossings of the ac supply, a zero voltage switch (SN72440) is used. This integrated circuit generates a pulse every time the ac waveform crosses zero volts. Most of the circuitry (resistors, capacitors, and diodes) that must be connected to the SN72440 has been left out for clarity. The reader is referred to an interface circuits data book for the complete hookup. The value of resistor R1 controls the output pulse width. We will use R1 equal to 15 kΩ, which gives a pulse width of 100 μs. As long as it is shorter than the gate trigger pulse width of the SCR (chosen at 150 μs), and long enough so that our program does not miss it, everything will work out well. This criterion will be explained in part 3.

A bidirectional buffer (the 74243) is necessary so that data can flow into or out of the computer. Normally, the buffer is set up to transmit data out of the computer. This is ensured by the 8.2 kΩ pull-up resistor holding the enable line high. When data is to be fed into the computer, the enable line will be pulled low by the negative-going pulse generated for the input port by the 7442 decoder.

The output port consists of a D-type flip-flop (the 7474), which serves as a latch to store the data generated by the program on data bus line $D0$. The output of the 7474 is the input to the motor drive circuit shown in Figure 6-13. It is also the pulse waveform shown in Figure 6-14.

3. The third and final part of the speed control task is the computer program. Although this program could run by itself, it might also be part of a larger program, which is in control of several processes simultaneously. Many times this type of program will be an interrupt routine, that is, one that is accessed by interrupting the computer either automatically or manually by an operator pressing a button.

In our case we will assume that the speed control program is the main program and that it gets interrupted by an external source. The interrupt can be performing one of several things: start the motor, stop the motor, increase the speed, or decrease the speed. In each case all that is done is to make the main program stop momentarily, jump to a small subroutine where some data is changed, and then jump back to the main program.

Before examining the program operation, we will perform a numerical analysis. When the motor starts up or changes speed, it will be done gradually. In this way an excessive starting current will be avoided. The speed will change after every 5-s interval until the final speed is reached. To determine when 5 s has elapsed, the program counts the pulses received from the 72440 (see Figure 6-15). Since the ac supply is 60 Hz and there are two zero crossings per cycle, the pulses out of the 72440 will be at a 120-Hz rate.

$$120 \frac{\text{pulses}}{\text{s}} \times 5 \text{ s} = 600 \text{ pulses}$$

$$(600)_{10} = (0258)_{16}$$

The hexidecimal (hex) number 0258 is entered into the program with the label PCNT. When 600 pulses are counted, the speed will be changed if it has not yet reached the final speed.

The speed is controlled by varying the time delay (t_1 in Figure 6-14), which in effect is varying the average (dc) armature voltage.

With zero time delay,

$$V_{dc} = 0.318 \times V_m = 0.318 \times 170 = 54 \text{ V}$$

This represents the maximum voltage, hence maximum speed, which can be attained with the half-wave configuration shown in Figure 6-13. In our task the decision was made to bring the motor from zero speed to maximum speed in seven steps. The seven steps correspond to time delays of 6 ms down to 0 in 1-ms steps.

Thus using the equation in Figure 6-6e for $t_1 = 6$ ms,

$$V_{dc} = \frac{170}{2\pi} \left(1 + \cos \frac{2\pi}{T} t_1 \right)$$

for 60 Hz, $T = \frac{1}{60}$, and

$$V_{dc} = \frac{170}{2\pi} (1 + \cos 377 t_1)$$

Substituting the various time delays selected into the equation above, we can calculate the voltage at each step. They are shown in Table 6-2. The initial time delay (6 ms in our case) is entered with the label T1INIT and the final time delay (0 in our case) is entered with the label T1FINL. A 1-ms delay is accomplished by going through a loop, which takes eight machine cycles, 125 ($7D) times. This will use 8×125, or 1000 machine cycles. Since our computer has a 1-MHz clock, each cycle takes 1 μs (1/1 MHz); hence 1000 cycles will take 1000×1 μs, or 1 ms. Therefore, to get an initial delay of 6 ms, T1INIT must be set to the hex equivalent of 750 (6×125), which is $02EE. Every time 5 s has elapsed, the present time delay T1PRES is reduced by $7D (1 ms) until T1FINAL is reached.

Table 6-2

t_1 (s)	V_{dc} (V)
6×10^{-3}	9.8
5×10^{-3}	18.7
4×10^{-3}	28.8
3×10^{-3}	38.6
2×10^{-3}	46.8
1×10^{-3}	52.2
0	54.1

Whenever an output pulse is to be generated, the contents of STCOM are examined. If they are found to equal $FF, we jump out of the program to STOP. In our case when the program goes to STOP, it jumps back to the monitor program and awaits an interrupt directing the computer to start the program again (JMP $0000).

The program operation can best be described by examining a flowchart (Figure 6-16) together with the program itself (Table 6-3). The program starts by initializing the following four items:

1. STCOM is cleared; this ensures that the command to stop the motor $FF is not present from a previous use.
2. The output at $8000 (input to the motor drive circuit) is set to zero.
3. The initial time delay (t_1) is read into the program from T1INIT. This sets the initial voltage or starting current of the motor. In our example the initial voltage is 9.8 V (see Table 6-2). If the armature resistance were 1 Ω, the starting current would be 9.8 A.
4. The pulse count is read into the program from PCNT. This number determines the interval between changes in voltage. It was arbitrarily set to 5 s in our case.

As we enter the program loop, the question is asked: Should we stop the motor? If yes, then branch to the monitor or perhaps another program. If no, then continue.

The next step is to monitor the input port ($1000) continuously until a 1 is detected in data bit D_0. This means that we are receiving a pulse and that the ac supply is crossing zero. After this the program waits for the time delay (t_1) and then proceeds to output a pulse at address $8000.

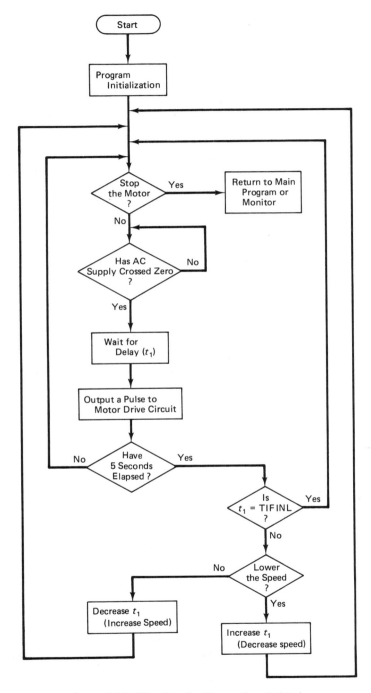

Figure 6-16 Flowchart for the speed control task.

Table 6-3

```
0000  7F 00 70              CLR     STCOM
0003  4F                    CLR   A
0004  B7 80 00              STA   A $8000
0007  FE 00 6E              LDX     T1INIT
000A  FF 00 6A              STX     T1PRES
000D  FE 00 6C              LDX     PCNT
0010  FF 00 71              STX     SAVEP
0013  F6 00 70      LOOP    LDA   B STCOM
0016  C1 FF                 CMP   B #$FF
0018  27 4D                 BEQ     STOP
001A  B6 10 00      ZCROSS  LDA   A $1000
001D  44                    LSR   A
001E  24 FA                 BCC     ZCROSS
0020  FE 00 6A              LDX     T1PRES
0023  08                    INX
0024  09            DELAY   DEX
0025  26 FD                 BNE     DELAY
0027  86 01                 LDA   A #$01
0029  B7 80 00              STA   A $8000
002C  CE 00 12              LDX     #$12
002F  09            PW      DEX
0030  26 FD                 BNE     PW
0032  4F                    CLR   A
0033  B7 80 00              STA   A $8000
0036  FE 00 71              LDX     SAVEP
0039  09                    DEX
003A  FF 00 71              STX     SAVEP
003D  26 D4                 BNE     LOOP
003F  FE 00 6C              LDX     PCNT
0042  FF 00 71              STX     SAVEP
0045  FE 00 6A              LDX     T1PRES
0048  BC 00 73              CPX     T1FINL
004B  27 C6                 BEQ     LOOP
004D  2D 0C                 BLT     SLOWER
004F  C6 7D                 LDA   B #$7D
0051  09            CHT1    DEX
0052  5A                    DEC   B
0053  26 FC                 BNE     CHT1
0055  FF 00 6A              STX     T1PRES
0058  7E 00 13              JMP     LOOP
005B  C6 7D         SLOWER  LDA   B #$7D
005D  08            INT1    INX
005E  5A                    DEC   B
005F  26 FC                 BNE     INT1
0061  FF 00 6A              STX     T1PRES
0064  7E 00 13              JMP     LOOP
0067  7E FC 00      STOP    JMP     MONITR
006A                T1PRES  RMB     2
006C                PCNT    RMB     2
006E                T1INIT  RMB     2
0070                STCOM   RMB     1
0071                SAVEP   RMB     2
0073                T1FINL  RMB     2
FC00                MONITR  EQU     $FC00
                            END
```

The pulse count is decremented and we ask the question: Have we counted 600 pulses? In other words, has 5 s elapsed? If not, we return to LOOP, check the stop command, and wait for the next zero crossing. If 5 s has elapsed, we must check the time delay. If it is equal to T1FINL (zero in our case), maintain the present delay T1PRES and go back to LOOP. If we have not yet reached the final delay, ask the question: Do we have to decrease t_1 (increase the speed) or increase t_1 (decrease the speed)? In either case we proceed to adjust t_1 by 1 ms and then jump back to LOOP.

The program continues until an interrupt causes $FF to appear in STCOM. We then jump out of the program loop, no more pulses are generated, and either the motor will stop or T1FINL is changed (calling for a new speed) and we then return to the start of our program at $0000. If the latter is the case, the motor would never really stop since the interrupt routine calling for a change in speed would take but a few microseconds.

The task just covered is about as simple as we could get. To get higher speeds, the full-wave drive circuit shown in Figure 6-17 could be used in place of Figure 6-13. This would double the maximum dc voltage available to the motor (compare Figure 6-6e to Figure 6-6f), which would enable us to run the motor at about twice the previous speed. The dc excitation for the field circuit has been left out of Figure 6-17 intentionally. It must be included unless the motor is of the permanent-magnet type.

To develop our task further and give it the capability of driving the motor in either direction would require the addition of switches (either semiconductor or relays) to reverse the polarity of the armature voltage. The configuration would be somewhat similar to the one in Figure 5-28, which was used for plugging. Here, too, the switch control would be optically coupled to the computer.

6-2.3.1 Closed-Loop Speed Control.

The speed control task just discussed was an example of open-loop control. Although it is simple to explain and build, it has one major drawback. As the load on the motor increases, the speed of the motor will decrease. Since there is nothing used that can sense the actual speed, neither the computer nor the drive circuitry is aware that the motor has slowed down. Under the increased load it will not run at the speed for which it has been programmed.

In this section we look at closed-loop speed control. We will see that for changes in speed, corrective action will be taken automatically. In this way the motor will run at the selected speed. We will look at two examples. In the first the computer will be **off-line.** An off-line computer control system is one in which the computer selects

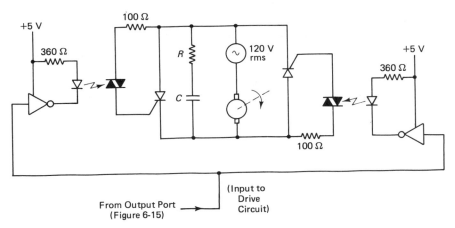

Figure 6-17   Full-wave drive circuit for the speed control task.

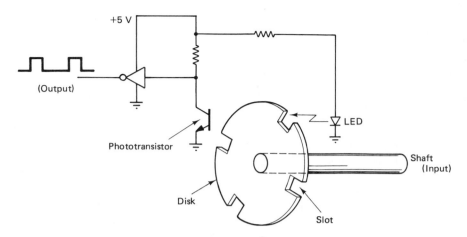

Figure 6-18 Simplified diagram of an incremental shaft encoder.

only the desired speed of the motor and produces the necessary input. The correction for speed changes is done with additional circuitry.

In the second example the computer will be **on-line.** An on-line computer control system is one in which the computer senses the actual speed and adjusts the input so that the desired speed is reached. In other words, the computer is involved in the corrective action.

Before we look at the off-line system, a special device will be discussed. An **incremental shaft encoder** is a device that has as its input a shaft that can rotate. The output of the encoder is an electrical signal. As the encoder shaft turns it produces a pulse waveform on its output. Depending on the construction of the encoder, it will produce a pulse every time its shaft turns through a fixed angle. A simplified picture of an incremental shaft encoder, which provides optoisolation as well, is shown in Figure 6-18. For the encoder shown, there are four slots spaced 90° apart on a disk. Whenever a slot passes the LED, light will shine through it. The passing light strikes the phototransistor. This turns the transistor on, driving its collector to zero volts. When this happens the inverter output jumps up to 5 V, producing a pulse. The frequency of the pulse train is a function of the shaft speed and the number of slots on the disk. Equations 6-1 relate the frequency (*f*) of the pulse train in hertz to the number of slots on the disk (*n*) and the speed of the shaft. The speed has units or rev/min and rad/s for the English and SI unit conventions, respectively.

(English)

$$f = \frac{Sn}{60} \qquad (6\text{-}1a)$$

(SI)

$$f = \frac{\omega n}{2\pi} \qquad (6\text{-}1b)$$

Example 6-2 (English)

An incremental shaft encoder has 40 equally spaced slots on its disk. If the shaft is rotating at 1800 rev/min, what will be the frequency of the output pulse train?

Solution

Direct substitution into Eq. 6-1a gives

$$f = \frac{1800 \text{ rev/min} \times 40 \text{ slots/rev}}{60 \text{ s/min}}$$

$$= 1200 \text{ Hz} = 1.2 \text{ kHz}$$

Example 6-3 (SI)

An incremental shaft encoder has 60 equally spaced slots on its disk. If the shaft is rotating at 200 rad/s, what will be the frequency of the output pulse train?

Solution

Direct substitution into Eq. 6-1b gives

$$f = \frac{200 \text{ rad/s} \times 60 \text{ slots/rev}}{2\pi \text{ rad/rev}}$$

$$= 1910.8 \text{ Hz} = 1.91 \text{ kHz}$$

By making the slots the same size as the space between them, the output waveform would be a square wave. Thus in Example 6-2, if the slots are 4.5° and the spacing between the slots is 4.5°, the output would be a square wave. For Example 6-3 the slots and spacing would be 3° to give a square-wave output.

Off-Line Control. There are many closed-loop speed controllers. The one described here will make use of a **phased-locked loop** (PLL). The PLL can best be described by examining Figure 6-19.

The phase-locked loop is used in many ways. They all however are based on the same principle. Two frequencies are compared, and one of them, the feedback frequency, is driven to be equal or very close to another (the input frequency). In our case

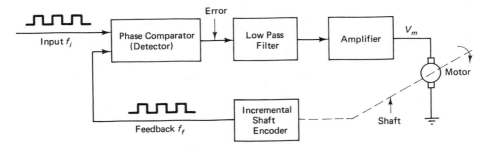

Figure 6-19 Simple phase-locked loop.

the input frequency is generated by the computer. The feedback frequency is generated by a shaft encoder that is driven by the motor. As we have just seen, the encoder frequency is proportional to its shaft speed (Eqs. 6-1). Therefore, by increasing the input frequency, the encoder frequency (hence motor speed) will increase accordingly.

The phase comparator is a circuit that detects a difference in phase between two periodic waveforms. Often, the two waveforms are sinusoidal; in our case they are square waves. The comparator output is made up of two signals. One of them is at a frequency $f_i + f_f$ and the other is at a frequency $f_i - f_f$. The magnitude of these signals is a function of the phase difference between them. The sum frequency $f_i + f_f$ will be blocked by the low-pass filter; however, the difference frequency will be passed converted to dc and amplified. The amplified error signal V_m will drive the motor, producing a feedback signal. As the amplifier gain is made larger, the error will decrease and f_f will become almost the same as f_i. However, for this system to operate there must be an error, even though it may be very small.

The computer, since it is off-line, is involved only in the generation of the input signal. Table 6-4 shows a program that could be used to control the motor's speed by varying the input frequency. Figure 6-20 is a flowchart for the program. The program generates a square wave whose frequency is controlled by the two-digit hexadecimal number stored at SPEED. The contents of SPEED can be entered and/or changed by an operator or another computer via the computer's interrupt system. The motor is stopped or started via a nonmaskable interrupt (NMI). The NMI can also be caused by an operator or another computer. When STCOM is loaded with $FF, the program

Table 6-4

```
0000  4F                          CLR  A
0001  B7 00 34        STA  A   STCOM
0004  86 70                      LDA  A   #$70
0006  B7 00 36                   STA  A   SPEED
0009  B6 00 34   START  LDA  A   STCOM
000C  81 FF                      CMP  A   #$FF
000E  27 F9                      BEQ      START
0010  86 01                      LDA  A   #$01
0012  B7 80 00                   STA  A   $8000
0015  F6 00 36                   LDA  B   SPEED
0018  5A          TON            DEC  B
0019  26 FD                      BNE      TON
001B  FE 00 33                   LDX      ONE
001E  09          DELAY          DEX
001F  26 FD                      BNE      DELAY
0021  86 00                      LDA  A   #$00
0023  B7 80 00                   STA  A   $8000
0026  F6 00 36                   LDA  B   SPEED
0029  5A          TOFF           DEC  B
002A  26 FD                      BNE      TOFF
002C  7E 00 09                   JMP      START
002F  73 00 34   CHANGE  COM     STCOM
0032  3B                         RTI
0033  01          ONE            FCB      01
0034              STCOM          RMB      2
0036              SPEED          RMB      1
00F7                             ORG      $00F7
00F7  7E 00 2F                   JMP      CHANGE
                                 END
```

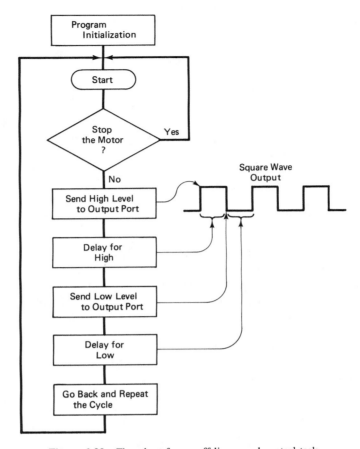

Figure 6-20 Flowchart for an off-line speed control task.

stops generating the square wave. The output, at $8000, will stay at zero volts. It is important to use a phase comparator that gives a zero volt output for a dc input. They do not all work the same way and, depending on their construction, could have a nonzero output for dc input.* This would create the need for additional hardware and/or software to stop the motor. When a NMI is encountered, program flow is interrupted, and we jump to an interrupt service subroutine (ISS). The ISS simply complements the contents of STCOM and then returns to the main program. Thus, if the motor is stopped (STCOM = $FF) and a NMI is encountered, STCOM will be changed to $00 and the motor will start to run.

If the load on the motor increases, causing it to slow down, there will be a decrease in feedback frequency. This will give rise to a larger error, which when amplified will make the motor go faster until it reaches the original speed. Thus excellent speed regulation is obtained as a result of the feedback being used (closed-loop control).

*The Motorola MC4044 phase-frequency detector is suitable for this application. Properly wired, it could provide a zero-volt output for an input frequency equal to zero.

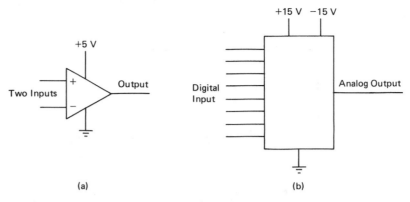

Figure 6-21 (a) Symbol for a comparator; (b) symbol for a digital to analog converter.

On-Line Control. In the speed control system examined here, the motor speed is sensed with a tachometer. The tachometer, whose shaft is connected to the motor shaft, provides a voltage proportional to its shaft speed. Before we look at the system in detail, three devices new to us will be discussed. They are: a comparator, a digital-to-analog converter (DAC), and the Motorola MC6820 peripheral interface adapter (PIA).

A comparator, whose symbol is shown in Figure 6-21a, is a circuit that compares two voltages. When the voltage at the positive terminal is greater than the one at the negative terminal, the output will be high (+5 V). If the voltage at the negative terminal becomes greater than or equal to the one at the positive terminal, the output will go low (0 V).

There are many different types of digital-to-analog converters. The one used here and shown in Figure 6-21b converts an 8-bit binary number into a proportional analog output from 0 to +5 V. The DATEL model DAC-HV8B-100 is well suited for this application.

Since the maximum input is 255 (i.e., **11111111** in binary) the gain of the DAC used is:

$$\text{gain} = \frac{\text{full-scale output}}{\text{maximum input}}$$

$$= \frac{5 \text{ V}}{255 \text{ steps}} = 0.0196 \text{ V/step}$$

This number is important since it and the tachometer gain will tell us what digital input is needed to run the motor at a given speed. This will be discussed in more detail shortly.

To explain the operation of the peripheral interface adapter properly would take a considerable amount of time. This would take away from the topic we are trying to cover here: on-line microcomputer control of a dc motor. We will therefore only cover it very briefly.

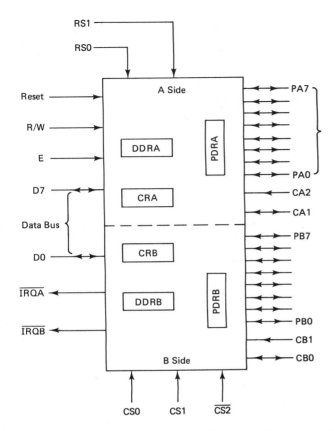

Figure 6-22 Circuit symbol for the PIA.

The PIA shown in Figure 6-22 can provide input and/or output ports for a microcomputer. There are 16 lines which are programmed to be any combination of inputs and outputs. In addition, there are four lines that can be used as interrupts. Two of these four lines can also be used as outputs or for handshaking. Furthermore, one of these two (CB2) can provide 1 mA of current at 1.5 V when used as an output. The CB2 line thus has the capability of driving the base of a transistor switch.

With the exception of some minor differences, such as the buffering of CB2, the PIA is split into two identical sides. The A side and B side each contain three 8-bit registers. The peripheral data register (PDRA and PDRB) is the actual input or output port connecting the computer's data bus to the outside world. The data direction register (DDRA and DDRB) bits determine whether each line coming out of the peripheral data register is an input or output (a **1** makes the corresponding line an output, while a **0** makes it an input). The control register (CRA and CRB) controls the behavior of the interrupt system. In addition, bit 2 of the control register helps select an internal register. If it is a **0,** a specific address determined by the chip select and register select lines will be the location of the DDR. If it is a **1,** the same address will be the location of the PDR. In this way only four addresses are needed for the six registers.

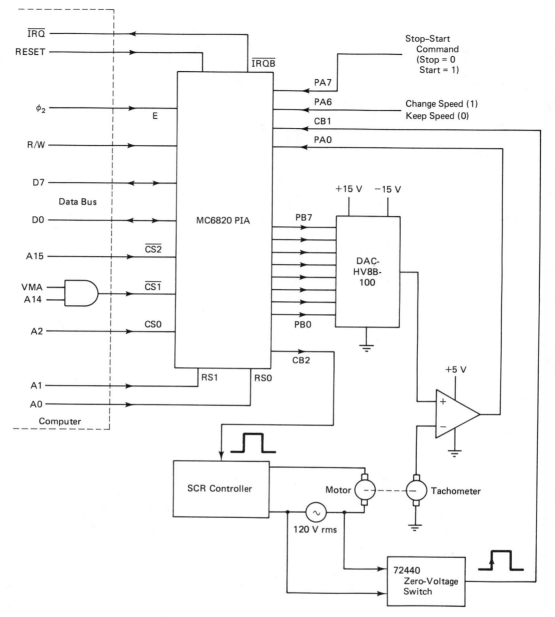

Figure 6-23 Diagram of an on-line speed control system.

Resetting the microcomputer will clear all the registers in the PIA. Thus, after a reset all 16 peripheral data lines will be inputs and since bit 2 of the control registers is a **0,** the addresses mentioned above will be the location of the DDRs.

In our application all peripheral data lines on the B side will be outputs. On the A side, although we will just be using three of the lines, all eight lines will be made inputs. Therefore, DDRA must contain $00 (all **0**'s) and DDRB must contain $FF (all **1**'s). The control register contents will be explained as we go along.

Figure 6-23 is a diagram of our on-line speed control system. Although the computer has been left out of the picture, the pertinent connections are shown. The box labeled "SCR controller" contains the full-wave drive circuit shown in Figure 6-17. Thus the dc voltage applied to the motor can vary from 0 V to 108 V (see Figure 6-6f when $t_1 = 0$).

The commands to start or stop the motor and to change speed are read from the two input lines PA7 and PA6, respectively. These signals could be coming from any of several places; such as an operator, another computer, a temperature or overload sensor, and so on. The source would depend on the particular application. The CB1 line is used as an interrupt. Whenever this line makes a low-to-high transition, the computer knows that the ac supply has crossed zero volts. The computer will then jump to an interrupt service subroutine (ISS) and wait for a time delay. After the time delay has elapsed, we return to the main program and set line CB2 high. This will fire the SCR.

Partial address decoding is used. With the connections shown to the chip and register select lines, the addresses in Table 6-5 are decoded for the six internal registers of the PIA. Since we are using partial address decoding there are actually many more addresses that could be used.

Table 6-5

$4004	Either DDRA or PDRA, depending on status of CRA bit 2
$4005	CRA
$4006	Either DDRB or PDRB, depending on status of CRB bit 2
$4007	CRB

Referring back to Figure 6-22, on the A side none of the interrupt lines (CA1 or CA2) are being used; therefore, our only concern is to set bit 2 to a **1.** This will enable us to read PDRA at $4004. The number $04 (**00000100**) when stored at $4005 will do this.

The B side is more involved. In addition to making bit 2 of CRB a **1** so that $4006 will be the address of PDRB, the following is needed. To recognize the interrupt on line CB1 on a low-to-high transition, bits 1 and 0 of CRB must both be a **1.** Bits 6 and 7 of CRB are internal interrupt flags and will be cleared (**0**). To make the output on line CB2 go high, bits 5, 4, and 3 of CRB must be **1 1 1,** respectively. To make CB2 go low, this must be changed to **1 1 0.** This gives rise to the control words shown in Table 6-6. Storing either of these at $4007 will make the B side of the PIA work as specified above.

Table 6-6

Binary number 7 6 5 4 3 2 1 0	Hex number	Output on CB2
0 0 1 1 0 1 1 1	$37	Low
0 0 1 1 1 1 1 1	$3F	High

The program is listed in Table 6-7. To help analyze the program we can refer to its flowchart, shown in Figure 6-24. The program starts with the initialization of the PIA and data, such as the selected speed (NEWSP) and the initial delay for t_1 (T1INIT). The value used here, $0400, gives an initial delay of 8.19 ms. Since the half-cycle of the supply is 8.33 ms ($\frac{1}{2}$ of 1/60 Hz) the initial motor voltage would be about zero. The delay depends on the clock speed. The delay routine, which is the ISS, takes eight machine cycles. With a 1-MHz clock speed the ISS will take 8 μs each time we go through it. Thus with T1INIT equal to $0400 (1024 in decimal) the initial delay will be 1024 $\times$ 8 μs, or 8.19 ms, as stated above.

Table 6-7

0000	FE 00 6D	START	LDX		T1INIT
0003	FF 00 69		STX		T1PRES
0006	B6 00 6C		LDA	A	NEWSP
0009	B7 00 6B		STA	A	SPCOM
000C	86 FF		LDA	A	#$FF
000E	B7 40 06		STA	A	$4006
0011	86 04		LDA	A	#$04
0013	B7 40 05		STA	A	$4005
0016	B6 00 6B		LDA	A	SPCOM
0019	B7 40 06		STA	A	$4006
001C	86 37	LOOP	LDA	A	#$37
001E	B7 40 07		STA	A	$4007
0021	B6 40 04	STOP	LDA	A	$4004
0024	48		ASL	A	
0025	24 FA		BCC		STOP
0027	48		ASL	A	
0028	24 06		BCC		SAME
002A	B6 00 6C		LDA	A	NEWSP
002D	B7 40 06		STA	A	$4006
0030	74 40 04	SAME	LSR		$4004
0033	24 06		BCC		SLOWER
0035	7A 00 69	FASTER	DEC		T1PRES
0038	7E 00 3E		JMP		CONT
003B	7C 00 69	SLOWER	INC		T1PRES
003E	FE 00 69	CONT	LDX		T1PRES
0041	BC 00 6D		CPX		T1INIT
0044	2A 08		BPL		UPLIM
0046	FE 00 69		LDX		T1PRES
0049	2B 07		BMI		LOLIM
004B	7E 00 53		JMP		WAIT
004E	09	UPLIM	DEX		
004F	7E 00 53		JMP		WAIT
0052	08	LOLIM	INX		
0053	0E	WAIT	CLI		
0054	3E		WAI		
0055	86 3F		LDA	A	#$3F
0057	B7 40 07		STA	A	$4007
005A	C6 06		LDA	B	#$06
005C	5A	PW	DEC	B	
005D	26 FD		BNE		PW
005F	7E 00 1C		JMP		LOOP
0062	FE 00 69	ISS	LDX		T1PRES
0065	09	DELAY	DEX		
0066	26 FD		BNE		DELAY
0068	3B		RTI		
0069		T1PRES	RMB		2
006B		SPCOM	RMB		1
006C	C0	NEWSP	FCB		$C0
006D	04 00	T1INIT	FDB		$0400
00F7			ORG		$00F7
00F7	7E 00 62		JMP		ISS
			END		

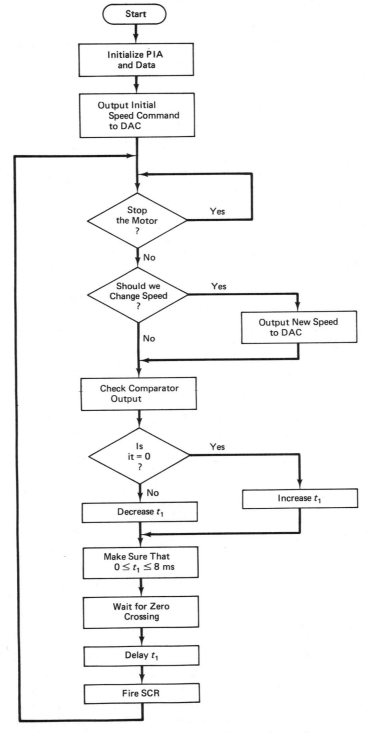

Figure 6-24 Flowchart for on-line speed control.

The following calculations are used to determine the contents of NEWSP. They are based on a tachometer gain of 20 mV/(rad/s) and the DAC's resolution, already calculated as 0.0196 V/step.

If the selected speed is 1800 rev/min, we must first convert it to rad/s,

$$1800 \frac{\text{rev}}{\text{min}} \times 2\pi \frac{\text{rad}}{\text{rev}} \times \frac{1}{60 \frac{\text{s}}{\text{min}}} = 188.5 \text{ rad/s}$$

Knowing the speed in rad/s, we can multiply it by the tachometer gain to find the output of the tachometer.

$$188.5 \text{ rad/s} \times \frac{20 \text{ mV}}{\text{rad/s}} = 3.77 \text{ V}$$

Dividing this voltage by the resolution of the DAC will tell us the necessary input to the DAC (in decimal).

$$\frac{3.77 \text{ V}}{0.0196 \text{ V/step}} = 192.3 \text{ steps}$$

Rounding this off to 192 so that we fall within the linearity of the converter and then converting to hexidecimal, we get

$$(192)_{10} = (1100\ 0000)_2 = (CO)_{16}$$

$$\text{NEWSP} = \$CO$$

Note that the maximum speed that could be commanded would be the speed corresponding to 5 V, the full-scale voltage of the DAC.

$$\omega_{max} = \frac{5 \text{ V}}{\text{tach gain}}$$

$$= \frac{5 \text{ V}}{0.0196 \text{ V/(rad/s)}}$$

$$= 255.1 \text{ rad/s} \quad (\text{SI})$$

or

$$S_{max} = 2436 \text{ rev/min} \quad (\text{English})$$

Once initialization has been completed, the program outputs the speed command to the DAC. As the program enters the main loop the CB2 output is made zero. This step is actually part of the PIA initialization. It must be placed within the loop since we are constantly making CB2 go high and then low. This is how we are generating the pulse to fire the SCR. The next step examines PA7. As long as it is **0**, the program stays in a loop, does not continue, thus never fires the SCR. If PA7 is a **1**, we break out of the STOP loop and continue. The program now examines PA6. If it is a **1**, a new speed command has been entered at NEWSP by an interrupt of some kind and the command to the DAC is changed. The program now examines PAO, which is the output of the comparator. If it is **0,** it means that the tachometer output has become greater than or equal to the DAC output. T1PRES is then increased by 1, which has the effect of slowing down the motor. If PAO was a **0,** we would have decreased T1PRES by 1, causing the speed to increase slightly. An increase or decrease of T1PRES by a count of 1 changes the delay (t_1) by about 8 μs for a 1-MHz clock speed. This, in turn, produces a very small change in speed. The next portion of the program ensures that the time delay stays within its usable range ($0 \leq t_1 \leq 8.19$ ms).

The program now sits and waits for an interrupt (WAI), in other words, a zero crossing of the ac supply. When the interrupt occurs, we go to the ISS and delay for t_1 seconds. After we return from the ISS, CB2 is made high and we go through a small delay (the PW loop) to generate the desired pulse width. The program then jumps back to the start of the main LOOP, where CB2 is set low. The whole process now repeats itself. The PW loop produces a pulse width of about 40 μs, which is more than enough to fire the SCRs.

Our purpose here has been to show how a microcomputer could be used to control a dc motor. In the real world, however, more thought must go into the on-line control example. Although the system would work, it is very possible that a noticeable oscillation in the final speed of the motor would result. Whether or not the oscillation is noticeable depends on the motor's time constant. This, in turn, depends on the gain of the motor and the total inertia and friction in the system. The main contributors of inertia and friction are the motor and the load it is driving. The analysis needed to take these factors into account is beyond the scope of our purpose and of this book*. It has therefore been omitted.

The examples discussed here are but a few. All the techniques of microcomputer control are beyond our imagination. It is hoped, however, that what has been presented illustrates the tremendous capabilities of this small device.

SYMBOLS INTRODUCED IN CHAPTER 6

Symbol	Definition	Units: English and SI
R_x	Variable-field circuit resistance	ohms
R_w	Shunt-field winding resistance	ohms
V_s	AC supply voltage	rms volts
V_G	SCR gate trigger voltage	volts
V_L	Load voltage	volts
V_{AK}	SCR anode-to-cathode voltage	volts
V_{GK}	SCR gate-to-cathode voltage	volts

*For a treatment of the dynamics of a motor and the effects of load friction and inertia on its time constant, see *Introduction to Feedback Control Systems,* Emanuel and Leff, McGraw-Hill, 1979.

Symbol	Definition	Units: English and SI
t_1	SCR firing time delay	seconds
T	Period of V_s waveform	seconds
I_G	SCR gate current	mA
V_m	Peak value of supply voltage	volts
V_{dc}	DC value of load voltage	volts
α_0	Firing delay angle of SCR	degrees or radians
α	Conduction angle of SCR	degrees or radians
$A0, A1,$ $\ldots, A15$	Address lines of a microcomputer	—
$D0, D1,$ $\ldots, D7$	Data lines of a microcomputer	—
f	Frequency of incremental shaft encoder output	hertz
n	Number of slots on shaft encoder disk	—
f_i	Phase-locked-loop input frequency	hertz
f_f	Phase-locked-loop feedback frequency	hertz

QUESTIONS

1. What is the basic speed of a dc motor?

2. What type of control will vary a motor's speed:
(a) Only above its basic speed?
(b) Only below its basic speed?
(c) Above and below its basic speed?

3. Define the following terms: Ward–Leonard system; rotary amplifier.

4. Draw the circuit symbol, labeling all terminals, of an SCR, DIAC, and TRIAC.

5. Define the term "dc voltage."

6. With reference to an SCR define the following terms; holding current; conduction angle; firing delay angle.

7. With reference to a TRIAC, what is the difference between four-quadrant and two-quadrant triggering?

8. Why is it sometimes necessary to isolate low-voltage-control circuitry from motors and other inductive loads?

9. What is an RC snubber circuit used for?

10. What is an incremental shaft encoder?

11. In a phase-locked loop, what is the purpose of the phase comparator and the low-pass filter?

12. In the on-line control system shown in Figure 6-23, what is the purpose of the tachometer and the digital-to-analog converter (DAC)?

PROBLEMS

(English and SI)

1. A motor's voltage is varied with a single SCR by varying the firing angle. A 120-V 60-Hz supply is used. If the firing time delay is 3.5 ms, find:
(a) The dc motor voltage
(b) The firing delay angle in radians and degrees
(c) The SCR conduction angle in degrees

2. Repeat Problem 1 if the supply is 230 V, 50 Hz.

3. A 120-V 60-Hz supply is full-wave rectified with two SCRs to vary the dc voltage across a load. Find:
(a) The firing time delay needed to get 20 V dc
(b) The corresponding firing delay angle
(c) The maximum dc voltage obtainable
(d) The SCR conduction angle in degrees to get a maximum dc voltage

4. Repeat Problem 3 if the supply is 230 V, 50 Hz.

(English)

5. An incremental shaft encoder has 180 equally spaced slots on its disk. Find the frequency of the output if the encoder shaft is rotating at:
(a) 60 rev/min
(b) 900 rev/min
(c) 2000 rev/min

6. An incremental shaft encoder has 60 equally spaced slots on its disk. How fast must its shaft rotate to produce an output frequency of:
(a) 60 Hz?
(b) 1000 Hz?
(c) 2400 Hz?

(SI)

7. An incremental shaft encoder has 120 equally spaced slots on its disk. Find the frequency of the output if the encoder shaft is rotating at:
(a) 5 rad/s
(b) 80 rad/s
(c) 170 rad/s

8. An incremental shaft encoder has 90 equally spaced slots on its disk. How fast must its shaft rotate to produce an output frequency of:
(a) 50 Hz?
(b) 400 Hz?
(c) 3000 Hz?

Chapter 7

Transformers

The transformer is a device that has found its way into almost every product manufactured today which uses electricity. It is used for impedance matching, electrical isolation, ac coupling, and most important for the raising or lowering of a voltage to a desired level. It is for this last use that we will study the transformer in this book. The transformer will work only for a changing electrical signal. It is therefore not used with constant voltages. It finds much use, however, in ac circuits. Before we examine the transformer we will first review some principles of ac circuit theory.

7-1   REVIEW OF SINGLE-PHASE AC CIRCUITS

There are many types of changing (alternating) signals. The only one we will concern ourselves with here is the sinusoidal function. Figure 7-1 shows two sinusoidal waveforms. In Section 2-1.2 we saw that a dc generator naturally produces a sinusoidal voltage (see Figures 2-7 and 2-9). It was with the use of brushes and a commutator that the sinusoidal voltage was converted inside the generator to a dc voltage. However, if a commutator is not used and instead we use **slip rings** as shown in Figure 7-2, the output of the generator will remain ac. This will be the case since the brushes do not switch, but remain in contact with the same conductor as it rotates in the magnetic field. Thus when conductor A passes the north pole, brush A will be positive and brush B will be negative. This is found by applying Fleming's right-hand rule. When the coil rotates 180°, brush A will become negative and brush B will now be positive. If the waveforms in Figure 7-1 represent the voltage across and the current through a circuit component, we can state the following facts:

    1. Both waveforms have the same period, hence the same frequency.
    2. Since the zero crossings and positive peaks do not occur at the same time, the

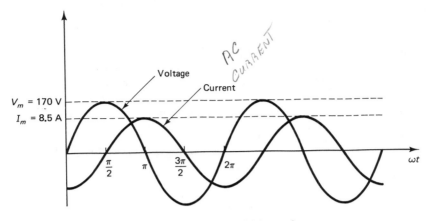

Figure 7-1 Two sinusoidal waveforms.

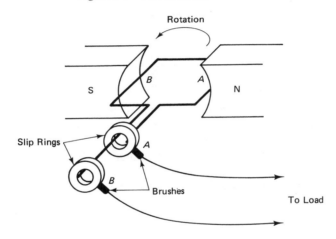

Figure 7-2 Slip rings used to bring ac out of a generator.

two waveforms are out of phase. Furthermore, the current is lagging the voltage by $\pi/2$ rad or 90°.

3. The effective (or rms) value of the waveforms* is given by Eqs. 7-1 and 7-2 where V_m and I_m are the maximum values.

$$V = \frac{V_m}{\sqrt{2}} \tag{7-1}$$

and

$$I = \frac{I_m}{\sqrt{2}} \tag{7-2}$$

*From this point on, all voltages and currents will be assumed rms. Anything else will be identified and given a subscript at the time it is covered.

Sec. 7-1 Review of Single-Phase AC Circuits **203**

4. Since the voltage and current are out of phase, we are led to the conclusion that the component in question is not purely resistive. It must contain a **reactive element** (capacitor and/or inductor). The fact that the current lags the voltage by exactly 90° tells us that the component is purely inductive. A pure inductor does not dissipate power; it merely stores energy in a magnetic field.

5. If we multiply the rms voltage by the rms current, we will obtain the **apparent power.** It is called this because it appears that a certain amount of power is being delivered to a component or load. It has units of volt-amperes (VA).

$$\text{apparent power} = P_a = VI \tag{7-3}$$

For our illustrative example (Figure 7-1)

$$V = \frac{V_m}{\sqrt{2}} = \frac{170 \text{ V}}{\sqrt{2}} = 120 \text{ V}$$

$$I = \frac{I_m}{\sqrt{2}} = \frac{8.5 \text{ A}}{\sqrt{2}} = 6 \text{ A}$$

and

$$P_a = VI = 120 \text{ V} \times 6 \text{ A}$$

$$= 720 \text{ VA}$$

We know from item 4, however, that in reality there is no **real power** being dissipated. Real power is the kind of power that ultimately does something for us (turn a motor; produce heat, cold, light; etc.). It is the real power that we pay for when we receive a bill from our utility company. To distinguish it from apparent power it has as its unit the watt and we will use the symbol P_w.

6. The **power factor** (PF) in an ac circuit is defined as the ratio of real to apparent power (Eq. 7-4):

$$\text{PF} = \frac{P_w}{P_a} \tag{7-4}$$

In our example,

$$\text{PF} = \frac{0 \text{ W}}{720 \text{ VA}} = 0$$

The power factor is also given by the cosine of the angle (θ) between the voltage and current being delivered to a load.

$$\text{PF} = \cos \theta \tag{7-5}$$

In Figure 7-1, $\theta = 90°$. Therefore, we can also find the power factor for our example using Eq. 7-5.

$$\text{PF} = \cos 90° = 0$$

7. In ac circuits the phase shift (θ) is caused by a reactive element present in the load. The load, made up of resistors, capacitors, and/or inductors, has an impedance

with a magnitude (Z) and an angle (θ) associated with it. This is the same angle (θ) that shows up as the phase shift between the voltage and current to the load. Therefore, if we know the angle of the impedance, we can calculate the power factor of the load using Eq. 7-5. In series circuits the angle of the impedance can be calculated using Eq. 7-9, where R is resistance and X_L and X_C are the inductive and capacitive reactance, respectively. X_L and X_C have the unit of ohms and depend on the magnitude of L, C, and the frequency f (Hz) of the applied voltage. The magnitude of the impedance can be calculated using Eq. 7-8.

$$X_L = 2\pi f L \tag{7-6}$$

$$X_C = \frac{1}{2\pi f C} \tag{7-7}$$

$$|Z| = \sqrt{R^2 + (X_L - X_C)^2} \tag{7-8}$$

$$\theta = \arctan \frac{X_L - X_C}{R} \tag{7-9}$$

If θ in Eq. 7-9 is positive, the impedance is inductive in nature, the current will lag the voltage, and the power factor is called a **lagging power factor.** If θ is negative, the impedance is capacitive in nature, the current will lead the voltage, and the power factor is called a **leading power factor.** When X_L equals X_C or neither are present in the circuit, θ will be zero and we have **unity power factor** (a purely resistive circuit).

8. As already stated, an inductor stores energy in a magnetic field. Similarly, a capacitor stores energy in an electric field. In an ac circuit this energy storage periodically builds up and breaks down. In this way energy flows from the source to the reactive element and then back from the reactive element to the source. The power necessary to build up the field is thus regained by the source when the field breaks down. A special name is given to this power since it does not represent real power. It is called **reactive power** and its unit is the var (volt-ampere-reactive). It is related to apparent power by Eq. 7-10, where θ again is the angle of the impedance.

$$P_r = P_a \sin \theta \tag{7-10}$$

or

$$P_r = VI \sin \theta$$

9. A helpful tool in analyzing the power flow in an ac circuit is the **power triangle.** It relates apparent power to real power and reactive power. It is shown in Figure 7-3. The triangle is shown for a lagging power factor. In other words, it is the picture that would be drawn if θ calculated in Eq. 7-9 were positive. The real power is always drawn on the reference axis. The reactive power is drawn at a right angle to real power. It is drawn down for lagging power factors and up for leading power factors. In addition to Eq. 7-10, Eqs. 7-11 and 7-12 can be derived from Figure 7-3 using trigonometry.

$$P_w = P_a \cos \theta = VI \cos \theta \tag{7-11}$$

$$P_a = \sqrt{P_w^2 + P_r^2} \tag{7-12}$$

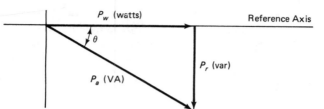

Reference Axis

P_w (watts)

θ

P_r (var)

P_a (VA)

Figure 7-3 Power triangle: shown for a lagging power factor.

Example 7-1 (The Series RL Circuit)

For the circuit shown in Figure 7-4, find:

(a) The total impedance

(b) The current

(c) The power factor (leading or lagging?)

(d) The apparent power, reactive power, and real power

(e) Draw the power triangle.

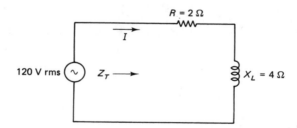

$R = 2\ \Omega$

I

120 V rms $Z_T \longrightarrow$

$X_L = 4\ \Omega$

Figure 7-4 Circuit for Example 7-1.

Solution

(a) Since we have a series circuit, $|Z|$ can be calculated using Eq. 7-8, where X_C is zero.

$$|Z| = \sqrt{2^2 + (4 - 0)^2}$$
$$= \sqrt{20} = 4.47\ \Omega$$

θ can be calculated using Eq. 7-9:

$$\theta = \arctan \frac{4}{2}$$

$$= 63.43°$$

(b) The rms current is found by applying Ohm's law.

$$I = \frac{\text{source voltage}}{Z}$$

$$= \frac{120 \text{ V}}{4.47 \text{ } \Omega}$$

$$= 26.85 \text{ A}$$

(c) Using Eq. 7-5 gives us

$$PF = \cos 63.43°$$

$$= 0.45 \text{ [lagging since } \theta \text{ calculated in part (a) was positive]}$$

(d) Using Eqs. 7-3, 7-10, and 7-11 yields

$$P_a = 120 \text{ V} \times 26.85 \text{ A}$$

$$= 3222 \text{ VA or } 3.222 \text{ kVA}$$

$$P_r = 3222 \sin 63.43°$$

$$= 2882 \text{ var or } 2.882 \text{ kvar}$$

$$P_w = 3222 \cos 63.43°$$

$$= 1441 \text{ W or } 1.441 \text{ kW}$$

Note that this could also have been found by finding the power dissipated in the resistor:

$$P_w = RI^2 = 2 \text{ } \Omega \times (26.85 \text{ A})^2$$

$$= 1442 \text{ W}$$

(e) The power triangle is shown in Figure 7-5.

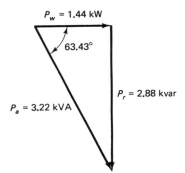

Figure 7-5 Power triangle for Example 7-1.

Example 7-2 (The Series RC Circuit)

For the circuit shown in Figure 7-6, find:

(a)   The total impedance

(b)   The current

(c)   The power factor (leading or lagging?)

(d)   The apparent power, reactive power, and real power

(e)   Draw the power triangle.

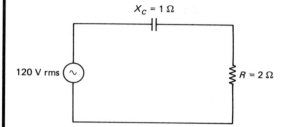

Figure 7-6   Circuit for Example 7-2.

Solution

(a)   Using Eq. 7-8 where X_L is zero,

$$|Z| = \sqrt{2^2 + (0 - 1)^2}$$
$$= 2.24 \ \Omega$$

From Eq. 7-9,

$$\theta = \arctan \frac{-1}{2}$$
$$= -26.57°$$

(b) $I = \dfrac{120 \ V}{2.24 \ \Omega}$

$\qquad = 53.57 \ A$

(c) $PF = \cos \theta = \cos (-26.57°)$

$\qquad = 0.89$ [leading since θ calculated in part (a) was negative]

(d) $P_a = VI$

$\qquad = 120 \ V \times 53.57 \ A$

$\qquad = 6.43 \ kVA$

$P_r = VI \sin \theta$

$\qquad = 6.43 \ k \sin (26.57°)$

$\qquad = 2.88 \ kvar$

$P_w = VI \cos \theta$

$$= 6.43\text{k} \cos 26.57°$$

$$= 5.75 \text{ kW}$$

As in Example 7-1, P_w can also be found by

$$P_w = RI^2 = 2 \text{ } \Omega \times (53.57 \text{ A})^2$$

$$= 5.74 \text{ kW}$$

(e) The power triangle is shown in Figure 7-7.

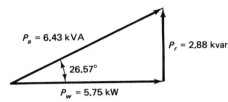

Figure 7-7 Power triangle for Example 7-2.

Example 7-3 *(The Series RLC Circuit)*

For the circuit shown in Figure 7-8, find:

(a) The total impedance

(b) The current

(c) The power factor (leading or lagging?)

(d) The apparent power, reactive power, and real power

(e) Draw the power triangle.

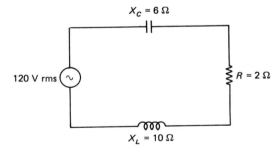

Figure 7-8 Circuit for Example 7-3.

Solution

(a) Using Eq. 7-8, we have

$$|Z| = \sqrt{2^2 + (10 - 6)^2}$$

$$= \sqrt{20} = 4.47 \text{ } \Omega$$

Using Eq. 7-9 gives us

$$\theta = \arctan \frac{10 - 6}{2} = \arctan \frac{4}{2}$$

$$= 63.43°$$

(b) $I = \dfrac{120 \text{ V}}{4.47 \ \Omega}$

$\qquad = 26.85 \text{ A}$

(c) $\text{PF} = \cos \theta = \cos 63.43°$

$\qquad = 0.45$ [lagging since θ calculated in part (a) is positive]

(d) $P_a = VI = 120 \text{ V} \times 26.85 \text{ A}$

$\qquad = 3.222 \text{ kVA}$

$P_r = VI \sin \theta$

$\qquad = 120 \text{ V} \times 26.85 \text{ A} \times \sin 63.43°$

$\qquad = 2.88 \text{ kvar}$

$P_w = VI \cos \theta$

$\qquad = 120 \text{ V} \times 26.85 \text{ A} \times \cos 63.43°$

$\qquad = 1.44 \text{ kW}$

(e) The power triangle is shown in Figure 7-9.

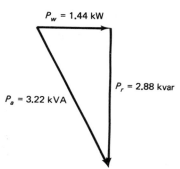

Figure 7-9 Power triangle for Example 7-3.

If we compare Example 7-3 with Ex. 7-1, we can make the following observations:

1. Although the components are different, the two circuits behave in the same way.
2. The real power was the same in both cases because the resistor and the current were the same in each example.
3. Even though a circuit contains both inductors and capacitors, it will behave like an inductive or capacitive circuit depending on the magnitude of X_L and X_C, and the circuit configuration.

Example 7-4

A single-phase 120-V source supplies power to three loads. Information on the loads is given in Table 7-1.

Table 7-1

Load	kVA	PF
A	15	0.9 lagging
B	5	0.9 leading
C	20	0.8 lagging

Find:

(a) The real power for each load

(b) The reactive power for each load

(c) The total real power

(d) The total reactive power

(e) The total apparent power

(f) The effective power factor of the system

Solution

(a) Load A: $P_w = P_a \cos \theta$

$$= 15 \text{ kVA } (0.9) = 13.5 \text{ kW}$$

Load B: $P_w = 5 \text{ kVA } (0.9) = 4.5 \text{ kW}$

Load C: $P_w = 20 \text{ kVA } (0.8) = 16 \text{ kW}$

(b) Load A: $P_r = P_a \sin \theta$

$$\theta = \arccos (0.9) = 25.84°$$

$$P_r = 15 \text{ kVA } \sin 25.84°$$

$$= 15 \text{ kVA } (0.44)$$

$$= 6.54 \text{ kvar (lagging)}$$

Load B: $P_r = P_a \sin \theta$

$$\theta = -25.84° \text{ (leading PF)}$$

$$P_r = 5 \text{ kVA } \sin (-25.84°)$$

$$= -2.18 \text{ kvar (leading)}$$

Load C: $P_r = P_a \sin \theta$

$$\theta = \arccos (0.8) = 36.87°$$

$$P_r = 20 \text{ kVA } \sin 36.87°$$

$$= 12 \text{ kvar (lagging)}$$

(c) Total P_w $= 13.5 \text{ kW} + 4.5 \text{ kW} + 16 \text{ kW}$

$$= 34 \text{ kW}$$

(d) Total P_r $= 6.54 \text{ kvar} - 2.18 \text{ kvar} + 12 \text{ kvar}$

                      $= 16.36 \text{ kvar (lagging)}$

(e) Total P_a from Eq. 7-12:

$$\text{total } P_a = \sqrt{34^2 + 16.36^2} \text{ kVA}$$

$$= \sqrt{1156 + 267.65} \text{ kVA}$$

$$= 37.85 \text{ kVA}$$

The power triangle is shown in Figure 7-10.

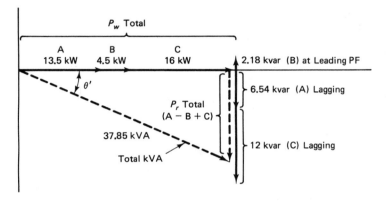

Figure 7-10   Power triangle for Example 7-4.

(f) Using Eq. 7-4, we can calculate the system power factor.

$$\text{PF} = \frac{P_w}{P_a} \text{ (using total quantities)}$$

$$= \frac{34 \text{ kW}}{37.85 \text{ kVA}}$$

$$= 0.9 \text{ lagging (and } \theta' = \arccos 0.9)$$

$$\theta' = 25.84° \text{ (system power factor angle)}$$

7-1.1 Measurement of Power

The measurement of power in dc circuits was very simple. All we had to do was measure the voltage across and current through a load, then multiply the two and obtain the power in watts. We have just seen that in ac circuits, multiplying voltage with current will give us apparent power, not real power. To obtain real power we would have to know the power factor (or phase shift between V and I).

The problem is solved by using a special instrument that measures watts. It is called a **wattmeter.** By sensing voltage and current at the same time, the wattmeter can determine the phase shift between them and thus determine real power. The

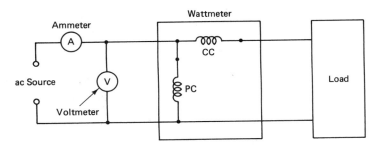

Figure 7-11 Typical wattmeter connection to a single-phase load.

voltage is sensed by a **potential coil** (PC) and the current is sensed by a **current coil** (CC). Since the current coil senses load current, it must be in series with the load. The potential coil is placed in parallel with the load so that it can sense the voltage. Thus the wattmeter has four terminals, two for the PC and two for the CC. A typical wattmeter connection is shown in Figure 7-11. Since the current coil will very often have large currents flowing through it, its terminals are made large and with an excellent conductor. In this way there will be very little contact resistance and a negligible voltage drop across the terminals. In high-power situations the current is sensed with a clamp-on probe. This eliminates the need for a potentially dangerous measurement, that is, opening up a circuit in which a large current normally flows.

Figure 7-11 also shows a voltmeter and ammeter connected. Using their readings, we could calculate apparent power (see Eq. 7-3). Knowing the real and apparent power, the power factor could be calculated (Eq. 7-4). Finally, once the power factor is known, we could calculate reactive power using Eq. 7-10. It should be pointed out that there are meters similar in construction to wattmeters that measure reactive power (var). They are called **varmeters.** Meters are also available which give a reading of power factor **(power factor meter).** All of these meters are available with analog or digital readouts. In addition, the digital instruments are available with a binary-coded-decimal (BCD) output so that they can be easily interfaced with computers.

Example 7-5

A circuit connected as Figure 7-11 has the following meter readings: voltmeter, 120 V; ammeter, 9.4 A; wattmeter, 650 W. Find:

(a) The apparent power

(b) The power factor

(c) The reactive power

Solution

(a) Using Eq. 7-3 gives us

$$P_a = VI = 120 \text{ V} \times 9.4 \text{ A}$$

$$= 1128 \text{ VA} = 1.128 \text{ kVA}$$

(b) Using Eq. 7-4 and noting that the wattmeter reading is P_w, we have

$$PF = \frac{P_w}{P_a}$$

$$= \frac{650 \text{ W}}{1128 \text{ VA}}$$

$$= 0.58$$

(c) Since PF = cos θ = 0.58:

$$\theta = \text{arccos } 0.58$$

$$= 54.81°$$

Using Eq. 7-10 yields

$$P_r = P_a \sin \theta$$

$$= 1128 \text{ VA sin } 54.81°$$

$$= 921.9 \text{ var}$$

7-1.1.1 Wattmeter Overloads.   One of the dangers in using wattmeters is that they can "burn up" even though they are measuring a very small power. The potential and current coils of a wattmeter each have a rating that should not be exceeded. The full-scale power reading (wattage range) of the meter is usually the product of the two coil ratings. In cases when the power factor is very low, the meter reading will be low. However, it is very possible that one of the coil ratings is being exceeded. If this condition continues, the coil whose rating is being exceeded will burn up.

Example 7-6

A wattmeter is used to measure power on a 120-V line. Its potential and current coils are rated 150 V and 10 A, respectively. The line voltage remains constant. Find:

(a) The wattage range of the meter

(b) The current in the coil when the meter reads full scale and the power factor is 1.0

(c) The current in the coil when the meter reads full scale and the power factor is 0.5

Solution

(a) Since nothing else is specified, the wattage range (full-scale reading) is

$$150 \text{ V} \times 10 \text{ A} = 1500 \text{ W} = 1.5 \text{ kW}$$

(b) Using Eq. 7-11 gives us

$$P_w = VI \cos \theta$$

$$1500 \text{ W} = 120 \text{ V} \times I \times 1.0$$

Solving for I, we have

$$I = \frac{1500 \text{ W}}{120 \text{ V}} = 12.5 \text{ A}$$

(c) Part (b) is repeated; however, this time the power factor is 0.5.

$$1500 \text{ W} = 120 \text{ V} \times I \times 0.5$$

$$I = \frac{1500}{120 \times 0.5}$$

$$= 25 \text{ A}$$

Note that in both cases [parts (b) and (c)] the current coil rating is exceeded even though the power reading is within the wattage range of the meter. The current in part (b) caused by the line voltage being lower than the potential coil rating is negligible. However, the current in part (c) caused by the low power factor is 2.5 times the coil rating and could possibly burn up the meter. Some wattmeters are built so that they can withstand this condition. They are called **low-power-factor wattmeters.** Even these meters can be damaged, however, when the power factor drops below 0.5.

7-2   BASIC TRANSFORMER THEORY

7-2.1 Transformer Construction

There are many sizes and shapes of transformers. They can be as small as the eraser on a pencil or bigger than a garbage truck. The size depends on the power that must be delivered by the transformer. They can have as few as four wires coming out of them or many more. We will look at the simplest type of transformer: a single-phase core-type transformer with two windings.

The transformer is made by winding two coils of wire around a core. The core, of uniform cross-sectional area, is made with laminated plates of a ferromagnetic material much like the materials studied in Chapter 1. A picture of one is shown in Figure 7-12 together with the circuit symbol. The input voltage is applied to the **primary** winding. The subscript 1 will be used for all primary quantities. Thus V_1, I_1, and N_1 represent the primary voltage, current, and the number of turns on the primary winding. The output voltage is taken from the **secondary** winding. The subscript 2

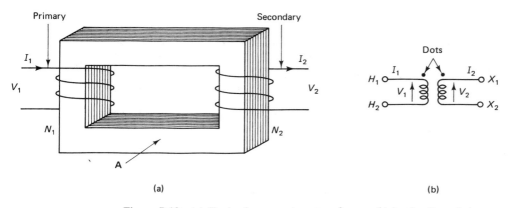

Figure 7-12   (a) Single-phase core-type transformer; (b) its circuit symbol.

will be used for all secondary quantities (V_2, I_2, and N_2). The dots shown on the schematic are used to indicate the phase (or polarity) relationship between primary and secondary voltages. They are to be interpreted in the following way. At some instant of time the dotted terminal on the primary winding will be positive with respect to the undotted terminal. At the same instant of time the dotted terminal on the secondary winding will also be positive with respect to its undotted terminal.

Some transformers will also have their terminals labeled H_1, H_2, H_3, . . . and X_1, X_2, X_3, ... Those terminals labeled with the letter H are connected to the high-voltage winding. The terminals marked with an X are connected to the low-voltage winding. In addition, like the dots, the H and X notation gives polarity information. If at any instant H_1 is positive with respect to H_2, X_1 will be positive with respect to X_2 at the same instant.

A **step-up transformer** is one in which the secondary (output) voltage is higher than the primary (input) voltage. If the secondary voltage is lower than that on the primary side, we call it a **step-down transformer.**

7-2.2 Ideal Transformer Relationships

Any device that transmits power from one point to another will have internal losses. In this sense the device is not ideal. The transformer is no exception and it, too, has internal losses. Transformers made today have efficiencies of 95 to 99%. With such high efficiency, it makes sense to analyze the transformer as an ideal device. A practical transformer will be discussed in Section 7-3.

In Chapter 1 we saw that a current flowing in a coil would produce an MMF (see Figure 1-1). The MMF in turn would create a flux, ϕ, whose magnitude depends on the magnetic circuit. If the current producing the MMF is an alternating one (ac), the flux will also be alternating. Figure 7-13 shows an ac source applied to the primary of a transformer. Most of the flux produced will exist in the core. It links the primary and secondary windings and is called the **mutual flux.** Some of the flux produced by the primary and secondary currents does not remain in the core, and leaks out of the core. This is called **leakage flux** and is shown as ϕ_1 and ϕ_2 in Figure 7-13. We will

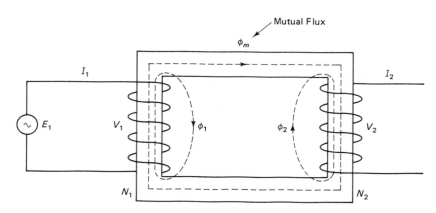

Figure 7-13 Alternating voltage applied to a transformer.

assume in this section that the leakage flux is negligible, therefore giving us 100% magnetic coupling. The relationship between the applied voltage E_1 and the maximum value of the mutual flux is given by

(English)

$$E_1 = 4.44fN_1\phi_m \times 10^{-8} \qquad (7\text{-}13\text{a})$$

(SI)

$$E_1 = 4.44fN_1\phi_m \qquad (7\text{-}13\text{b})$$

In both equations E_1 is the rms value of the voltage applied to the primary winding, f is the frequency of the voltage E_1 in hertz, and N_1 is the number of turns on the primary winding. In Eq. 7-13a, ϕ_m is the peak value of the mutual flux produced by the applied voltage in lines, and in Eq. 7-13b, the unit for ϕ_m is the weber (Wb). Note that when the applied voltage is directly across the primary winding (i.e., no series resistance), E_1 will equal V_1 (the voltage induced in the primary winding) and the two can be interchanged in Eqs. 7-13. Furthermore, Eqs. 7-13 can be applied to the secondary winding. When this is done, the subscript 2 is used in place of 1. In this case ϕ_m is the source and V_2 will be the voltage induced in the secondary winding. If there is no secondary current (no load), V_2 will be measurable at the secondary terminals. We will expand on this when the practical transformer is discussed in Section 7-3.

Solving Eq. 7-13b for ϕ_m, we get

(SI)

$$\phi_m = \frac{V_1}{4.44fN_1} \qquad (7\text{-}14)$$

A similar equation can be written for the secondary winding:

(SI)

$$\phi_m = \frac{V_2}{4.44fN_2} \qquad (7\text{-}15)$$

By equating Eq. 7-14 to Eq. 7-15, and simplifying, we obtain the following basic transformer equation:

$$\frac{V_1}{V_2} = \frac{N_1}{N_2} \qquad (7\text{-}16)$$

Equation 7-16 states that the ratio of primary voltage to secondary voltage is equal to the ratio of primary turns to secondary turns. Any units can be used for V_1 and V_2 as long as they are both the same, that is, both kV rms, peak-to-peak (p-p), mV rms, or similar units. The turns ratio of a transformer is defined as the number of primary turns divided by the number of secondary turns. It is given by Eq. 7-17 and the letter a is used to denote it.

$$\text{turns ratio} = a = \frac{N_1}{N_2} \qquad (7\text{-}17)$$

Example 7-7

It is desired to build a step-down transformer. For a primary voltage of 120 V it should produce 24 V at its secondary. The primary is to have 400 turns. Find:

(a) The secondary turns required

(b) The turns ratio

Solution

(a) Using Eq. 7-16, we have

$$\frac{120 \text{ V}}{24 \text{ V}} = \frac{400 \text{ turns}}{N_2}$$

Solving for N_2 gives us

$$N_2 = 400 \times \frac{24}{120}$$

$$= 80 \text{ turns}$$

(b) Using Eq. 7-17 yields

$$a = \frac{N_1}{N_2} = \frac{400 \text{ turns}}{80 \text{ turns}}$$

$$= 5$$

Example 7-8 (English)

If in Example 7-7 the applied voltage is 60 Hz, find the peak mutual flux linking the primary and secondary coils.

Solution

Equation 7-13a is used, but first we will rearrange it to solve for ϕ_m.

$$\phi_m = \frac{E_1}{4.44 f N_1 \times 10^{-8}}$$

$$= \frac{120}{4.44 \times 60 \times 400 \times 10^{-8}}$$

$$= 112{,}613 \text{ lines}$$

Example 7-9 (SI)

If in Example 7-7 the applied voltage is 50 Hz, find the peak mutual flux linking the primary and secondary coils.

Solution

Equation 7-14 is used.

$$\phi_m = \frac{120}{4.44 \times 50 \times 400}$$
$$= 1.35 \times 10^{-3} \text{ Wb}$$

7-3 PRACTICAL SINGLE-PHASE TRANSFORMER

We will see in this section that Eq. 7-16 is not exact when we deal with a practical transformer under load conditions. This is due to the fact that the transformer windings are not ideal and have a small amount of resistance. Also, in addition to the leakage flux, there are core losses within the transformer. These are the hysteresis and eddy current losses that were discussed in Section 4-5.1. Since the transformer has a magnetic core in which a flux is changing, it too will have these losses.

7-3.1 Transformer Ratings

A transformer will have written on it four numbers which are its ratings. They are:

1. The rated apparent power of the secondary winding. It is expressed in VA or kVA and represents the maximum power that should be drawn from the secondary. Apparent power is used rather than real power since we can have very little real power delivered to a load and still be exceeding the voltage and/or current rating. We saw in Section 7-1.1.1 that this occurs when the power factor is very low.
2. The primary voltage for which it has been designed.
3. The secondary voltage that would appear across a load when rated voltage is applied to the primary and rated VA are delivered to the load. We shall see that at no-load (0 VA) conditions, the secondary voltage will be higher than this number.
4. The frequency for which the transformer has been designed. We shall see that a transformer rated for 60 Hz cannot be used at lower frequencies; however, it could be used at higher frequencies.

Thus if a transformer is rated 120/1200 V, 6 kVA, 60 Hz, it means the following: when 120 V is applied to the primary, the secondary voltage will be 1200 V when the transformer is supplying 6 kVA to a load. Furthermore, if we assume 100% efficiency, the kVA delivered to the load by the secondary must equal the kVA supplied to the primary. Using this assumption, we can calculate the rated currents for both primary and secondary windings. These represent the maximum currents that should be allowed to flow in the windings. A fuse or circuit breaker is used in one of the windings to protect the transformer from excessive currents.

$$I_{1\text{rated}} = \frac{\text{rated VA}}{V_{1\text{rated}}} \tag{7-18}$$

$$I_{2\text{rated}} = \frac{\text{rated VA}}{V_{2\text{rated}}} \tag{7-19}$$

In our case,

$$I_{1\text{rated}} = \frac{6000 \text{ VA}}{120 \text{ V}} = 50 \text{ A}$$

$$I_{2\text{rated}} = \frac{6000 \text{ VA}}{1200 \text{ V}} = 5 \text{ A}$$

Here the secondary would most likely be protected by a 5-A fuse.

Note that the currents are in the opposite ratio from that of the voltages. The voltage was steppped up by a factor of 10, whereas the current was stepped down by a factor of 10. This leads to another useful relationship, which is an extension of Eq. 7-16.

$$\frac{V_1}{V_2} = \frac{I_2}{I_1} = a \tag{7-20}$$

It should be noted that Eq. 7-20 is useful only at large loads. At no load, for example, the secondary current is zero, whereas the primary current is not. A small primary current called the **magnetizing current** is present. It is this current that produces the MMF, hence the flux in the core.

7-3.2 Basic Relationships

To understand transformer behavior more thoroughly, we will analyze it with reference to Figure 7-13. To visualize the current and voltage relationships, the vector diagrams in Figure 7-14 are drawn. When a voltage E_1 is applied to the primary and the secondary is open circuited (no load), the transformer appears as an inductor. If we assume for the moment that the inductor (transformer) is ideal, a small current I_M (magnetization current) will flow in the primary ($I_1 = I_M$), setting up a magnetic field. This current and the flux it produces will lag E_1 by 90°. Furthermore, a voltage V_1 is induced in the primary winding which is equal to E_1 but 180° out of phase with it. This condition is shown in Figure 7-14a. Since the angle between primary voltage and current is 90°, the power factor (cos θ) will be zero. Thus no power is delivered to the transformer.

If a wattmeter were connected to the primary, however, we would read some amount of power being delivered to the transformer. What, then, is wrong with the analysis and diagram drawn in Figure 7-14a?

Since an alternating flux exists, the core will have eddy current and hysteresis losses (see Section 4-5.1). These core losses represent real power (watts) lost in the transformer. If real power is being supplied to the transformer, there must be a component of current in the primary which is in phase with E_1. It is shown in Figure 7-14b as I_C (current providing core loss). The actual primary current (I_1) is not I_M as before, but the vector sum of I_M and I_C. The angle between I_1 and E_1 is no longer 90°, but rather a little less (θ_0). Furthermore, cos θ_0 is called the transformer's **no-load power factor.**

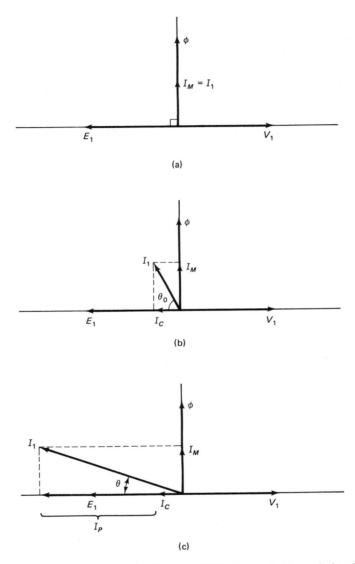

Figure 7-14 Transformer vector diagrams: (a) ideal no load; (b) practical no load; (c) practical supplying load.

If a resistive load is connected to the transformer's secondary, power will be delivered to it. This power must in turn be supplied to the primary by the source, E_1. There will now be an additional component of current (I_p) in phase with E_1 which is supplying real power to the transformer. The transformer in turn delivers this power to the load. Figure 7-14c represents this condition. Note now that I_1 is the vector sum of I_M and the current in phase with $E_1(I_C + I_p)$. In addition, θ is getting smaller, which means that the power factor is approaching unity.

Keep in mind that the load introduced here was resistive. If it had been a reactive (i.e., a capacitive or inductive) load, I_M would have changed, which in turn would have changed the power factor. A capacitive load would decrease I_M (increasing the power factor), whereas an inductive load would increase I_M (decreasing the power factor).

Example 7-10

A transformer rated 120/600 V, 60 Hz, 2 kVA has its primary energized with rated voltage while the secondary is open. A wattmeter connected to the primary measures 10 W and an ammeter measures 0.5 A in the primary. Find:

(a) The no-load power factor

(b) The magnetization current

(c) The current supplying core loss

(d) The power factor if a resistor is connected to the secondary and the wattmeter now reads 1.5 kW.

Solution

(a) Using Eq. 7-11, we have

$$P_w = VI \cos \theta_0$$

$$\cos \theta_0 = \frac{P_w}{VI}$$

$$= \frac{10 \text{ W}}{120 \text{ V} \times 0.5 \text{ A}}$$

$$= 0.167$$

(b) Applying trigonometry to Figure 7-14b yields

$$I_M = I_1 \sin \theta_0$$

θ_0 is found from part (a): $\theta_0 = \arccos 0.167$

$$= 80.39°$$

$$I_M = 0.5 \text{ A} \times \sin 80.39°$$

$$= 0.493 \text{ A}$$

(c) Similarly, from Figure 7-14b,

$$I_C = I_1 \cos \theta_0$$

$$= 0.5 \text{ A} \times 0.167$$

$$= 0.0833 \text{ A}$$

(d) Since the load is resistive, the increase in primary current will be in phase with the applied voltage. Referring to Figure 7-14c,

$$I_p + I_C = \frac{P_W}{E_1} = \frac{1500}{120} = 12.5 \text{ A}$$

I_M is still 0.493 A. Thus

$$\theta = \arctan \frac{I_M}{I_p + I_C} = \arctan \frac{0.493}{12.5}$$

$$= 2.26°$$

and

$$\text{PF} = \cos \theta = 0.999 \approx 1$$

7-3.2.1 Reflecting Impedance.
When a transformer under load is to be analyzed, we are faced with the solution of two separate circuits. These are the primary circuit and the secondary circuit. The two are not independent of each other. To be able to solve for the currents and voltages, we must reflect all quantities from primary to secondary, or vice versa. This reduces the problem to the analysis of one circuit.

Consider the circuit of Figure 7-15a.

$$V_2 = Z_L I_2 \tag{7-21}$$

From Eq. 7-16 we can write

$$V_1 = V_2 \times \frac{N_1}{N_2} \tag{7-22}$$

Substituting Eq. 7-21 into Eq. 7-22, we have

$$V_1 = (Z_L I_2) \times \frac{N_1}{N_2} \tag{7-23}$$

From Eq. 7-20, $I_2 = aI_1$. Substituting this into Eq. 7-23 and substituting a for N_1/N_2, we get

$$V_1 = Z_L(aI_1) \times a$$

or

$$V_1 = a^2 Z_L I_1$$

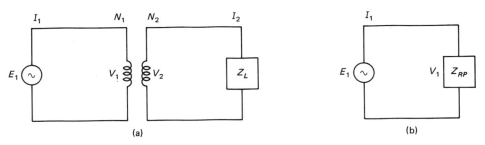

(a)   (b)

Figure 7-15   Transformer connected to impedance: (a) actual circuit; (b) primary circuit with equivalent reflected impedance.

which gives us

$$\frac{V_1}{I_1} = a^2 Z_L \tag{7-24}$$

However, if we look at Figure 7-15b, the reflected impedance Z_R is given by

$$Z_{RP} = \frac{V_1}{I_1}$$

Combining this with Eq. 7-24, we obtain the equation for reflecting impedance from secondary to primary:

$$Z_{RP} = a^2 Z_L \tag{7-25}$$

When reflecting an impedance from primary to secondary, we divide by a^2:

$$Z_{RS} = \frac{Z_1}{a^2} \tag{7-26}$$

In cases where admittance must be reflected, the formulas are just the opposite. From secondary to primary,

$$Y_{RP} = \frac{Y_L}{a^2} \tag{7-27}$$

and from primary to secondary,

$$Y_{RS} = a^2 Y_1 \tag{7-28}$$

Example 7-11

A 120/12.6-V 60-Hz 25-VA transformer is connected as shown in Figure 7-16a. Find the rated primary and secondary currents. Also find the actual currents that are flowing and the percent of rated current that they represent.

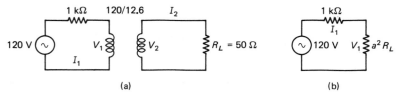

(a) (b)

Figure 7-16 (a) Circuit for example 7-11; (b) circuit for example 7-11 with all quantities reflected to primary.

Solution

The rated currents are obtained from the transformer ratings. Using Eq. 7-18, we have

$$\text{rated } I_1 = \frac{25 \text{ VA}}{120 \text{ V}} = 0.21 \text{ A}$$

Using Eq. 7-19 gives us

$$\text{rated } I_2 = \frac{25 \text{ VA}}{12.6 \text{ V}} = 1.98 \text{ A} \approx 2 \text{ A}$$

The turns ratio is given by Eq. 7-16:

$$a = \frac{V_1}{V_2} = \frac{120 \text{ V}}{12.6 \text{ V}} = 9.52$$

The reflected resistance in Figure 7-16b is

$$a^2 R_L = (9.52)^2 \times 50 \ \Omega = 4535 \ \Omega$$

Applying Ohm's law to Figure 7-16b yields

$$I_1 = \frac{120 \text{ V}}{(1000 + 4535) \ \Omega}$$

$$= 0.02 \text{ A}$$

$$V_1 = I_1 \times a^2 R_L$$

$$= 0.02 \text{ A} \times 4535 \ \Omega = 98.3 \text{ V}$$

The secondary voltage can be found with Eq. 7-16:

$$V_2 = \frac{V_1}{a}$$

$$= \frac{98.3 \text{ V}}{9.52} = 10.3 \text{ V}$$

and using Ohm's law again, we get

$$I_2 = \frac{V_2}{R_L} = \frac{10.3 \text{ V}}{50 \ \Omega}$$

$$= 0.21 \text{ A}$$

To find the percent of rated quantities, divide I_1 and I_2 by the rated currents and multiply by 100%:

$$\frac{0.02 \text{ A}}{0.21 \text{ A}} \times 100\% = 9.5\%$$

and

$$\frac{0.21 \text{ A}}{2 \text{ A}} \times 100\% = 10.5\%$$

7-3.2.2 Transformer Equivalent Circuit. To be able to predict a transformer's behavior accurately when it is supplying load, we need an equivalent circuit for it just as we did for the motor and generator. Example 7-11 represented an approximate solution. To solve a problem properly, quantities such as the resistance and leakage flux of the windings should be included. Figure 7-17 is a simplified equivalent circuit

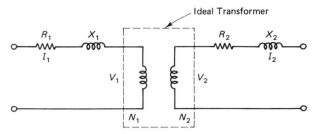

Figure 7-17 Simplified equivalent circuit of a transformer.

for a transformer. Resistors R_1 and R_2 represent the resistance of the primary and secondary windings, respectively. These quantities can be measured quite easily with an ohmmeter. The inductors, shown as reactances X_1 and X_2, represent the leakage flux in the primary and secondary windings. The leakage flux (see Figure 7-13) is a magnetic field developed which does not contribute to the mutual flux linking the coils. Its repesentation as a pure inductive reactance is therefore a reasonable one.

To simplify our equivalent circuit further, we can reflect all quantities to either the primary or secondary using Eq. 7-25 or Eq. 7-26. The reduced circuits are shown in Figure 7-18. Note that in each version of the equivalent circuit V_t and E_1 are the same but V_1 and V_2 are different; furthermore, that portion within the dashed line now represents an ideal transformer. It is ideal in the sense that there is 100% coupling between the two windings. It still has a core loss, which we will investigate shortly. The reflected quantities in Figure 7-18 are given by Eqs. 7-29 to 7-32, where $a = N_1/N_2$.

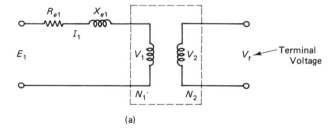

(a)

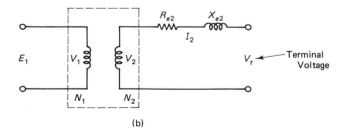

(b)

Figure 7-18 Reduced equivalent circuit of a transformer: (a) quantities reflected to primary; (b) quantities reflected to secondary.

$$R_{e1} = R_1 + a^2R_2 \qquad (7\text{-}29)$$

$$X_{e1} = X_1 + a^2X_2 \qquad (7\text{-}30)$$

$$R_{e2} = \frac{R_1}{a^2} + R_2 \qquad (7\text{-}31)$$

$$X_{e2} = \frac{X_1}{a^2} + X_2 \qquad (7\text{-}32)$$

In addition, the equivalent impedance seen looking into either primary or secondary is given by Z_{e1} and Z_{e2}, respectively. These are expressed by Eqs. 7-33 and 7-34 and were derived using Eq. 7-8.

$$Z_{e1} = \sqrt{R_{e1}^2 + X_{e1}^2} \qquad (7\text{-}33)$$

$$Z_{e2} = \sqrt{R_{e2}^2 + X_{e2}^2} \qquad (7\text{-}34)$$

The equivalent parameters in Figure 7-18 can be found with a simple technique called the **short-circuit test.** In this test one of the windings is short circuited, usually the low-voltage side. The other winding has connected to it a voltmeter, an ammeter, a wattmeter, and a variable voltage supply, as shown in Figure 7-19. The voltage is increased until rated current exists in the windings. The voltage needed is small, only about 5% of the rated voltage for that winding. Since the voltage is so small, the core loss is negligible. This is so because the core losses depend on flux and flux in turn depends on voltage. Furthermore, since rated current is flowing in the windings, the wattmeter reading represents the transformer's rated copper losses. A transformer's copper losses are the I^2R losses in its windings. Referring to Figure 7-18, the copper loss is given by Eq. 7-35 or Eq. 7-36.

$$P_{\text{Cu}} = R_{e1}I_1^2 \qquad (7\text{-}35)$$

$$P_{\text{Cu}} = R_{e2}I_2^2 \qquad (7\text{-}36)$$

When the test is applied, three readings are recorded: W, A, and V. Equation 7-35 is then used to solve for R_{e1}. Note that $P_{\text{Cu}} = W$ and $I_1 = A$. By dividing V by A (Ohm's law) we obtain the transformer's impedance looking into the primary, Z_{e1}. Knowing this and R_{e1}, we can solve for X_{e1} using Eq. 7-33.

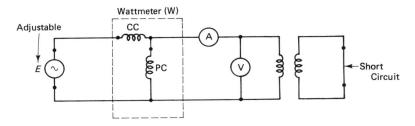

Figure 7-19 Circuit diagram for short-circuit tests.

Example 7-12

A 120/24-V 60-Hz 240-VA step-down transformer has a short-circuit test done on it. The measurements are made on the primary (high-voltage side) and the secondary (low-voltage side) is short circuited. The wattmeter reads 3.2 W, the voltmeter reads 2.8 V, and the ammeter reads rated current. Find the equivalent circuit parameters as seen on the primary and secondary.

Solution

First, we must solve for the rated primary current using Eq. 7-18 and find the ammeter reading.

$$I_{1r} = \frac{240 \text{ VA}}{120 \text{ V}} = 2 \text{ A}$$

Using Eq. 7-35 gives us

$$R_{e1} = \frac{P_{\text{Cu}}}{I_1^2} = \frac{W}{I_{1r}^2}$$

$$= \frac{3.2 \text{ W}}{(2 \text{ A})^2}$$

$$= 0.8 \text{ } \Omega$$

Now applying Ohm's law, we can find Z_{e1}.

$$Z_{e1} = \frac{2.8 \text{ V}}{2 \text{ A}} = 1.4 \text{ } \Omega$$

From Eq. 7-33,

$$X_{e1} = \sqrt{Z_{e1}^2 - R_{e1}^2}$$

$$= \sqrt{1.96 - 0.64} = 1.1 \text{ } \Omega$$

Using Eq. 7-26, we have

$$R_{e2} = \frac{R_{e1}}{a^2}$$

where $a = V_1/V_2 = 120 \text{ V}/24 \text{ V} = 5$; thus

$$R_{e2} = \frac{0.8}{25} = 0.032 \text{ } \Omega$$

and

$$X_{e2} = \frac{X_{e1}}{a^2} = \frac{1.1}{25} = 0.044 \text{ } \Omega$$

Figure 7-20 shows two equivalent circuits for this transformer.

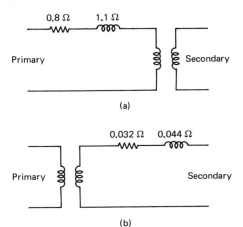

Figure 7-20 Equivalent circuits for the transformer of Example 7-12: (a) parameters reflected to primary; (b) parameters reflected to secondary.

7-3.2.3 Physical Limitations.

When hysteresis was discussed in Section 1-2.1, we saw that a magnetic material was limited in the amount of flux it could hold by its cross-sectional area. Different materials have different maximum flux densities. In Examples 1-11 and 1-12 we saw that this affected the size of the magnetic material. The minimum cross-sectional area of a transformer's core is determined by this parameter. The other factors that affect the size of a transformer's core are the voltage rating, number of winding turns, and the frequency of the applied voltage. Examples 7-13 and 7-14 will illustrate these relationships. Note that the size (thickness or gauge) of the wire used for the windings is determined by the current rating of the transformer.

Example 7-13 (English)

A 440/40-V step-down transformer has 550 turns on its primary. It is made with a material whose maximum flux density is 40,000 lines/in^2. Find the minimum cross-sectional area needed for its core when:

(a) It is to be used for voltages at 60 Hz

(b) It is to be used for voltages at 400 Hz

Solution

(a) Using Eq. 7-13a, we can find ϕ_m, the peak flux linking the primary and secondary windings.

$$\phi_m = \frac{E_1}{4.44 f N_1 \times 10^{-8}}$$

$$= \frac{440}{4.44 \times 60 \times 550 \times 10^{-8}} = 300.3 \text{ kilolines}$$

From Eq. 1-2a, we have

$$A_{\min} = \frac{\phi_m}{B_{\max}}$$

$$= 300.3\,\text{kilolines}/40\,\text{kilolines/in}^2$$

$$= 7.5\,\text{in}^2$$

(b) Repeating part (a), however, using 400 Hz gives us

$$\phi_m = \frac{440}{4.44 \times 400 \times 550 \times 10^{-8}}$$

$$= 45\,\text{kilolines}$$

and

$$A_{\min} = \frac{45\,\text{kilolines}}{40\,\text{kilolines/in}^2}$$

$$= 1.1\,\text{in}^2$$

Example 7-14 (SI)

A 440/40-V step-down transformer has 550 turns on its primary. It is made with a material whose maximum flux density is 1.2 T. Find the minimum cross-sectional area needed for its core when:

(a) It is to be used for voltages at 50 Hz

(b) It is to be used for voltages at 400 Hz

Solution

(a) Using Eq. 7-13b, we can find ϕ_m, the peak flux linking the primary and secondary windings.

$$\phi_m = \frac{E_1}{4.44fN_1}$$

$$= \frac{440}{4.44 \times 50 \times 550}$$

$$= 3.6 \times 10^{-3}\,\text{Wb}$$

From Eq. 1-2b, we have

$$A_{\min} = \frac{\phi_m}{B_{\max}}$$

$$= \frac{3.6 \times 10^{-3}\,\text{Wb}}{1.2\,\text{T}}$$

$$= 3 \times 10^{-3}\,\text{m}^2$$

(b) Repeating part (a), however, using 400 Hz,

$$\phi_m = \frac{440}{4.44 \times 400 \times 550}$$

$$= 0.45 \times 10^{-3} \text{ Wb}$$

and

$$\mathbf{A}_{\min} = \frac{0.45 \times 10^{-3} \text{ Wb}}{1.2 \text{ T}}$$

$$= 0.375 \times 10^{-3} \text{ m}^2$$

The preceding examples illustrate how the size of a transformer can be significantly reduced by using it with higher frequencies. It is for this reason that generators on aircraft develop power at 400 Hz rather than at 50 or 60 Hz. By doing this, the weight of the electromagnetic equipment on an aircraft is drastically reduced. Furthermore, it should be pointed out that a transformer designed for use at 50 or 60 Hz can always be used as rated at higher frequencies. On the other hand, one designed for 400 Hz would not work properly at lower frequencies. If it was used at a lower frequency, the core would saturate and the secondary voltage would not resemble or be proportional to the primary voltage.

In a similar way, operating a transformer at voltages greater than its rating will produce nonlinear effects. From Eqs. 7-13 we can see that ϕ_m is proportional to E_1. Thus if E_1 is too large, ϕ_m will be greater than that designed for and the core will saturate. Figure 7-21 illustrates this behavior with a transformer having a turns ratio

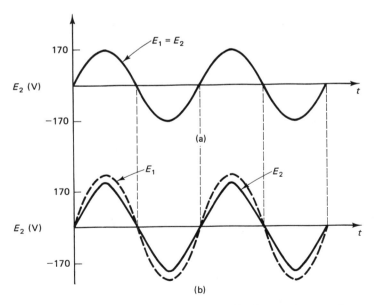

Figure 7-21 Secondary voltages of a transformer: (a) primary voltage equal to rated voltage; (b) primary voltage is 50% greater than rated voltage.

of 1. Its primary is rated 120 V (170 V peak). It is interesting to note that not only have we lost the 1:1 ratio, but also the secondary voltage now looks like a triangular waveform.

7-3.3 Voltage Regulation

Since the controlled output of a transformer is voltage (the current is a function of the load), its regulation has the same meaning as the dc generator (see Section 4-4). It is easy to visualize if we look at Figure 7-18b. In this configuration the primary winding voltage V_1 is equal to the applied voltage E_1. The secondary winding voltage V_2 will be V_1 divided by the turns ratio. Under no-load conditions I_2 is zero; thus there is no voltage drop across R_{e2} and X_{e2}. Therefore, at no load V_t will equal V_2. The voltage V_2 represents the transformer's no-load voltage (V_{NL}) and is generally an unknown quantity that we must solve for. As the transformer begins to supply load (current), E_1, V_1, and V_2 remain the same. The terminal voltage V_t will start to decrease since there will now be a voltage drop across R_{e2} and X_{e2} within the transformer. When the transformer is supplying rated current (full-load condition), the terminal voltage will be the transformer's full-load voltage (V_{FL}). This voltage is usually known and is given by the transformer's ratings. Thus in Examples 7-12 and 7-13, the full-load voltages are 24 V and 40 V, respectively.

Voltage regulation has the same definition as before and Eq. 4-6 is rewritten here as Eq. 7-37 for convenience.

$$\frac{\% \text{ voltage}}{\text{regulation}} = \frac{V_{NL} - V_{FL}}{V_{FL}} \times 100 \qquad (7\text{-}37)$$

Theoretically, predicting a transformer's voltage regulation involves three steps:

1. Perform the short-circuit test and determine the equivalent circuit parameters R_{e2} and X_{e2}.
2. Solve the circuit in Figure 7-18b for V_2 (this is V_{NL}), noting that V_t is the transformer's rated voltage and I_2 is its rated current.
3. Apply Eq. 7-37, where V_{FL} is the transformer's rated voltage and V_{NL} is V_2 found in step 2.

Example 7-15

Find the percent voltage regulation of the transformer in Example 7-12 when it supplies rated load at unity power factor.

Solution

Since we already know the equivalent circuit from Example 7-12, step 1 has been done and we proceed to draw the secondary circuit. The parameters X_{e2} and R_{e2} are from Figure 7-20b. Also, since the load is at unity power factor, it is purely resistive in nature (see item 7 in Section 7-1). The circuit is shown in Figure 7-22. At rated load the current I_2

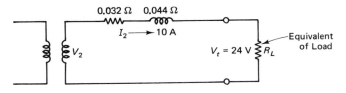

Figure 7-22 Circuit for Example 7-15.

is rated current (10 A) and the voltage V_t is rated voltage. Kirchhoff's voltage law is used to solve for V_2, keeping in mind that the voltages must be added as vectors.

$$V_2^2 = (0.032 \times 10 + 24)^2 + (0.044 \times 10)^2$$

$$= 591.46 + 0.19 = 591.65$$

$$V_2 = 24.32 \text{ V}$$

Now using Eq. 7-37 gives us

$$\% \text{ V.R.} = \frac{24.32 - 24}{24} \times 100 = 1.33\%$$

A problem like Example 7-15 becomes slightly more complicated when the load is not at unity power factor. We saw in Section 7-1 that a load which is inductive in nature will produce a lagging power factor and one that is capacitive will produce a leading power factor. The problem can be simplified by writing a general equation that can always be used. It is written as Eq. 7-38, where $\cos \theta$ is the power factor. When the power factor is lagging, the positive sign is used and when it is leading the negative sign is used.

$$V_{\text{NL}} = V_2 = \sqrt{(R_{e2}I_{2r} + V_{\text{FL}} \cos \theta)^2 + (X_{e2}I_{2r} \pm V_{\text{FL}} \sin \theta)^2} \qquad (7\text{-}38)$$

(+ for lagging PF, − for leading PF)

Example 7-16

A 2400/120-V 6-kVA transformer has the following parameters found from a short-circuit test: $R_{e2} = 0.12 \ \Omega$ and $X_{e2} = 0.22 \ \Omega$. Find the percent voltage regulation when it supplies rated load at:

(a) PF = 1

(b) PF = 0.85 lagging

(c) PF = 0.85 leading

Solution

First find the rated secondary current,

$$I_{2r} = \frac{6 \text{ kVA}}{120 \text{ V}} = 50 \text{ A}$$

(a) At PF = 1, cos θ = 1 and sin θ = 0. Using Eq. 7-38, we have

$$V_{NL} = \sqrt{(0.12 \times 50 + 120 \times 1)^2 + (0.22 \times 50 + 120 \times 0)^2}$$

$$= 126.48 \text{ V}$$

From Eq. 7-37,

$$\% \text{ V.R.} = \frac{126.48 - 120}{120} \times 100 = 5.4\%$$

(b) Again using Eq. 7-38 but with PF = 0.85 lagging, cos θ = 0.85 and sin θ = 0.53. Then

$$V_{NL} = \sqrt{(0.12 \times 50 + 120 \times 0.85)^2 + (0.22 \times 50 + 120 \times 0.53)^2}$$

$$= 131.26 \text{ V}$$

$$\% \text{ V.R.} = \frac{131.26 - 120}{120} \times 100 = 9.38\%$$

(c) Using Eq. 7-38 again but with PF = 0.85 leading yields

$$V_{NL} = \sqrt{(0.12 \times 50 + 120 \times 0.85)^2 + (0.22 \times 50 - 120 \times 0.53)^2}$$

$$= 120.13 \text{ V}$$

$$\% \text{ V.R.} = \frac{120.13 - 120}{120} \times 100 = 0.11\%$$

It is interesting to note that the best regulation in Example 7-16 was achieved with the leading power factor [part (c)]. This happened because the capacitive load counteracted the inductive nature of the transformer.

7-3.4 Efficiency

To calculate the efficiency of a transformer, we apply the general formula used many times before:

$$\eta\% = \frac{P_o}{P_i} \times 100 \tag{7-39}$$

In this equation P_o and P_i represent the output power and input power, respectively, in watts or kilowatts. Usually, the output power is known and the input power is found by adding the output power to the losses of the transformer. This was identical to the procedure for the dc generator (see Eqs. 4-8 and 4-9). Equation 4-8 is rewritten here as

$$P_i = P_o + \text{total losses} \tag{7-40}$$

There are two losses in the transformer which must be considered. One of them, the rated copper loss, can be found by doing a short-circuit test (Section 7-3.2.2). The other is the transformer's rated core loss. The core loss can be found by another test, called the **open-circuit test.** In this test rated voltage is applied to one winding,

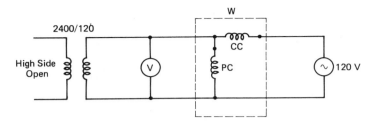

Figure 7-23 Typical open-circuit test connection.

usually the low side, while the other is left open. A wattmeter on the energized winding gives a reading that represents the transformer's core loss. Although this is an approximation, it is a good one. Since core losses depend on flux and flux depends on voltage, the core loss read by the wattmeter will be a rated loss as long as rated voltage is applied. As the load is varied, this loss will not change significantly. Furthermore, since one winding is open, the current in the energized winding will be very small and the copper loss in that winding will be negligible. A typical connection is shown in Figure 7-23.

By applying these tests, we can find the transformer's losses. The core loss will be assumed constant; however, the copper loss will vary as the current changes. Since the copper loss varies as the square of the current, the copper loss at half rated load will be one-fourth of the rated value. Note that if a transformer is rated 440/220 V, 10 kVA and it is supplying half-rated load, it is supplying 5 kVA. The voltage across the load is still about 220 V; however, the current to the load has halved. The power it is supplying depends on the power factor of the load.

Example 7-17

A 120/440-V 3-kVA transformer supplies rated load at unity power factor. Find the power output and the secondary current.

Solution

From Eq. 7-11, where $\cos \theta = 1$,

$$P_o = P_a \cos \theta = 3 \text{ kVA} \times 1$$

$$= 3 \text{ kW}$$

From Eq. 7-19,

$$I_{2r} = \frac{3 \text{ kVA}}{440 \text{ V}}$$

$$= 6.82 \text{ A}$$

Example 7-18

The transformer of Example 7-17 is supplying half-rated load at a power factor of 0.9 lagging. Find the power output and the secondary current.

Solution

We will again use Eq. 7-11; however, this time $\cos \theta = 0.9$ and P_a is one-half of rated output.

$$P_a = \tfrac{1}{2}(3 \text{ kVA}) = 1.5 \text{ kVA}$$

$$P_o = P_a \cos \theta = 1.5 \text{ kVA} \times 0.9$$

$$= 1.35 \text{ kW}$$

At half-rated load, the current will be one-half of rated current.

$$I_2 = \tfrac{1}{2}I_{2r} = 3.41 \text{ A}$$

The equation for efficiency can now be written as follows, where P_{Cu} and P_{core} are the copper and core losses, respectively:

$$\eta\% = \frac{P_o}{P_o + P_{Cu} + P_{core}} \times 100 \qquad (7\text{-}41)$$

In Eq. 7-41 all quantities must be in watts or kilowatts. The efficiency can also be written in terms of the VA output of the transformer.

$$\eta\% = \frac{\text{VA}_{out} \times \text{PF}}{\text{VA}_{out} \times \text{PF} + P_{Cu} + P_{core}} \times 100 \qquad (7\text{-}42)$$

Example 7-19

A 10-kVA 220/600-V 60-Hz transformer has open-circuit and short-circuit tests performed. Find its efficiency for the following conditions:

(a) Full load, PF = 1

(b) Half load, PF = 1

(c) Full load, PF = 0.8 lagging

(d) Half load, PF = 0.8 lagging

(e) 125% of full load, PF = 1

The results of the tests are as follows:

Open-circuit test (high side open)	Short-circuit test (low side shorted)
V = 220 V	V = 35 V
I = 1 A	I = rated current
W = 120 W	W = 200 W

Solution

From the tests, the rated copper loss is 200 W and the rated core loss is 120 W.

(a) Using Eq. 7-42, we have

$$\eta = \frac{10{,}000 \times 1}{10{,}000 \times 1 + 200 + 120} \times 100$$

$$= 96.9\%$$

(b) We again use Eq. 7-42; however, this time the VA_{out} is 5000 since we are at half load. The PF is still unity, but P_{Cu} will be one-fourth of its rated value.

$$\eta = \frac{5000 \times 1}{5000 \times 1 + 200(\frac{1}{2})^2 + 120} \times 100$$

$$= \frac{5000}{5000 + 50 + 120} \times 100$$

$$= 96.7\%$$

(c) In this case we are at full load, but the power factor has dropped to 0.8.

$$\eta = \frac{10{,}000 \times 0.8}{10{,}000 \times 0.8 + 200 + 120} \times 100$$

$$= 96.15\%$$

(d) At one-half load and PF = 0.8 lagging, we get

$$\eta = \frac{5000 \times 0.8}{5000 \times 0.8 + 200(\frac{1}{2})^2 + 120} \times 100$$

$$= \frac{4000}{4000 + 50 + 120} \times 100$$

$$= 95.9\%$$

(e) Finally, when the load is increased beyond rated load, we have the following calculation. Note that the copper loss in the denominator now increases with the square of the load increase.

$$VA_{out} = 1.25 \times 10{,}000 = 12{,}500 \text{ VA}$$

$$\eta = \frac{12{,}500 \times 1}{12{,}500 \times 1 + 200(1.25)^2 + 120} \times 100$$

$$= \frac{12{,}500}{12{,}500 + 312.5 + 120} \times 100$$

$$= 96.66\%$$

A couple of points should be made about efficiency and the variables affecting it. A transformer's efficiency increases (gets better) as the load it supplies gets closer to rated load. Above full load the efficiency begins to decrease. In addition, as the power factor decreases, so will the efficiency. Thus we can conclude that the maximum efficiency occurs at full load and unity power factor. Furthermore, we should note that whether the power factor is leading or lagging will have no effect on efficiency. In other words, the results in parts (c) and (d) in Example 7-19 would have been the same if the power factor had been 0.8 leading.

7-3.5 Multiple-Winding Transformers

Very often many different voltage transformations are needed in a piece of equipment. Rather than have a separate transformer for each one, a single transformer with several windings on the secondary is used. In this way, one single-phase voltage can be stepped up or down to different levels. Analyzing this type of transformer can be done by reflecting from primary to secondary, or vice versa, as before. We will do the latter since the former technique requires the solution of several circuits (one for each secondary winding).

Example 7-20

A four-winding transformer is shown in Figure 7-24a. Secondary winding X is connected to a purely inductive load, Y is connected to a purely capacitive load, and Z is connected to a purely resistive load. Find:

(a) The voltage and current of each winding.

(b) The power factor at which the transformer is operating.

(c) The VA being supplied by each winding.

Solution

(a) The first step is to find the turns ratio for each winding. Using this, each secondary voltage can be found. In addition, each load can be reflected to the primary, as shown in Figure 7-24b.

Winding X:

$$\text{turns ratio} = \frac{200}{400} = \frac{1}{2}$$

$$V_X = E_1 \div \tfrac{1}{2} = 240 \text{ V}$$

Winding Y:

$$\text{turns ratio} = \frac{200}{50} = 4$$

$$V_Y = E_1 \div 4 = 30 \text{ V}$$

Winding Z:

$$\text{turns ratio} = \frac{200}{100} = 2$$

$$V_Z = E_1 \div 2 = 60 \text{ V}$$

Looking at Figure 7-24b, we find that

$$I_L = \frac{120 \text{ V}}{15 \ \Omega} = 8 \text{ A}$$

$$I_C = \frac{120 \text{ V}}{32 \ \Omega} = 3.75 \text{ A} \quad \text{and} \quad I_R = \frac{120 \text{ V}}{40 \ \Omega} = 3 \text{ A}$$

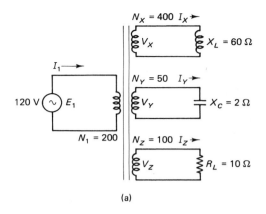

(a)

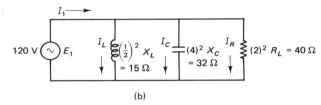

(b)

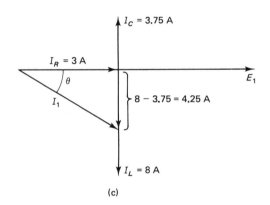

(c)

Figure 7-24 (a) Circuit for Example 7-20; (b) equivalent primary circuit; (c) vector addition of primary currents.

To find the primary current, I_1, we apply Kirchhoff's current law, remembering that we must add the currents as vectors. The vector addition is shown in Figure 7-24c. Note that I_R is in phase with E_1, I_C leads E_1 by 90°, and I_L lags E_1 by 90°.

$$I_1 = \sqrt{(3)^2 + (8 - 3.75)^2}$$
$$= 5.2 \text{ A}$$

This is the current that would be measured if an ammeter were placed in the primary of Figure 7-24a. The secondary currents are found by applying Eq. 7-20 to I_L, I_C, and I_R, respectively.

$$I_X = I_L \times \tfrac{1}{2} = 4 \text{ A}$$

$$I_Y = I_C \times 4 = 15 \text{ A}$$

$$I_Z = I_R \times 2 = 6 \text{ A}$$

(b) The power factor is given by the cosine of the angle between I_1 and E_1 in Figure 7-24c. Using trigonometry, we have

$$\theta = \arctan \frac{4.25}{3}$$

$$= 54.78°$$

$$\text{PF} = \cos 54.78° = 0.58$$

(c) The VA supplied by each winding is found by applying Eq. 7-3.

Winding X:

$$P_a = V_X I_X = 240 \text{ V} \times 4 \text{ A} = 960 \text{ VA}$$

Winding Y:

$$P_a = V_Y I_Y = 30 \text{ V} \times 15 \text{ A} = 450 \text{ VA}$$

Winding Z:

$$P_a = V_Z I_Z = 60 \text{ V} \times 6 \text{ A} = 360 \text{ VA}$$

Primary Winding:

$$P_a = E_1 I_1 = 120 \text{ V} \times 5.2 \text{ A} = 624 \text{ VA}$$

It should be noted that since the windings supply different VA, they would each be constructed of a different gauge wire.

A variation of the multiple-winding transformer is the **tapped transformer.** This type can have several secondary voltages with only one winding. An example of a **center-tapped transformer** is shown in Figure 7-25.

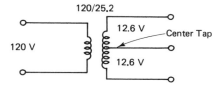

Figure 7-25 Center-tapped transformer.

7-3.6 Autotransformer

In the transformers studied so far, power was transferred from primary to secondary magnetically. There was no electrical connection between input and output. By electrically connecting the primary and secondary windings, we can effectively increase the power capability of a transformer. Some power is transferred magnetically as before; however, now we transfer an additional amount of power by direct electrical conduction. This leads to a more economical and efficient transformer. What is lost

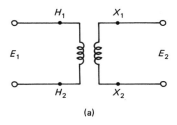

(a)

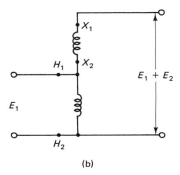

(b)

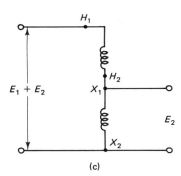

(c)

Figure 7-26 (a) Single-phase transformer; (b) wired as a step up autotransformer; (c) wired as a step down autotransformer.

in the process is the electrical isolation between primary and secondary that existed before. Care must be taken to connect the two windings properly with regard to their polarity markings. Figure 7-26 shows two typical connections. In Figure 7-26b we have a step-up autotransformer, whereas in Figure 7-26c it is wired to step down.

There are obviously other possibilities. For example, the input in Figure 7-26b could have been applied to the X_1–X_2 terminals. The autotransformer would then be stepping up from E_2 to $E_1 + E_2$. In addition, other transformations are possible by connecting the windings with opposite polarity. For example, if X_1 and X_2 were interchanged in Figure 7-26c, the autotransformer would be stepping down from $E_1 - E_2$ to E_2.

The autotransformer's rating is obtained by maintaining each winding's VA rating as specified for the original single-phase transformer.

Example 7-21

A 220/440-V 3-kVA transformer is used to build a 440/660-V step-up autotransformer.

(a) Draw a diagram of the connection showing proper polarity markings.

(b) Calculate all currents under rated conditions and the kVA rating of the auto-transformer.

Solution

(a) See Figure 7-27.

(b) To calculate the currents, we must first find the rated currents for the original transformer from its ratings. From Eq. 7-18,

$$I_{1r} = \frac{3000 \text{ VA}}{220 \text{ V}} = 13.64 \text{ A}$$

and from Eq. 7-19.

$$I_{2r} = \frac{3000 \text{ VA}}{440 \text{ V}} = 6.82 \text{ A}$$

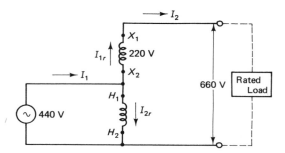

Figure 7-27 Solution to part (a) of Example 7-21.

Referring to Figure 7-27, we see that the autotransformer's secondary current (I_2) is the same as the original transformer's low-voltage winding current, I_{1r}. Thus the rated or full-load I_2 is given by the rated winding current I_{1r}.

$$I_2 = I_{1r} = 13.64 \text{ A}$$

The kVA out of the autotransformer is obtained by multiplying secondary voltage by secondary current.

$$660 \text{ V} \times 13.64 \text{ A} = 9002 \text{ VA} \approx 9 \text{ kVA}$$

The kVA into the autotransformer should equal that coming out. This assumes 100% efficiency which is fairly reasonable. Thus

$$9002 \text{ VA} = 440 \text{ V} \times I_1$$

$$I_1 = \frac{9002}{440} = 20.46 \text{ A}$$

Applying Kirchhoff's current law to the node at H_1–X_2, we get as a check:

$$I_{2r} = I_1 - I_{1r} = 20.46 - 13.64$$

$$= 6.82 \text{ A (as originally calculated)}$$

In Example 7-21 we can note the following with reference to the circuit in Figure 7-27. Of the 9 kVA ($440 \times I_1$) going into the autotransformer, 3 kVA ($440 \times I_{2r}$) is delivered to the output by **induction.** The remaining 6 kVA [$440 \text{ V} \times (I_1 - I_{2r})$] is delivered to the output by **conduction.**

Example 7-22

A 600/240-V 7.5 kVA transformer is used to build a 600/360-V step-down autotransformer.

(a) Draw a diagram of the connection showing proper polarity markings.

(b) Calculate all currents under rated conditions and the kVA rating of the autotransformer.

Solution

(a) See Figure 7-28. Note that the polarities are opposing ($600 \text{ V} - 240 \text{ V} = 360 \text{ V}$).

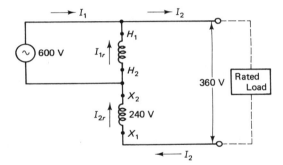

Figure 7-28 Solution to part (a) of Example 7-22.

(b) To calculate the currents, first find the rated currents of the original transformer.

$$I_{1r} = \frac{7500 \text{ VA}}{600 \text{ V}} = 12.5 \text{ A}$$

$$I_{2r} = \frac{7500 \text{ VA}}{240 \text{ V}} = 31.25 \text{ A}$$

Referring to Figure 7-28, we see that

$$I_2 = I_{2r} = 31.25 \text{ A}$$

Thus the kVA out is

$$360 \text{ V} \times I_2 = 360 \text{ V} \times 31.25 \text{ A} = 11.25 \text{ kVA}$$

The kVA into the autotransformer will be the same; therefore,

$$11.25 \text{ kVA} = 600 \text{ V} \times I_1$$

$$I_1 = \frac{11.25 \text{ kVA}}{600 \text{ V}} = 18.75 \text{ A}$$

Applying Kirchhoff's current law to the node H_1 as a check:

$$I_1 + I_{1r} = I_2$$

$$I_{1r} = I_2 - I_1 = 31.25 \text{ A} - 18.75 \text{ A}$$

$$= 12.5 \text{ A (as originally calculated)}$$

Note that in Example 7-22, of the 11.25 kVA being transferred from input to output, 7.5 kVA (600 V $\times$ I_{1r}) is done by induction. The remaining 3.75 kVA [600 V $\times$ ($I_1 - I_{1r}$)] is transferred by conduction.

7-4   REVIEW OF THREE-PHASE AC THEORY

Up to this point we have been dealing with only one ac voltage at any given time. We call this voltage single-phase ac. The electrical equipment that we use in our homes is built to operate with this type of voltage. It is for this reason that the power supplied to our homes by the electrical utilities is in the form of single-phase power. There are many advantages, however, in using more than one voltage simultaneously. These voltages would have the same magnitude but differ in phase; that is, their positive or negative peaks would occur at different times. If we could look at them simultaneously on an oscilloscope, we would see a picture similar to that shown in Figure 7-29.

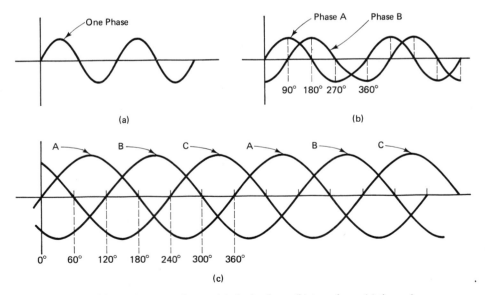

Figure 7-29   Various ac voltages: (a) single phase; (b) two phase; (c) three phase.

When there is more than one phase in a voltage system, it is called a **polyphase system.** The most common polyphase system is the three-phase system shown in Figure 7-29c. The way the three voltages are labeled determines the **phase sequence.** The waveforms travel from right to left. Therefore, if we watch the 0° axis, the positive peak of phase A would cross first, then phase B, and finally phase C. After that the cycle repeats. We call the order in which the waves follow each other the phase sequence. For Figure 7-29c the sequence is ABC, BCA, or CAB; all three mean exactly the same sequence. By switching any two of the three waveforms (A with B, B with C, or A with C) we produce a second phase sequence, ACB (or CBA or BAC). With three phases there are only two possible sequences.

The advantages of three-phase power are the following:

1. More power can be transmitted for a given cable size and cost.
2. It provides a smoother flow of power than the pulsating power flow in a single-phase system.
3. It allows us to use three-phase machines, which can run more smoothly and are cheaper to build for a given size than is a single-phase machine.

7-4.1 Numerical Relationships

Depending on the source of power, the three voltages are transmitted on either three or four wires. In the three-wire system, the voltage between any two lines is called the **line voltage** (V_L). The four-wire system is basically the same except that a fourth wire, called the **neutral,** is added. In this case the voltage between any line and the neutral is called a **phase voltage** (V_p). The voltage between any two lines is the line voltage as before. Figure 7-30 will help clarify these definitions.

For the two cases shown in Figure 7-30, the following should be noted:

1. The three line voltages (V_{AB}, V_{BC}, and V_{CA}) are equal in magnitude but differ in phase, as shown in Figure 7-29c.
2. The three phase voltages (V_{AN}, V_{BN}, and V_{CN}) are equal in magnitude but differ in phase, as shown in Figure 7-29c.
3. If we apply Kirchhoff's voltage law to Figure 7-30b, we can write the following equation:

$$V_{AB} = V_{AN} + V_{NB}$$

However, keeping in mind the fact that V_{AN} and V_{NB} are out of phase 60° (see Figure 7-29c), we realize that they must be added as vectors. If we did that, we would come up with a basic relationship given as Eq. 7-43:

$$V_L = \sqrt{3}\, V_p \qquad (7\text{-}43)$$

Although a three-phase system can be connected to any type of load, we will examine what is called a **balanced three-phase load.** This type of load is made up of three identical impedances connected in either of the two ways shown in Figure 7-31. Their appearance gives them the names **delta** (Δ) and **wye** (Y).

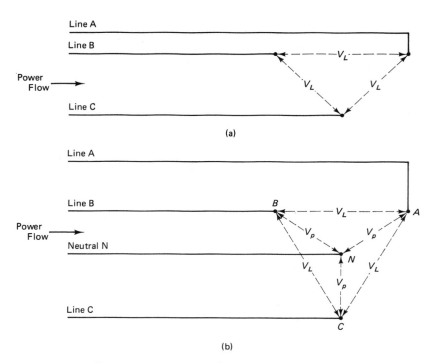

Figure 7-30 (a) Three-phase three-wire system; (b) three-phase four-wire system.

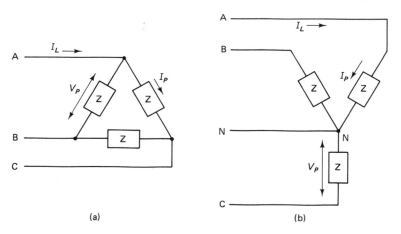

Figure 7-31 Balanced three-phase loads: (a) in a delta configuration; (b) in a wye configuration.

Although a neutral line is shown in Figure 7-31b, it actually does not serve us a purpose. This is so because in a balanced system the current in the neutral will be zero. In a practical balanced system, however, the neutral line is monitored with a circuit breaker. In the event of a failure in one of the phases, the balance is broken, a current will appear in the neutral and the breaker will trip.

In the delta configuration (Figure 7-31a) the voltage across one phase of the load (V_p) is equal to the line voltage. The current in each phase of the load (I_p) can be found by dividing V_p (or V_L in this case) by Z. Because of the 120° phase difference, the line current I_L is given by Eq. 7-44.

$$\Delta \text{ load} \quad \begin{cases} I_L = \sqrt{3}\, I_p & (7\text{-}44) \\[2mm] V_L = V_p & (7\text{-}45) \end{cases}$$

In the wye configuration (Figure 7-31b) the line current is equal to the phase current. In this case, however, the voltage across one phase of the load (V_p) is equal to the line voltage divided by $\sqrt{3}$ (see Eq. 7-43).

$$\text{Y load} \quad \begin{cases} I_p = I_L & (7\text{-}46) \\[2mm] V_p = \dfrac{V_L}{\sqrt{3}} & (7\text{-}47) \end{cases}$$

The apparent power in each phase of the load is $V_p \times I_p$. For the entire load the total apparent power would be three times this.

$$P_a = 3V_p I_p \tag{7-48}$$

Substituting Eqs. 7-46 and 7-47 into Eq. 7-48, we get the more commonly used expression for apparent power. Equation 7-49, which is in terms of the line quantities, is valid whether the load is delta or wye connected.

$$P_a = \sqrt{3}\, V_L I_L \tag{7-49}$$

Furthermore, we can solve for the reactive and real power delivered to the load. Note the similarity between Eqs. 7-50 and 7-51 and those for single-phase ac (Eqs. 7-10 and 7-11).

$$P_r = \sqrt{3}\, V_L I_L \sin\theta \tag{7-50}$$

$$P_w = \sqrt{3}\, V_L I_L \cos\theta \tag{7-51}$$

Note, as before, that $\cos\theta$ is the power factor, and θ is the angle of the impedance Z in Figure 7-31.

Three-phase power is almost always specified in terms of the line quantities. Therefore, a 440-V 10-kVA rated system is one whose line voltage is 440 V and which delivers a total of 10 kVA. From this information the rated line current can be found using Eq. 7-49.

Example 7-23

A 230-V three-phase system supplies 2 kVA at unity power factor to a balanced wye-connected load. Find all the currents and voltages. In addition, find an equivalent representation of the load.

Solution

The first step is to draw a picture of the system indicating what is given and what must be found. In Figure 7-32 the load has been represented by three **resistors**, R_{eq}. This is

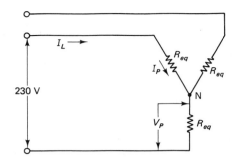

Figure 7-32 Diagram for Example 7-23.

so because we were told that load is supplied at unity power factor. It may very well be made up of capacitors and inductors; however, since the power factor is unity, it can be represented as purely resistive. The line voltage is given as 230 V. Using this and the 2 kVA supplied with Eq. 7-49, we can solve for the line current.

$$I_L = \frac{2000 \text{ VA}}{\sqrt{3} \times 230 \text{ V}} = 5.02 \text{ A}$$

For the wye-connected load,

$$I_p = I_L = 5.02 \text{ A}$$

and

$$V_p = \frac{V_L}{\sqrt{3}} = \frac{230 \text{ V}}{\sqrt{3}}$$

$$= 132.8 \text{ V}$$

Using Ohm's law, we have

$$R_{eq} = \frac{V_p}{I_p} = \frac{132.8 \text{ V}}{5.02 \text{ A}}$$

$$= 26.45 \ \Omega$$

Example 7-24

A 440-V 50-kVA three-phase system supplies rated load at a lagging power factor of 0.8 to an industrial facility. Find:

(a) The full-load line current

(b) The real power supplied

(c) The reactive power supplied

(d) Under what conditions could maximum power be supplied?

Solution

(a) From Eq. 7-49,

$$I_L = \frac{50,000 \text{ VA}}{\sqrt{3} \times 440 \text{ V}}$$

$$= 65.6 \text{ A}$$

(b) From Eq. 7-51 and noting that cos θ = PF = 0.8, we have

$$P_w = \sqrt{3} \times 440 \text{ V} \times 65.6 \text{ A} \times 0.8$$

$$= 39{,}995 \text{ W} \approx 40 \text{ kW}$$

(c) First find θ.

$$\theta = \arccos 0.8 = 36.87°$$

$$\sin \theta = 0.6$$

Using Eq. 7-50 gives us

$$P_r = \sqrt{3} \times 440 \text{ V} \times 65.6 \text{ A} \times 0.6$$

$$= 29{,}996 \text{ var} \approx 30 \text{ kvar}$$

(d) To get the conditions for maximum power supplied, we examine Eq. 7-51. It is clear from this equation that V_L and I_L can only be as large as their rated values. The only variable left is cos θ, the power factor. This has a maximum value of 1. Thus maximum power can be delivered when the power factor is unity.

$$\text{max. } P_w = \sqrt{3} \times 440 \text{ V} \times 65.6 \text{ A} \times 1.0$$

$$= 50 \text{ kW}$$

This is the same as the kVA rating expressed in kW. In other words, if 50 kVA is supplied at unity power factor, 50 kW is being supplied.

After going through Example 7-24, we can make the following observations. When a utility supplies power to a user, it will only get paid for the real power (kW) that it supplies. Thus at low power factors, the utility's lines can be operating at rated conditions and yet it will not be making as much money as it could if the power factor was higher.

In a similar analysis, the industrial facility of Example 7-24 may have a total power need of 45 kW. However, since the facility is operating at a power factor of 0.8, it can only draw 40 kW from the lines. Thus the facility cannot operate all of its equipment unless the power factor increases to 0.9 (50 kVA × 0.9 = 45 kW).

In light of these observations, it is to everyone's advantage (supplier and user) to operate at power factors as close to unity as possible. In Section 10-6 we will see how the power factor can be changed. The technique is called **power factor correction.**

7-4.2 Power Measurement in Three-Phase Systems

There is a basic theorem called **Blondel's theorem,** which states: In a polyphase system with N wires, the total power can be measured with $N - 1$ wattmeters. Thus in a three-phase three-wire system, two $(3 - 1)$ wattmeters are needed to measure the total power. In a three-phase four-wire system, three $(4 - 1)$ wattmeters are needed. However, if the load is balanced, one wire (the neutral) can be eliminated and we can measure the power with two wattmeters. Figure 7-33 shows the connection using two and three wattmeters. In either case shown in Figure 7-33, the total power is the

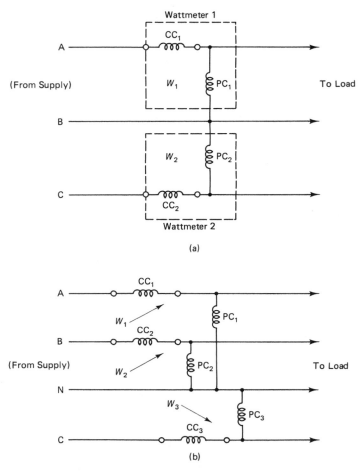

Figure 7-33 Measurement of power in (a) a three-wire system; (b) a four-wire system.

algebraic sum of the wattmeter readings. Since we will be concerned here only with balanced loads, the **two-wattmeter** technique will be covered in detail.

When the power factor of the load is unity, the meter readings will be positive and equal to each other. The total power will be their sum, $W_1 + W_2$. As the power factor changes in a lagging direction, one of the readings, say W_1, will be larger. This reading is always positive. The smaller reading, W_2 in our case, must be tested to see if it is positive or negative. A negative reading would indicate a flow of power from load to supply. However since the positive reading W_1 is greater than the negative reading W_2, the net power, $W_1 - W_2$, is still positive. As long as it is positive, the net flow of power is from the supply to the load.

If W_2 had been the larger reading, it would have been the positive reading and W_1 would have to be tested for polarity. The test for polarity is as follows. Refer to Figure 7-33a.

1. If W_2 is the smaller reading, disconnect PC_2 from line B. Touch the disconnected end of PC_2 to line A. If W_2 reads upscale, its original reading was positive. However, if W_2 now reads downscale, its original reading was negative.

2. If W_1 is the smaller reading, disconnect PC_1 from line B. Touch the disconnected end of PC_1 to line C. If W_1 reads upscale, its original reading was positive. If, however, W_2 now reads downscale, its original reading was negative.

7-5   THREE-PHASE TRANSFORMERS

Just as we needed transformers when working with single-phase voltages, we also need them when working with three-phase voltages. One possibility would be to use three separate but identical single-phase transformers. It is initially cheaper, though, to use one polyphase (three-phase, in this case) transformer. In the event of a failure, however, the entire three-phase transformer must be replaced, whereas if three separate single-phase units were used, we would only have to replace one of them. Thus the choice of which scheme to use would depend on the particular application. In either case we analyze three-phase transformation in the same way. A simplified picture of a **core-type** three-phase transformer is shown in Figure 7-34.

To help clarify our discussion, the primary windings are labeled with uppercase letters (A, B, C) and the secondary windings are labeled with lowercase letters (a, b, c). Note that since the three primary voltages are 120° out of phase, the three

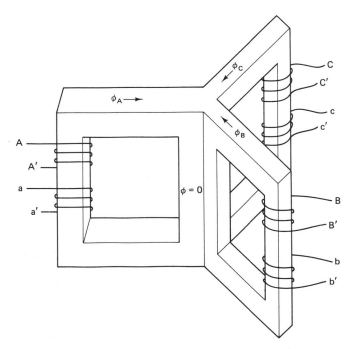

Figure 7-34   Simplified diagram of a core-type three-phase transformer.

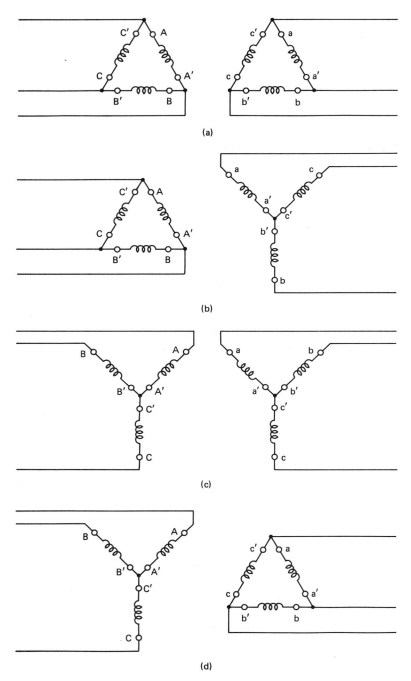

Figure 7-35 Four basic three-phase transformer connections: (a) Δ-Δ; (b) Δ-Y; (c) Y-Y; (d) Y-Δ.

components of flux (ϕ_A, ϕ_B, and ϕ_C) will also be 120° out of phase. The three add up vectorally in their common path, and the flux in the middle leg will be zero. The middle leg is therefore omitted in the construction of the transformer, since it is not needed as a flux path.

By connecting the primary and secondary windings together in different ways, we can produce different characteristics for the transformer of Figure 7-34. However, regardless of how they are connected, the total kVA that can be handled will not change. It should be pointed out that the three primary windings are identical, as are the three secondary windings.

7-5.1 Basic Transformer Connections

The primary windings in Figure 7-34 can be interconnected in two basic ways. By connecting A', B', and C' together and the three phase voltages to A, B, and C, the primary will be in a Y connection (see Figure 7-35c). On the other hand, if we connect A' to B, B' to C, and C' to A, the primary will be in a Δ connection (see Figure 7-35a). In a similar way the secondary windings can be connected in either Δ or Y. Each of the connections has advantages over the other. As we examine the various connections, the advantages will be pointed out. There are four possible connections and they are shown in Figure 7-35. Note that the primary windings, when connected to a three-phase system, can be treated just like the balanced three-phase loads covered in Section 7-4.1. Therefore, the equations used in that section can all be used for the three-phase transformer. Furthermore, the rated kVA of a three-phase transformer is three times the rating of each primary or secondary winding.

Example 7-25

A three-phase transformer supplies a rated load of 25 kVA at 208 V to a balanced load. Its primary is connected to a 1200 V supply. Find the kVA rating and turns ratio of each transformer and the primary and secondary currents and voltages if it is wired:

(a) Y-Y

(b) Δ-Δ

Solution

(a) Each of the three transformers must be rated at one-third of the total:

$$\frac{25 \text{ kVA}}{3} = 8.333 \text{ kVA}$$

The solution of the currents and voltages is fairly easy when Figure 7-36 is drawn. The primary current will be equal to the high-side line current. Since the transformer is supplying 25 kVA, it must also be drawing 25 kVA. From Eq. 7-49,

$$I_1 = I_{L1} = \frac{25,000 \text{ VA}}{\sqrt{3} \times 1200 \text{ V}} = 12.03 \text{ A}$$

From Eq. 7-47,

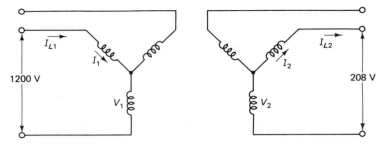

Figure 7-36 Diagram for part (a) of Example 7-25.

$$V_1 = \frac{1200 \text{ V}}{\sqrt{3}} = 693 \text{ V}$$

On the secondary we can solve for the quantities in the same way.

$$I_2 = I_{L2} = \frac{25{,}000 \text{ VA}}{\sqrt{3} \times 208 \text{ V}} = 69.4 \text{ A}$$

$$V_2 = \frac{208 \text{ V}}{\sqrt{3}} = 120 \text{ V}$$

The individual transformer turns ratio can be found using Eq. 7-20.

$$a = \frac{V_1}{V_2} = \frac{693 \text{ V}}{120 \text{ V}} = 5.78$$

Note that this is the same as the ratio of the line voltages 1200/208.

(b) Each transformer is still rated 8.333 kVA; this does not change with the type of connection. We now draw Figure 7-37 for the Δ-Δ connection. From Figure 7-37 it is clear that

$$V_1 = 1200 \text{ V}$$

and

$$V_2 = 208 \text{ V}$$

The high-side and low-side line currents I_{L1} and I_{L2} are the same as part (a) since the kVA and line voltages are the same. From Eq. 7-44,

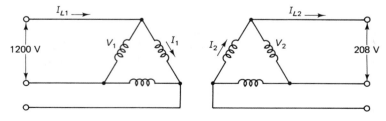

Figure 7-37 Diagram for part (b) of Example 7-25.

$$I_1 = \frac{I_{L1}}{\sqrt{3}} = \frac{12.03 \text{ A}}{\sqrt{3}} = 6.95 \text{ A}$$

$$I_2 = \frac{I_{L2}}{\sqrt{3}} = \frac{69.4 \text{ A}}{\sqrt{3}} = 40 \text{ A}$$

The turns ratio is given by

$$a = \frac{V_1}{V_2} = \frac{1200 \text{ V}}{208 \text{ V}} = 5.78$$

As a check in both parts, if we multiply V_1 by I_1 or V_2 by I_2 we should get the individual transformers rating. For part (a),

$$V_1 \times I_1 = 693 \text{ V} \times 12.03 \text{ A} \approx 8333 \text{ VA}$$

and for part (b),

$$V_1 \times I_1 = 1200 \text{ V} \times 6.95 \text{ A} \approx 8333 \text{ VA}$$

The difference is the error introduced by rounding off the currents and voltages.

Example 7-26

A three-phase transformer steps up 1200 V to 13,200 V for transmission. It is rated 100 kVA. Find the rating of each transformer, the primary and secondary voltages and currents, and the turns ratio if it supplies rated load.

(a) It is Δ-Y connected

(b) It is Y-Δ connected.

Solution

(a) Each transformer is rated at 100 kVA/3 = 33.33 kVA.

A diagram for part (a) is shown in Figure 7-38. Using Eq. 7-49, we get the two line currents:

$$I_{L1} = \frac{100,000 \text{ VA}}{\sqrt{3} \times 1200 \text{ V}} = 48.11 \text{ A}$$

$$I_{L2} = \frac{100,000 \text{ VA}}{\sqrt{3} \times 13,200 \text{ V}} = 4.37 \text{ A}$$

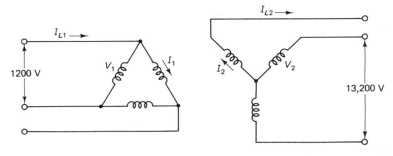

Figure 7-38 Diagram for part (a) of Example 7-26.

From Figure 7-38,

$$I_2 = I_{L2} = 4.37 \text{ A}$$

$$I_1 = \frac{I_{L1}}{\sqrt{3}} = \frac{48.11 \text{ A}}{\sqrt{3}} = 27.78 \text{ A}$$

$$V_1 = 1200 \text{ V}$$

$$V_2 = \frac{13{,}200 \text{ V}}{\sqrt{3}} = 7621 \text{ V}$$

The turns ratio is

$$a = \frac{V_1}{V_2} = \frac{1200 \text{ V}}{7621 \text{ V}} = 0.16$$

Note this is not the same as the ratio of the line voltages as in Example 7-25.

(b) Each transformer is still rated

$$\frac{100 \text{ kVA}}{3} = 33.33 \text{ kVA}$$

The picture now looks as shown in Figure 7-39. Using Eq. 7-49, the line currents are

$$I_{L1} = \frac{100{,}000 \text{ VA}}{\sqrt{3} \times 1200 \text{ V}} = 48.11 \text{ A}$$

$$I_{L2} = \frac{100{,}000 \text{ VA}}{\sqrt{3} \times 13{,}200 \text{ V}} = 4.37 \text{ A}$$

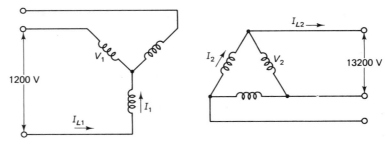

Figure 7-39 Diagram for part (b) of Example 7-26.

Note they are the same as in part (a). From Figure 7-39,

$$I_1 = I_{L1} = 48.11 \text{ A}$$

$$I_2 = \frac{I_{L2}}{\sqrt{3}} = \frac{4.37 \text{ A}}{\sqrt{3}} = 2.53 \text{ A}$$

$$V_2 = 13{,}200 \text{ V}$$

$$V_1 = \frac{1200 \text{ V}}{\sqrt{3}} = 692.8 \text{ V}$$

The turns ratio is

$$a = \frac{V_1}{V_2} = \frac{692.8 \text{ V}}{13,200 \text{ V}}$$

$$= 0.0525$$

Again it is different than the ratio of the line voltages, which is

$$\frac{1200 \text{ V}}{13,200 \text{ V}} = 0.091$$

As a check, each transformer's rating is:

Part (a): $V_1 \times I_1 = 1200 \text{ V} \times 27.78 \text{ A} = 33.3 \text{ kVA}$

Part (b): $V_1 \times I_1 = 692.8 \text{ V} \times 48.11 \text{ A} = 33.3 \text{ kVA}$

By examining the results of Examples 7-25 and 7-26, we can make the following observations:

1. The transformer's winding in the Δ configuration has the full line voltage across it. In the Y configuration the winding voltage is 57% $(1/\sqrt{3})$ of the line voltage. Thus in high-voltage low-kVA transformations, the Y-Y configuration is better than the Δ-Δ; whereas in low voltage high-kVA transformations, the Δ-Δ is preferred.
2. The winding current in the Δ configuration is 57% of the line current. In the Y configuration the winding current is equal to the line current. In high-kVA systems we can take advantage of this by combining the two configurations (Δ-Y or Y-Δ), thus reducing the cost of the transformer. The Δ-Y is used as a step-up transformer and would probably be found at the utility end of a transmission line. The Y-Δ is used as a step-down transformer and would probably be found at the customer end of the transmission line.

In addition to the differences mentioned above, we can note the following. The Y has an advantage in that a neutral line can be used with it. Furthermore, since no current flows during a no-load condition, a phase can be replaced without fear of interrupting a large current. The Δ, on the other hand, since it is a closed circuit, can have current flowing in the windings, even at no load. This would occur if one of the windings was connected backwards. A simple test can be made and is shown in Figure 7-40. Before connecting point X to Y, the voltage between them is measured. In a properly connected Δ, the voltmeter should read zero (or close to it). A small voltage may be present if the three transformations are not identical. In the event that one winding is connected backwards, the voltmeter reading will be about twice the winding voltage. By reversing the individual windings, the correct connection can be found when the voltmeter reads zero.

An advantage of the Δ configuration is that it still provides three-phase power even if one of the windings is removed for repair. This configuration is sometimes used intentionally and is called an **open delta** or V connection. It should be noted,

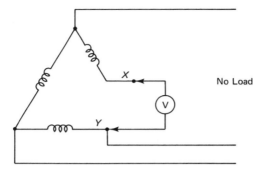

Figure 7-40 Test for proper phasing in a Δ.

however, that there is a loss in their power capability. In other words, two 10-kVA transformers when connected open delta (V-V) can supply only 17.3 kVA, not 20 kVA as might be expected. Example 7-27 will illustrate this.

Example 7-27

Two 230/600-V 10-kVA transformers are connected open delta. Find:

(a) The rated primary and secondary currents

(b) The total kVA that can be supplied without exceeding the transformer ratings

(c) The total kVA that can be supplied if a third transformer, identical to the others, is connected to form a Δ-Δ

Solution

A diagram of the connection is shown in Figure 7-41.

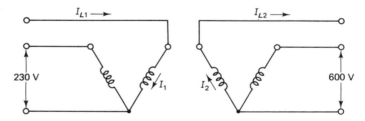

Figure 7-41 Diagram for Exercise 7-27.

(a) Using Eqs. 7-18 and 7-19, the rated currents are

$$I_1 = \frac{10{,}000 \text{ VA}}{230 \text{ V}} = 43.5 \text{ A}$$

$$I_2 = \frac{10{,}000 \text{ VA}}{600 \text{ V}} = 16.67 \text{ A}$$

Note from Figure 7-41 that these are also the line currents.

$$I_{L1} = I_1 = 43.5 \text{ A}$$

$$I_{L2} = I_2 = 16.67 \text{ A}$$

(b) Since we are supplying three-phase power, we must still use Eq. 7-49 to find the total kVA supplied.

$$P_a = \sqrt{3} \times 600 \text{ V} \times 16.67 \text{ A}$$

$$= 17{,}320 \text{ VA} = 17.32 \text{ kVA}$$

This represents 86% of the total if the two transformers supplied single-phase power.

$$\frac{17.32}{20} \times 100\% = 86.6\%$$

(c) If a third transformer is added to form a Δ-Δ configuration, the full-load kVA becomes

$$3 \times 10 \text{ kVA} = 30 \text{ kVA}$$

Note that by adding one 10-kVA transformer, we have increased our capability by 12.7 kVA.

A special variation of the Δ configuration is a **four-wire delta,** as shown in Figure 7-42. This configuration is commonly used to bring power from the transmission lines into our homes. The transformer is mounted at the power lines. The three wires labeled A, B, and N are brought into our homes. The N wire comes off a **center tap** of one secondary winding. We therefore have two voltages available in our homes: the voltage from A to B (240 V) and the voltage from A to N or B to N (120 V). Note that the 120-V power is in phase with the 240-V power. Only one phase comes into our homes. If we were to look inside the fuse or circuit breaker panel in our home, we would see that the A and B wires are colored red and black, respectively, and the N wire is white and serves as the neutral within the house.

Industrial facilities will have all four wires (A, B, C, and N) brought into them by the utility. This is done so that they have the common 120-V single-phase power available and the 240-V three-phase power for their heavy machinery.

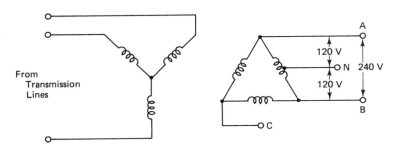

Figure 7-42 Four-wire delta.

SYMBOLS INTRODUCED IN CHAPTER 7

Symbol	Definition	Units: English and SI
V	Sinusoidal voltage	rms volts
I	Sinusoidal current	rms amps
V_m	Maximum or peak value of V	volts
I_m	Maximum or peak value of I	amperes
P_a	Apparent power	VA
P_w	Real power	W
P_r	Reactive power	var
PF	Power factor	—
f	Frequency of sinusoidal voltage	hertz
θ	Power factor angle	degrees
L	Inductance	henries
X_L	Inductive reactance	ohms
C	Capacitance	farads
X_C	Capacitive reactance	ohms
$|Z|$	Magnitude of complex impedance	ohms
ϕ_1	Primary winding leakage flux	lines or webers
ϕ_2	Secondary winding leakage flux	lines or webers
ϕ_m	Peak value of mutual flux linking primary and secondary windings	lines or webers
N_1	Number of turns on primary winding	—
N_2	Number of turns on secondary winding	—
E_1	Voltage applied to primary winding	volts
V_1	Voltage induced in primary winding	volts
V_2	Voltage induced in secondary winding	volts
a	Transformer turns ratio	—
I_1	Primary current	amperes
I_2	Secondary current	amperes
I_{1r}	Rated primary current	amperes
I_{2r}	Rated secondary current	amperes
V_{1r}	Rated primary voltage	volts
V_{2r}	Rated secondary voltage	volts
I_M	Primary-winding magnetizing current	amperes
I_c	Current providing core losses	amperes
I_p	Current supplying real power which is eventually delivered to a load	amperes
θ_0	No-load power factor angle	degrees
Z_L	Complex load impedance	ohms
Z_R	Reflected impedance	ohms
Z_{RP}	Secondary impedance reflected to primary	ohms
Z_{RS}	Primary impedance reflected to secondary	ohms
Y_L	Load admittance	siemen (S)
Y_1	Primary admittance	siemen
Y_{RP}	Secondary admittance reflected to primary	siemen
Y_{RS}	Primary admittance reflected to secondary	siemen
R_1	Equivalent resistance of primary winding	ohms
R_2	Equivalent resistance of secondary winding	ohms
X_1	Equivalent reactance of primary winding	ohms

Symbol	Definition	Units: English and SI
X_2	Equivalent reactance of secondary winding	ohms
Z_1	Equivalent impedance of primary winding	ohms
Z_2	Equivalent impedance of secondary winding	ohms
R_{e1}	Total transformer resistance looking into primary	ohms
R_{e2}	Total transformer resistance looking into secondary	ohms
X_{e1}	Total transformer reactance looking into primary	ohms
X_{e2}	Total transformer reactance looking into secondary	ohms
P_{Cu}	Transformer copper losses	watts
P_{core}	Transformer core losses	watts
V_L	Three-phase line-to-line voltage	volts
V_p	Three-phase line-to-neutral (or phase) voltage	volts
I_L	Three-phase line current	amperes
I_p	Three-phase "phase" current	amperes
R_{eq}	Equivalent resistance of a three-phase load	ohms
I_{L1}	Primary-side line current of a three-phase transformer	amperes
I_{L2}	Secondary-side line current of a three-phase transformer	amperes

QUESTIONS

1. What is the difference between slip rings and brushes? When would each be used?

2. Define the following terms: reactive element; apparent power; real power; reactive power; power factor.

3. Can the power factor ever be larger than unity? Explain.

4. How can we tell the difference between the potential coil and current coil terminals on a wattmeter?

5. Is it possible to damage a wattmeter while measuring power within its wattage range? Explain.

6. Define the following terms: primary winding; secondary winding; step-down and step-up transformers; mutual flux; leakage flux.

7. How could we obtain, with measurements and calculations, the no-load power factor of a transformer?

8. Describe the short-circuit test, open-circuit test, and what they are used for.

9. Describe in words what it means if the voltage regulation is positive, zero, or negative.

10. Can a transformer's efficiency be greater than 100%? Explain.

11. What is the main disadvantage of an autotransformer?

12. How many possible phase sequences are there in a three-phase system? How could we change the sequence?

PROBLEMS

1. In Figure 7-43, $R = 10\ \Omega$, $X_L = 15\ \Omega$, $X_C = 0\ \Omega$, and $V = 230$ V rms. Find:
(a) Z
(b) I
(c) PF
(d) P_a, P_w, P_r
(e) Draw the power triangle.

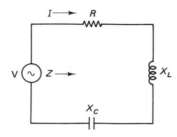

Figure 7-43

2. In Figure 7-43, $R = 20\ \Omega$, $X_L = 0\ \Omega$, $X_C = 30\ \Omega$, and $V = 230$ V rms. Find:
(a) Z
(b) I
(c) PF
(d) P_a, P_w, P_r
(e) Draw the power triangle.

3. In Figure 7-43, $R = 40\ \Omega$, $X_L = 15\ \Omega$, $X_C = 25\ \Omega$, and $V = 230$ V rms. Find:
(a) Z
(b) I
(c) PF
(d) P_a, P_w, P_r
(e) Draw the power triangle.

4. In Figure 7-44, $R = 6$ kΩ, $L = 1.27$ H, and $C = 0.01\ \mu$F. Find:
(a) Z
(b) I
(c) PF
(d) P_a, P_w, P_r
(e) Draw the power triangle.

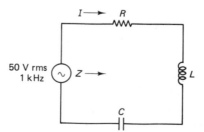

Figure 7-44

5. In Figure 7-44, $R = 7$ kΩ, $L = 4.77$ H, and $C = 0.0265$ μF. Find:

(a) Z

(b) I

(c) PF

(d) P_a, P_w, P_r

(e) Draw the power triangle.

6. The four loads described in Table 7-2 are connected to a single-phase 230-V source. Find:

(a) P_w for each load

(b) P_r for each load

(c) P_w total

(d) P_r total

(e) P_a total

(f) The effective power factor for the entire system

Table 7-2

Load	kVA	PF
A	5	1.0
B	10	0.85 lagging
C	12	0.80 lagging
D	10	0.6 leading

7. In Problem 6, what would the effective power factor of the system become if load D were removed?

8. A load was tested as shown in Figure 7-11. The meter readings were 230 V, 12.8 A, and 2.75 kW. Find the load:

(a) kVA

(b) pf

(c) kvar

9. A wattmeter is used to measure power on a 230-V line. Its potential and current coils are rated 300 V and 10 A, respectively. Find:

(a) The wattage range of the meter

(b) The maximum power that can be read before a coil rating is exceeded

10. The wattmeter of Problem 9 is used to measure power on a 120-V line. The load has a constant PF of 0.7. What is the maximum power that can be read before the current coil rating is exceeded?

11. A transformer with a rated primary voltage of 230 V has 500 primary turns. Find the secondary turns required and the turns ratio if the desired secondary voltage is:

(a) 1150 V

(b) 460 V

(c) 1200 V

(d) 12 V

(e) 60 V

12. Repeat Problem 11 if the rated primary voltage is 120 V.

13. A transformer is rated 230/460 V, 10 kVA. Find the rated:
(a) Primary current
(b) Secondary current

14. A transformer is rated 120/12.6 V, 300 VA. Find the rated:
(a) Primary current
(b) Secondary current

15. A transformer is rated 18,000/220,000 V, 60,000 kVA. Find the rated:
(a) Primary current
(b) Secondary current

16. A 230/460-V 5-kVA transformer has rated voltage applied to the primary with the secondary open (no load). A wattmeter connected to the primary measures 12 W and an ammeter measures 0.3 A in the primary. Find:
(a) The no-load power factor of the transformer
(b) The magnetization current
(c) The current supplying core loss
(d) The transformer's power factor if a unity power factor load is connected to the secondary and the wattmeter reads 2 kW

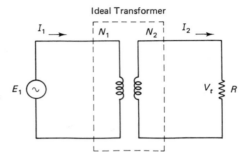

Figure 7-45

17. In Figure 7-45, $E_1 = 230$ V, $N_1 = 200$, $N_2 = 40$, and $R = 2$ Ω. Find:
(a) The turns ratio
(b) V_t
(c) I_2
(d) I_1

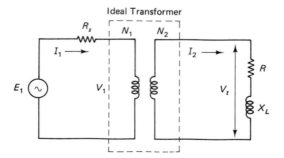

Figure 7-46

18. In Figure 7-45, $E_1 = 120$ V, $N_1 = 400$, $N_2 = 80$, and $R = 1$ kΩ. Find:
(a) The turns ratio
(b) V_t
(c) I_2
(d) I_1

19. In Figure 7-45, $E_1 = 230$ V, $V_t = 920$ V, and $I_1 = 20$ A. Find:
(a) The turns ratio
(b) I_2
(c) R

20. In Figure 7-46, $E_1 = 230$ V, $N_1 = 320$, $N_2 = 40$, $R = 3$ Ω, $X_L = 12.5$ Ω, and $R_s = 408$ Ω. Find:
(a) I_1 **(c)** I_2
(b) V_1 **(d)** V_t

21. If the transformer in Problem 20 is rated 60 VA, what is the maximum value E_1 can have before the transformer rating is exceeded?

22. In Figure 7-46, $E_1 = 120$ V, $N_1 = 100$, $N_2 = 200$, $R_s = 2.75$ Ω, and $R = 5$ Ω. If the primary current is measured and found to be 24 A, find X_L.

23. If the transformer in Problem 22 is rated 3 kVA, what percent of rated current is flowing in the primary?

24. A 230/28-V 400-Hz 0.5-kVA step-down transformer has a short-circuit test done on it. The secondary is shorted and the measurements are made on the high side. The wattmeter reads 18 W, the voltmeter reads 12 V, and the ammeter reads rated current. Find:
(a) R_{e1}
(b) X_{e1}
(c) R_{e2}
(d) X_{e2}

25. A 1200/230-V 50-Hz 17.5-kVA step-down transformer has a short-circuit test done on it. The primary is shorted and the measurements are made on the low side. If $W = 510$ W, $V = 15$ V, and I is rated current, find:
(a) R_{e2}
(b) X_{e2}
(c) R_{e1}
(d) X_{e1}

26. A 120/1200-V 60-Hz 5-kVA step-up transformer has a short-circuit test done on it. The secondary is shorted and measurements are made on the low side. If $W = 240$ W, $V = 9$ V, and I is rated current, find:
(a) R_{e1}
(b) X_{e1}
(c) R_{e2}
(d) X_{e2}

27. The transformer in Figure 7-47 is rated 230/460 V, 50 Hz, 1 kVA and has $R_{e2} = 0.04$ Ω and $X_{e2} = 0.1$ Ω. Find its percent voltage regulation if it supplies rated load at unity power factor.

28. Repeat Problem 27 with the power factor 0.8 lagging.

29. The transformer in Figure 7-47 is rated 120/12.6 V, 60 Hz, 250 VA and has $R_{e2} = 0.005$ Ω and $X_{e2} = 0.008$ Ω. Find its percent voltage regulation if it supplies rated load at unity power factor.

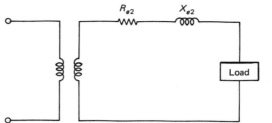

R_{e2} X_{e2}

Load

Figure 7-47

30. Repeat Problem 29 if the power factor is 0.9 leading.

31. The transformer in Figure 7-47 is rated 4600/230 V, 50 Hz, 12 kVA. A short-circuit test is performed and its equivalent parameters are found: $R_{e2} = 0.15\ \Omega$ and $X_{e2} = 0.2\ \Omega$. Find the percent voltage regulation when it supplies rated load at:
(a) PF = 1
(b) PF = 0.8 lagging
(c) PF = 0.85 leading

32. A 230/120-V 10-kVA transformer supplies rated load at unity power factor. Find:
(a) The power output
(b) The secondary current

33. Repeat Problem 32 if the transformer is supplying half-rated load at a power factor of 0.85 lagging.

34. A 115/230-V 4-kVA transformer has open- and short-circuit tests performed. The results are given in Table 7-3.

Table 7-3

Open-circuit test (high side open)	Short-circuit test (high side shorted)
$V = 115$ V	$V = 7$ V
$I = 0.5$ A	I = rated current
$W = 220$ W	$W = 100$ W

Find its efficiency under the following conditions:
(a) Full load, PF = 1
(b) Full load, PF = 0.75 lagging
(c) Half load, PF = 1
(d) Half load, PF = 0.75 lagging
(e) One-fourth load, PF = 0.5 lagging

35. During an open-circuit test a wattmeter reads 600 W and in a short-circuit test it reads 900 W. If the transformer is rated 50 kVA, find its efficiency under the following conditions:
(a) Full load, PF = 1
(b) Full load, PF = 0.9 lagging
(c) Half load, PF = 0.9 lagging
(d) Half load, PF = 0.6 lagging
(e) 120% of full load, PF = 0.8 lagging

36. In Figure 7-48, $E_1 = 220$ V, $N_1 = 400$, $N_X = 200$, $N_Y = 100$, $N_Z = 1000$, $R = 10\ \Omega$, $X_L = 2\ \Omega$, and $X_C = 50\ \Omega$. Find:

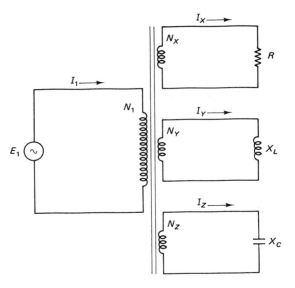

Figure 7-48

(a) I_X
(b) I_Y
(c) I_Z
(d) I_1
(e) The power factor
(f) The VA rating needed for winding Z

37. In Figure 7-48, $E_1 = 32$ V, $R = \infty$ (open circuit), $X_L = 1\ \Omega$, and $X_C = 100\ \Omega$. The winding turns are given in Problem 36. Assuming an ideal transformer, find:
(a) The primary current
(b) The power factor

38. In Figure 7-48, $E_1 = 120$ V, $N_1 = 100$, $N_X = 50$, $N_Y = 1000$, $N_Z = 500$, $R = 2\ \Omega$, $X_L = 1200\ \Omega$, and $X_C = 150\ \Omega$. Find:
(a) I_1
(b) I_X
(c) I_Y
(d) I_Z
(e) The VA rating needed for the primary winding
(f) The VA rating needed for winding Y

39. A 120/240-V 6-kVA transformer is used to build a 120/360-V step-up autotransformer. Draw a diagram of the connection, showing proper polarity markings, and find at full load:
(a) The primary current
(b) The secondary current
(c) The kVA of the autotransformer

40. A 240/28-V 1-kVA transformer is used to build a 212/28-V step-down autotransformer. Draw a diagram of the connection showing proper polarity markings and find at full load:
(a) The primary current
(b) The secondary current
(c) The kVA of the autotransformer

41. A 110/440-V 50-kVA transformer is used to build a 440/550-V step-up autotransformer. Draw a diagram of the connection showing proper polarity markings and find at full load:
(a) The primary current
(b) The secondary current
(c) The kVA of the autotransformer

42. In Problem 41, how much of the total kVA is transferred by:
(a) Induction?
(b) Conduction?

43. A 220/10-V 5-kVA transformer is used to build a 220/230-V step-up autotransformer. For the autotransformer, find:
(a) The kVA rating
(b) The kVA transferred by induction
(c) The kVA transferred by conduction

44. A 110/230-V 20-kVA transformer is used to build a 230/120-V step-down auto-transformer. Draw a diagram of the connection showing proper polarity markings and find at full load:
(a) The primary current
(b) The secondary current
(c) The kVA rating of the autotransformer

45. For the autotransformer designed in Problem 44, how much kVA is transferred by:
(a) Induction?
(b) Conduction?

46. A 10-kVA transformer is tested. The wattmeter in an open-circuit test reads 180 W. In a short-circuit test the wattmeter reads 230 W. Find the efficiency of the transformer at:
(a) Half load, PF = 0.7 lagging
(b) Full load, PF = 1
(c) 125% of rated load, PF = 0.8 lagging

47. A 120-V three-phase system supplies 60 kVA to a balanced Δ load. At the load find:
(a) The line voltage
(b) The line current
(c) The phase voltage
(d) The phase current

48. Repeat Problem 47 if the load is Y-connected.

49. A 600-V three-phase system supplies 10 kVA to a balanced Δ load. At the load find:
(a) The line voltage
(b) The line current
(c) The phase voltage
(d) The phase current

50. Repeat Problem 49 if the load is Y-connected.

51. A 230-V three-phase system supplies 25 kVA at a 0.85 lagging power factor to a factory. Find:
(a) The line current
(b) The real power supplied
(c) The reactive power supplied

52. Repeat Problem 51 with the power factor raised to unity. Note that 25 kVA is still supplied.

53. A 400-V three-phase system supplies 30 kW at a power factor of 0.8 lagging to a factory. Find:

(a) The apparent power being drawn from the supply
(b) The line current
(c) The reactive power being drawn from the supply

54. Repeat Problem 53 with the power factor raised to unity. The power consumed by the factory is still 30 kW.

55. A three-phase transformer wired Δ-Δ supplies 60 kVA at 400 V to a balanced load. It is supplied by a 13,200-V supply. Find:
(a) The required kVA rating for each transformer
(b) The transformer turns ratio
(c) The high-side line current
(d) The primary phase current
(e) The low-side line current
(f) The secondary phase current

56. Repeat Problem 55 with the transformer wired Y-Δ.

57. A three-phase transformer wired Y-Y steps up 480 V to 13,200 V for transmission. A total of 2500 kVA is transmitted. Find:
(a) The required rating of each transformer
(b) The low-side line current
(c) The primary phase current
(d) The high-side line current
(e) The secondary phase current
(f) The transformer turns ratio

58. Repeat Problem 57 with the transformer wired Δ-Y.

59. Three identical transformers are connected in a Y-Δ configuration. They are used to step down 400,000 V to 24,000 V. The total three-phase load supplied is 350,000 kVA. Find the required:
(a) Primary voltage rating
(b) Primary current rating
(c) Secondary voltage rating
(d) Secondary current rating
(e) Turns ratio

60. Repeat Problem 59 with the transformer wired Δ-Y.

61. A three-phase transformer supplies 50 kW at 210 V to a balanced load. The load power factor is 0.9 lagging. The transformer wired Y-Δ is fed by a 600-V system. Find:
(a) The high-side line current
(b) The low-side line current
(c) The required turns ratio
(d) The kVA rating for each transformer

62. A Δ-Y transformer steps up 150 V to 2600 V and delivers 5000 kVA. Find:
(a) The kVA rating of each transformer
(b) The primary winding current rating
(c) The primary winding voltage rating
(d) The secondary winding current rating
(e) The secondary winding voltage rating

63. Repeat Problem 62 with the transformer wired Y-Δ.

64. Two 120/330-V 20-kVA transformers are connected open delta. Find:
(a) The rated primary current
(b) The rated secondary current

(c) The total kVA that can be supplied without exceeding the ratings

(d) The total kVA that can be supplied if a third identical transformer is added, forming a Δ-Δ connection.

65. A Δ-Δ transformer steps down 480 V to 240 V and can supply 50 kVA. Find the kVA that can still be supplied (without exceeding ratings) if one transformer is removed and operation is made open delta.

(English)

66. A 400-Hz aircraft transformer whose primary rated 28 V has 200 primary turns.
(a) Find the peak mutual flux in the core.
(b) Repeat part (a) if the input frequency is 60 Hz instead of 400 Hz.

67. A 115/25-V 60-Hz transformer is to be built. If the peak mutual flux desired is 250 kilolines, find:
(a) The primary turns required
(b) The secondary turns required

68. A 28/120-V step-up transformer has 70 primary turns. It is made with a material whose maximum flux density is 120 kilolines/in². Find the minimum cross-sectional area needed for its core when:
(a) It is to be used at 60 Hz
(b) It is to be used at 400 Hz

69. A 28/14-V 400-Hz transformer is to be built with 200 secondary turns. If a core cross-sectional area of 5 in² is desired, find the maximum flux density needed for the core material.

70. Repeat Problem 69 if the transformer is to be used at 60 Hz.

(SI)

71. A 400-Hz aircraft transformer whose primary is rated 28 V has 320 primary turns.
(a) Find the peak mutual flux in the core.
(b) Repeat part (a) if the input frequency is 50 Hz instead of 400 Hz.

72. A 230/25-V 50-Hz transformer is to be built. If the peak mutual flux desired is 0.003 Wb, find:
(a) The primary turns required
(b) The secondary turns required

73. A 28/230-V step-up transformer has 50 primary turns. It is made with a material whose maximum flux density is 2 T. Find the minimum cross-sectional area needed for its core when:
(a) It is to be used at 50 Hz
(b) It is to be used at 400 Hz

74. A 28/7-V 400-Hz transformer is to be built with 150 secondary turns. If a core cross-sectional area of 3.5×10^{-3} m² is desired, find the maximum flux density needed for the core material.

75. Repeat Problem 74 if the transformer is to be used at 50 Hz.

Chapter 8

The Synchronous Alternator

In Chapter 2 (Faraday's law) we saw that if a conductor moved through a magnetic field cutting the lines of flux, a voltage would be induced in it. If we tried doing the opposite, that is, move the magnetic field past a stationary conductor so that the lines of flux cut it, there would again be a voltage induced in the conductor. It is this principle upon which the alternator works. We will see in this chapter that an alternator can generate very large voltages compared to a dc generator. This is possible because the power is generated in a stationary winding. As a result of this we are not forced to connect a moving part to the outside world. In other words, brushes and slip rings are not needed.

8-1   BASIC CONSTRUCTION

The alternator consists of rotating part (**rotor**) and the stationary mainframe part (**stator**). The two parts perform a function opposite to that in the dc generator. Here, the rotor provides the magnetic field, which is made to rotate by turning the alternator shaft. The **armature** analog is the winding on the stator. It is from this stationary winding that the generated power is taken. Figure 8-1 shows a simplified diagram of a single-phase two-pole alternator.

The field winding is placed in slots on the rotor. It is energized by an external dc voltage brought into the machine with slip rings or brushes. The field excitation comes from a variable dc source called an **exciter**. It is by varying the exciter voltage that we control the voltage generated in the stator. The exciter can be either an electronic supply or a dc generator mounted right on the alternator shaft. The rotor shown in Figure 8-1 is called a **round** or **cylindrical rotor.** It is a two-pole rotor because the field winding produces two rotating field poles (one north and one south).

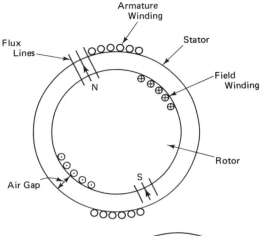

Figure 8-1 Single-phase two-pole alternator.

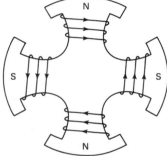

Figure 8-2 Four-pole rotor with salient pole construction.

The round rotor is used almost entirely for two-pole alternators and is well suited for high rotational speeds. In alternators with more than two poles, the **salient pole** construction shown in Figure 8-2 is usually used.

The stator winding is placed in slots on the main frame. As the field rotates, the flux lines cut the conductors, which are distributed so as to induce a sinusoidal voltage in the stator. In Figure 8-1, as the rotor turns 360 mechanical degrees (one revolution), there will be one cycle (360 electrical degrees) of an ac voltage induced in the stator. Since there is only one winding on the stator, only one varying voltage (single phase) is produced. If in Figure 8-1 the rotor had four poles instead of two, we would still have a single-phase voltage generated. This time, however, there would be two cycles (720 electrical degrees) of the generated voltage for every revolution (360 mechanical degrees) of the rotor.

8-2 FREQUENCY RELATIONSHIP

From the preceding section it should be clear that the frequency of the generated voltage is a function of both the number of poles and the speed of the rotor. Furthermore, for a two-pole field and rotor speed of 1 rev/s the generated voltage would be a 1 Hz sinusoid. The frequency can be related to the speed in rev/s and the number of poles by Eq. 8-1.

$$f(\text{Hz}) = \frac{\text{number of poles}}{2} \times \frac{\text{rev}}{\text{s}} \tag{8-1}$$

By converting units in Eq. 8-1, we can derive Eqs. 8-2, which solve for the generated frequency:

(English)

$$f = \frac{S\mathbf{P}}{120} \tag{8-2a}$$

(SI)

$$f = \frac{\omega\mathbf{P}}{4\pi} \tag{8-2b}$$

In Eqs. 8-2, $\mathbf{P}$ is the number of rotor poles, f is the frequency in hertz, S is the rotor speed in rev/min, and ω is the rotor speed in rad/s.

Example 8-1 (English)

An alternator is to generate 60 Hz.

(a) How fast must its shaft turn if it has four poles?

(b) If it is driven at 600 rev/min, how many poles must it have?

Solution

(a) Using Eq. 8-2a to solve for S, we have

$$S = 120 \times \frac{60 \text{ Hz}}{4 \text{ poles}}$$

$$= 1800 \text{ rev/min}$$

(b) Again using Eq. 8-2a, however, rearranging to solve for $\mathbf{P}$,

$$\mathbf{P} = 120\frac{f}{s}$$

$$\mathbf{P} = 120 \times \frac{60 \text{ Hz}}{600 \text{ rev/min}} = 12 \text{ poles}$$

Example 8-2 (SI)

An alternator is to generate 50 Hz.

(a) How fast must its shaft turn if it has six poles?

(b) If it is driven at 80 rad/s, how many poles must it have?

Solution

(a) Using Eq. 8-2b to solve for ω, we have

$$\omega = 4 \times \pi \times \frac{50 \text{ Hz}}{6 \text{ poles}}$$

$$= 104.7 \text{ rad/s}$$

(b) Again using Eq. 8-2b, however, rearranging to solve for **P**,

$$\mathbf{P} = \frac{4\pi f}{\omega}$$

$$= \frac{4 \times \pi \times 50 \text{ Hz}}{80 \text{ rad/s}} = 7.85 \text{ poles}$$

This is not possible; in fact, we must always have an even number of poles. Therefore, eight poles must be used and the speed would be adjusted to the following to produce the desired 50 Hz. From Eq. 8-2b,

$$\omega = 4 \times \pi \times \frac{50 \text{ Hz}}{8 \text{ poles}}$$

$$= 78.54 \text{ rad/s}$$

8-3   THE GENERATED VOLTAGE

The armature winding in Figure 8-1 will generate a voltage proportional to the number of poles **P**, flux per pole ϕ, speed, and number of conductors z in the stator winding. From Eqs. 2-2 the following equations can be derived, where E_p is the generated rms voltage for the single-phase stator winding in Figure 8-1. As before, S and ω have units of rev/min and rad/s, respectively.

(English)

$$E_p = 0.0185z\phi\mathbf{P}S \times 10^{-8} \qquad\qquad (8\text{-}3a)$$

(SI)

$$E_p = \frac{1.11z\phi\mathbf{P}\omega}{2\pi} \qquad\qquad (8\text{-}3b)$$

Substituting Eqs. 8-2 into 8-3, we can derive an expression for E_p which is now a function of the generated frequency.

(English)

$$E_p = 2.22z\phi f \times 10^{-8} \qquad\qquad (8\text{-}4a)$$

(SI)

$$E_p = 2.22z\phi f \qquad\qquad (8\text{-}4b)$$

Example 8-3 (English)

A single-phase aircraft alternator generates 400 Hz. The armature has 100 turns (200 conductors). Find the flux per pole necessary to generate 60 V rms.

Solution

Equation 8-4a is rearranged to solve for ϕ.

$$\phi = \frac{E_p}{2.22zf \times 10^{-8}}$$

$$= \frac{60}{2.22 \times 200 \times 400 \times 10^{-8}}$$

$$= 33,784 \text{ lines} \approx 33.78 \text{ kilolines}$$

Example 8-4 (SI)

A single-phase aircraft alternator generates 400 Hz. The armature has 120 turns (240 conductors). Find the flux per pole necessary to generate 54 V rms.

Solution

Equation 8-4b is rearranged to solve for ϕ.

$$\phi = \frac{E_p}{2.22zf}$$

$$= \frac{54}{2.22 \times 240 \times 400}$$

$$= 0.253 \times 10^{-3} \text{ Wb}$$

8-4 THE THREE-PHASE ALTERNATOR

Polyphase alternators are very common. We will examine the **three-phase alternator.** This type has three windings mounted on the stator as shown in Figure 8-3. Each of these windings will generate an ac voltage given by Eqs. 8-3. As long as each winding (A, B, and C) has the same number of conductors (z), the voltages generated by each will have the same magnitude. However, since the windings are placed 120° apart on the stator, their respective voltages will be out of phase by 120° just like those in Figure 7-29c. For the direction of rotation and at the instant shown in Figure 8-3, winding A will generate its peak voltage with A positive with respect to A'. This can be verified by applying Fleming's right-hand rule. When the rotor has turned 120° deg from the position shown, a peak voltage will be generated in winding C, with C positive with respect to C'. After another 120° rotation, winding B will be peaking with B positive with respect to B'. Finally, the rotor returns to its starting position and the three voltages will repeat the cycle they have just gone through. Note that the phase sequence for this case is ACB. If we were to reverse the direction of rotation (making it clockwise), the phase sequence would change to ABC.

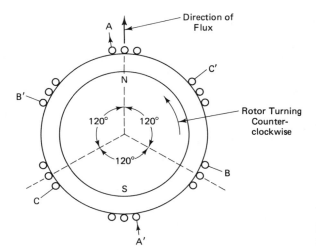

Figure 8-3 Diagram of a two-pole three-phase alternator.

8-4.1 Alternator Connections

By connecting the three windings in different ways we can obtain different characteristics for a given alternator. If A′, C′, and B′ are connected together and A, B, and C are brought out as the stator terminals, the alternator will be Y-connected, as shown in Figure 8-4a. The Δ-connected alternator is shown in Figure 8-4b. This is accomplished by connecting A′ to C, C′ to B, and B′ to A. These connection points are brought out of the alternator as the stator terminals.

In Figure 8-4a the alternator terminal voltage (V_t) is equal to $\sqrt{3} \times E_p$, whereas the line current will equal the current in the winding. For the Δ connection in Figure 8-4b, the terminal voltage (V_t) will equal E_p; now, however, the line current will be $\sqrt{3}$ times the current in the winding. In each case the same kVA are generated by the alternator. Thus to obtain a given amount of kVA at a high voltage and low current, the alternator should be Y connected. Connecting it in a Δ would produce the same kVA, however, at a low voltage and high current.

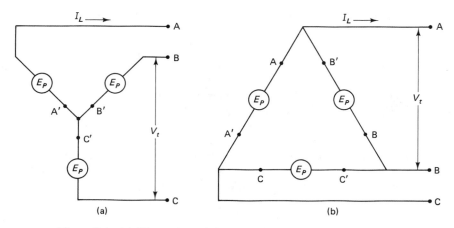

Figure 8-4 (a) Wye-connected alternator; (b) delta-connected alternator.

8-4.2 Alternator Ratings

As with the three-phase transformer, the alternator is rated in terms of its full-load line quantities. In other words, an alternator rated at 10 kVA, 208 V, rated speed, will have 208 V across its line terminals when it is delivering 10 kVA. Under these conditions the line current is the rated current and can be solved for using Eq. 7-49. Note that under no-load conditions the line current and kVA are zero. The terminal voltage, however, will usually be greater than 208 V. We will go over this point in detail in Section 8-6. The rated speed is the speed it must be driven at to produce the desired frequency, voltage, and kVA.

8-5 ALTERNATOR EQUIVALENT CIRCUIT

The stator of an alternator is made up of wire placed in slots in a metal frame. Each stator winding must therefore have resistance and inductance. If we were to draw a circuit for one phase of a three-phase stator winding, it would look like Figure 8-5.

In this equivalent circuit r_p is the resistance of the winding per phase and x_p is the **synchronous reactance per phase** due to the winding inductance. Since most alternators are designed to deliver large currents, the wire used is very thick and the resistance r_p will be quite small compared to the reactance x_p. Typically, x_p is 10 or more times the value of r_p. The combined effect of the reactance and resistance of a winding produces what we call the **synchronous impedance per phase** Z_s.

$$Z_s = \sqrt{r_p^2 + x_p^2} \qquad (8-5)$$

This equation and Figure 8-5 should be compared to Eq. 7-34 and Figure 7-18b for the transformer. They are very similar and we use similar tests to determine x_p and r_p for the alternator.

8-5.1 Resistance Measurement

The resistance is usually determined by applying a dc voltage to two of the stator terminals. By measuring the current and applying Ohm's law, we can determine a value of resistance r_a. This resistance is now used to find r_p depending on how the alternator is wired.

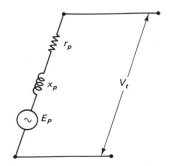

Figure 8-5 Equivalent circuit of one phase of a three-phase alternator.

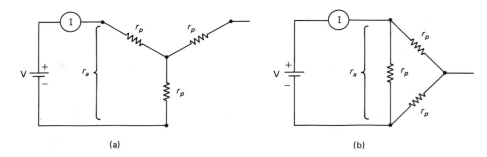

Figure 8-6 Circuits for determining resistance per phase of an alternator: (a) Y-connected armature; (b) Δ-connected armature.

In Figure 8-6a, r_a is equal to 2 times r_p. On the other hand, for the Δ connection in Figure 8-6b, r_a is equal to r_p in parallel with $2r_p$. Using these basic relations, Eqs. 8-6 can be derived, where r_a is equal to the voltage V divided by the current I in Figure 8-6.

$$\left.\begin{aligned} \text{Y connection:} \quad r_p' &= \frac{r_a}{2} \\[2ex] \Delta \text{ connection:} \quad r_p' &= \frac{3r_a}{2} \end{aligned}\right\} \tag{8-6}$$

For reasons whose explanation is beyond the scope of this text, the effective resistance per phase is about 40% larger than the value determined by this test. Therefore, once calculated, it is multiplied by 1.4 to give the effective resistance per phase.

$$r_p = 1.4 r_p' \tag{8-7}$$

8-5.2 Reactance Measurement

Two tests are needed to calculate the synchronous reactance. Both of them can be performed using the circuit shown in Figure 8-7. It does not matter whether the stator is Δ or Y connected, the procedure is the same. The test is shown here for the Y connection.

In the first test, called the **short-circuit test,** the alternator starts out with no field excitation ($I_f = 0$) running at rated speed. The switches are closed and the excitation is increased until we read rated current in the three ammeters. The field current I_f is recorded. The three readings (I_1, I_2, and I_3) represent the short-circuit stator current. Since we are Y connected, the currents measured by these three meters represent the rated phase current. Although the three readings I_1, I_2, and I_3 should be about the same, they can be averaged and we will call it the short-circuit phase current (I_{sc}).

In the second of these tests, called the **open-circuit test,** the three switches are kept open. With the alternator running at rated speed, the field excitation I_f is increased to the same value as that recorded in the short-circuit test. The open-circuit voltage

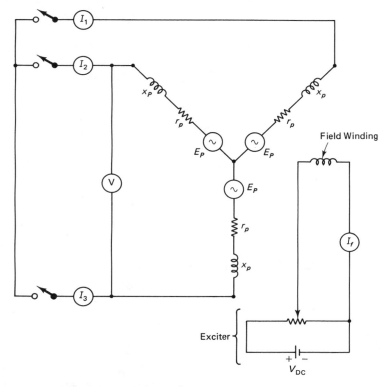

Figure 8-7 Circuit for determining an alternator's synchronous impedance.

measured by the voltmeter is $\sqrt{3}$ times the generated phase voltage under rated conditions. From this we can calculate V_{oc} (the open-circuit phase voltage).

The synchronous impedance is now calculated by dividing the open-circuit phase voltage by the short-circuit phase current.

$$Z_s = \frac{V_{oc}}{I_{sc}} \tag{8-8}$$

Taking the value of Z_s calculated from Eq. 8-8, the value of r_p calculated from Eq. 8-7, and substituting them into Eq. 8-5, we can solve for the synchronous reactance per phase x_p.

$$x_p = \sqrt{Z_s^2 - r_p^2} \tag{8-9}$$

Note that since we made the tests at rated speed and excitation, the reactance obtained by Eq. 8-9 is not due only to the winding inductance. Also included in x_p are the effects of armature reaction at rated conditions.

Example 8-5

A three-phase synchronous alternator rated 50 kVA, 220 V, 60 Hz is Y connected. It is tested as in Figures 8-6a and 8-7 to determine its per phase parameters. The test results are given below. Find r_p, Z_s, and x_p.

Resistance test: $V = 2$ V, $I = 22$ A

Short circuit test: $I_1 = I_2 = I_3 =$ rated current

$$I_f = 22 \text{ A}$$

Open circuit test: $I_f = 22$ A

$$V = 95 \text{ V}$$

Solution

From the results of the resistance test and Eq. 8-6 for the Y-connected armature,

$$r_a = \frac{V}{I} = \frac{2 \text{ V}}{22 \text{ A}} = 0.091 \ \Omega$$

$$r_p' = \frac{0.091 \ \Omega}{2} = 0.045 \ \Omega$$

From Eq. 8-7, finding the effective resistance per phase, we have

$$r_p = 1.4 \times 0.045 \ \Omega = 0.064 \ \Omega$$

The open-circuit test gives us V_{oc}; however, the measured voltage (95 V) is the open-circuit line voltage and we want the phase voltage. Since we are Y connected,

$$V_{oc} = \frac{95 \text{ V}}{\sqrt{3}} = 54.8 \text{ V}$$

In the short-circuit test I_{sc} measured was the rated line current. Since we are Y connected, the phase current equals the line current. Thus from Eq. 7-49,

$$I_{sc} = \frac{50,000 \text{ VA}}{\sqrt{3} \times 220 \text{ V}}$$

$$= 131.22 \text{ A}$$

The synchronous impedance is (using Eq. 8-8)

$$Z_s = \frac{54.8 \text{ V}}{131.22 \text{ A}} = 0.42 \ \Omega$$

Finally, using Eq. 8-9 to solve for the effective reactance per phase, we have

$$x_p = \sqrt{(0.42)^2 - (0.064)^2}$$

$$= 0.419 \approx 0.42 \ \Omega$$

Note that since x_p is much greater than r_p, the reactance per phase x_p is in effect the synchronous impedance Z_s.

8-6   VOLTAGE REGULATION

In Section 8-5 we looked at the alternator's equivalent circuit. Noting its similarity to the transformer, we should have very little difficulty in predicting an alternator's

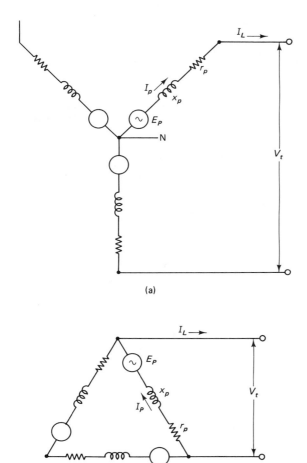

(a)

(b)

Figure 8-8 Two alternator armatures: (a) Y-connected; (b) Δ-connected.

regulation. The same basic formula (Eq. 7-37) is used; however, we must be careful not to confuse the phase voltages with line voltages. Either phase or line quantities can be used. Since our equivalent circuit is in terms of phase quantities, we will use phase voltages in determining voltage regulation. Figure 8-8 will help simplify our discussion.

Under full-load conditions the line current I_L is the rated current and the line voltage V_t is the rated (or full-load) voltage of the alternator. In the Δ connection the full-load phase voltage is the same as the line quantity; however, in the Y connection the full-load phase voltage is the line quantity divided by $\sqrt{3}$. Therefore, the full-load phase voltage can be calculated from the ratings. Using it and the full-load current, we can solve for the generated phase voltage (E_p) using ac circuit analysis.

Under no-load conditions the currents are zero and there will be no voltage drop across x_p and r_p. Therefore, by applying Kirchhoff's voltage law, the no-load phase voltage will equal the generated phase voltage E_p.

As an aid in solving for E_p, Eq. 8-10, which is very similar to Eq. 7-38 for the transformer, is given.

$$E_p = \sqrt{(I_p r_p + V_{FL} \cos \theta)^2 + (I_p x_p \pm V_{FL} \sin \theta)^2} \tag{8-10}$$

(+ for lagging PF, − for leading PF)

In Eq. 8-10, I_p is the rated phase current; V_{FL} is the rated phase voltage; r_p and x_p are the resistance and reactance per phase, respectively; and $\cos \theta$ is the power factor of the load.

Once we know E_p, we can solve for the voltage regulation using Eq. 8-11, noting that the no-load voltage (V_{NL}) is the quantity E_p just solved for.

$$\frac{\% \text{ voltage}}{\text{regulation}} = \frac{V_{NL} - V_{FL}}{V_{FL}} \times 100 \tag{8-11}$$

Example 8-6

A 1000-kVA 1200-V three-phase alternator is Y connected. Its resistance per phase is 0.12 Ω and the reactance per phase is 1.5 Ω. Find its voltage regulation if it supplies rated load at:

(a) PF = 1

(b) PF = 0.9 lagging

(c) PF = 0.9 leading

(d) What would the no-load line voltage be in part (c)?

Solution

(a) First find the rated line current using Eq. 7-49.

$$I_L = \frac{1,000,000 \text{ VA}}{\sqrt{3} \times 1200 \text{ V}}$$

$$= 481 \text{ A}$$

Since the alternator is Y connected,

$$I_p = I_L = 481 \text{ A}$$

Next find the rated phase voltage,

$$V_{FL} = \frac{1200 \text{ V}}{\sqrt{3}} = 693 \text{ V}$$

If PF = 1,

$$\cos \theta = 1$$

$$\theta = 0°$$

$$\sin \theta = 0$$

Solve for E_p using Eq. 8-10.

$$E_p = \sqrt{(481 \times 0.12 + 693 \times 1)^2 + (481 \times 1.5 + 693 \times 0)^2}$$

$$= \sqrt{(750.72)^2 + (721.5)^2} = 1041.2 \text{ V}$$

Thus

$$V_{NL} = E_p = 1041.2 \text{ V}$$

$$V_{FL} = 693 \text{ V}$$

and

$$\% \text{ V.R.} = \frac{1041.2 \text{ V} - 693 \text{ V}}{693 \text{ V}} \times 100$$

$$\text{V.R.} = 50.2\%$$

(b) V_{FL} and I_p are the same as part (a), except that the power factor has changed.

$$\text{PF} = 0.9 \text{ lagging}$$

$$\cos \theta = 0.9$$

$$\theta = 25.84°$$

$$\sin \theta = 0.44$$

From Eq. 8-10 and using the positive sign for lagging power factor.

$$E_p = \sqrt{(481 \times 0.12 + 693 \times 0.9)^2 + (481 \times 1.5 + 693 \times 0.44)^2}$$

$$= 1232 \text{ V}$$

$$\% \text{ V.R.} = \frac{1232 \text{ V} - 693 \text{ V}}{693 \text{ V}} \times 100$$

$$\text{V.R.} = 77.78\%$$

(c) As in part (b), V_{FL} and I_p are still the same. The power factor is still 0.9; however, now it is leading. Therefore, the minus sign is used in Eq. 8-10.

$$E_p = \sqrt{(481 \times 0.12 + 693 \times 0.9)^2 + (481 \times 1.5 - 693 \times 0.44)^2}$$

$$= 798.67 \text{ V}$$

$$\% \text{ V.R.} = \frac{798.67 \text{ V} - 693 \text{ V}}{693 \text{ V}} \times 100$$

$$\text{V.R.} = 15.2\%$$

It should be noted that the leading power factor load (capacitive-type load) tends to counteract the lagging (inductive) behavior of the stator.

(d) The no-load line voltage, also called the generated line voltage, is just the no-load phase voltage E_p converted to a line quantity.

$$E_{gL} = \sqrt{3}E_p = \sqrt{3} \times 798.67 \text{ V}$$

$$= 1383.3 \text{ V}$$

This is the voltage that would be measured from line to line at no-load conditions.

Example 8-7

Repeat Example 8-6, but this time assume that the alternator is Δ connected. The per phase parameters and the ratings are the same.

Solution

(a) First find the rated line current using Eq. 7-49. As in Example 8-6,

$$I_L = \frac{1,000,000 \text{ VA}}{\sqrt{3} \times 1200 \text{ V}}$$

$$= 481 \text{ A}$$

Now, however, the armature is Δ connected. Therefore,

$$I_p = \frac{I_L}{\sqrt{3}} = \frac{481 \text{ A}}{\sqrt{3}} = 277.7 \text{ A}$$

In the Δ connection the phase voltage equals the line voltage.

$$V_{FL} = 1200 \text{ V}$$

From Eq. 8-10.

$$E_p = \sqrt{(277.7 \times 0.12 + 1200 \times 1)^2 + (277.7 \times 1.5 + 0)^2}$$

$$= 1301.8 \text{ V}$$

Thus

$$V_{NL} = E_p = 1301.8 \text{ V}$$

$$V_{FL} = 1200 \text{ V}$$

and

$$\% \text{ V.R.} = \frac{1301.8 \text{ V} - 1200 \text{ V}}{1200 \text{ V}} \times 100$$

$$\text{V.R.} = 8.4\%$$

(b) $V_{FL} = 1200$ V and $I_p = 277.7$ A, as before. From Eq. 8-10,

$$E_p = \sqrt{(277.7 \times 0.12 + 1200 \times 0.9)^2 + (277.7 \times 1.5 + 1200 \times 0.44)^2}$$

$$= 1460 \text{ V}$$

$$\% \text{ V.R.} = \frac{1460 \text{ V} - 1200 \text{ V}}{1200 \text{ V}} \times 100$$

$$\text{V.R.} = 21.67\%$$

(c) The power factor is now leading. From Eq. 8-10,

$$E_p = \sqrt{(277.7 \times 0.12 + 1200 \times 0.9)^2 + (277.7 \times 1.5 - 1200 \times 0.44)^2}$$

$$= 1119 \text{ V}$$

$$\% \text{ V.R.} = \frac{1119 \text{ V} - 1200 \text{ V}}{1200 \text{ V}} \times 100$$

$$\text{V.R.} = -6.75\%$$

(d) Since the stator is Δ connected, the no-load line voltage equals the phase voltage.

$$E_{gL} = E_p = 1119 \text{ V}$$

There are a couple of interesting points to be made about the preceding examples. In Example 8-7, part (d), under no-load conditions the line voltage is 1119 V. When the alternator is fully loaded, this line voltage actually increases to 1200 V, hence the negative voltage regulation. In this case the capacitive load (leading PF) actually supplies some of the lagging reactive power (var) absorbed by the stator. This lessens the total kVA that must be generated by the alternator. Since the current remains at its full-load value, the generated voltage decreases.

In each case where E_p is different, what has changed is the field excitation (I_f) necessary to supply rated kVA to the different loads. This is actually done by an operator or an automatic control system.

Although the voltage regulation for the Δ connection was much better than that for the Y connection, a generalization should not be made. The examples did not really use identical machines. In fact, if the stator windings were identical, the rated voltage for the Δ would have been 693 V instead of the 1200 V used in Example 8-7. Thus, in effect, the two alternators used were of totally different design. In other words, the regulation is a function of the design and not the winding type (Δ or Y).

8-7   EFFICIENCY

The efficiency of an alternator is found in the same way as it was for the dc generator. The input power is determined by adding the power output to the internal losses of the alternator. The power output is a function of the power factor and the kVA supplied to a load. The internal losses, which are similar to those for a dc generator (mechanical, core, copper, and field losses), can be found by performing some simple tests. Once the losses are known, Eq. 8-12 is used to solve for the efficiency. Note the similarity of this equation to Eq. 7-42 for the transformer.

$$\eta(\%) = \frac{\text{VA}_{\text{out}} \times \text{PF}}{\text{VA}_{\text{out}} \times \text{PF} + P_{\text{Cu}} + P_{\text{core}} + P_{\text{field}} + P_{\text{mech}}} \times 100 \qquad (8\text{-}12)$$

8-7.1 Tests for Determining Alternator Losses

To test an alternator accurately, we must drive (turn) it with a dc motor that is calibrated. In other words, all the losses (hence efficiency) are known for the motor at every load condition. The arrangement would look as shown in Figure 8-9. Note that the power input of the alternator is equal to the power output of the motor. As long as the motor's efficiency and input are known, we can calculate the motor's output

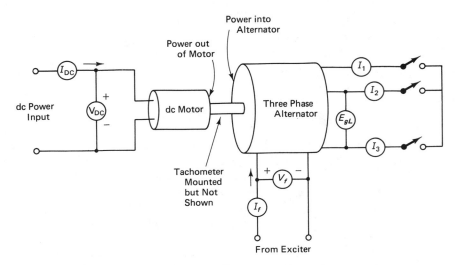

Figure 8-9 Arrangement to test a three-phase alternator.

(alternator input) at all times. From the figure we can see that eight meters are needed: two dc voltmeters, two dc ammeters, one ac voltmeter, and three ac ammeters. In addition, a tachometer is needed to monitor the alternator's speed.

The power input of the motor is given by $V_{dc} \times I_{dc}$. Multiplying this power by the motor's efficiency (η) gives us its output, or power input of the alternator. We will call the alternator input P_i.

$$P_i = V_{dc}I_{dc}\eta \tag{8-13}$$

To have some degree of accuracy we should also know the generated line voltage (E_{gL}) for the load the alternator will be supplying. This was the quantity calculated in part (d) of Examples 8-6 and 8-7. Although in those examples E_{gL} was smaller than rated voltage, it is usually larger than rated voltage (i.e., most loads are inductive, lagging pf). Three separate tests are needed. They are outlined below.

1. With no excitation supplied and all switches open, the alternator is driven at its rated speed. Since there is no voltage generated (no field excitation), there will be no core loss or power out of the alternator. In addition, since the switches are open, there is no copper loss in the stator. Hence the power out of the motor (P_{i1}) for this test is needed to overcome the mechanical losses of the alternator.

$$P_{i1} = P_{mech}(\text{alternator mechanical loss}) \tag{8-14}$$

2. With the alternator still running from test 1, increase the excitation until the generated line voltage (E_{gL}) is read across the lines. This is usually larger than its rated voltage. As before, since the switches are open, there is no stator copper loss and no power output from the alternator. However, since E_{gL} is at its normal value, the flux is rated and the core losses will be rated losses. Therefore, the

power out of the motor (P_{i2}) for this test is needed to overcome both mechanical and core losses of the alternator.

$$P_{i2} = P_{core} + P_{mech}$$

Substituting Eq. 8-14 into the equation above and solving for P_{core}, we get

$$P_{core} = P_{i2} - P_{i1} \qquad (8\text{-}15)$$

In addition, during this test the field meters give us the rated field loss.

$$P_{field} = V_f I_f \qquad (8\text{-}16)$$

3. With no excitation ($I_f = 0$) and the switches closed (short-circuit test), the alternator is brought up to rated speed. The excitation is slowly increased until rated current is measured in the lines. The excitation needed for this is minimal; thus the flux is very small and the core loss is negligible. However, since rated current is in the stator, the alternator has its rated copper loss present. Hence the power out of the motor (P_{i3}) for this test is needed to overcome both mechanical and stator copper losses in the alternator.

$$P_{i3} = P_{mech} + P_{Cu}$$

Substituting Eq. 8-14 into the equation above and solving for P_{Cu}, we get

$$P_{Cu} = P_{i3} - P_{i1} \qquad (8\text{-}17)$$

Example 8-8

A 208-V 3-kVA 60-Hz three-phase alternator is tested in the lab to determine its losses and efficiency. All tests are at rated speed. The test setup is shown in Figure 8-9 and the results are given in Table 8-1. Assume that the motor's efficiency is constant at 75%. Find the alternator's efficiency at:

(a) Full load, PF = 1

(b) Full load, PF = 0.9 lagging

(c) Half load, PF = 0.8 lagging

Table 8-1

Meter	Test 1	Test 2	Test 3
V_{dc} (V)	120	120	120
I_{dc} (A)	0.9	2.0	2.4
V_f (V)	0	5	1
I_f (A)	0	7.2	1.4
E_{gL} (V)	≈ 0	240	0
I_1 (A)	0	0	8.1
I_2 (A)	0	0	8.5
I_3 (A)	0	0	8.3

Solution

Using Eq. 8-13 in each case to calculate P_i, we get

$$\text{Test 1:} \quad P_{i1} = 120 \text{ V} \times 0.9 \text{ A} \times 0.75 = 81 \text{ W}$$

$$\text{Test 2:} \quad P_{i2} = 120 \text{ V} \times 2 \text{ A} \times 0.75 = 180 \text{ W}$$

$$\text{Test 3:} \quad P_{i3} = 120 \text{ V} \times 2.4 \text{ A} \times 0.75 = 216 \text{ W}$$

From Eq. 8-14,

$$P_{\text{mech}} = 81 \text{ W}$$

From Eq. 8-15,

$$P_{\text{core}} = 180 \text{ W} - 81 \text{ W} = 99 \text{ W}$$

From Eq. 8-17,

$$P_{\text{Cu}} = 216 \text{ W} - 81 \text{ W} = 135 \text{ W}$$

From Eq. 8-16 and test 2,

$$P_{\text{field}} = 5 \text{ V} \times 7.2 \text{ A} = 36 \text{ W}$$

The efficiency of the alternator is calculated using Eq. 8-12.

(a) $\quad \eta = \dfrac{3000 \times 1.0}{3000 \times 1.0 + 135 + 99 + 36 + 81} \times 100$

$\qquad = 89.53\%$

(b) $\quad \eta = \dfrac{3000 \times 0.9}{3000 \times 0.9 + 135 + 99 + 36 + 81} \times 100$

$\qquad = 88.5\%$

(c) In this part since we are at half load, the stator current will be half of its rated value. The copper losses will then be one-fourth of their rated value.

$$P_{\text{Cu}}(\text{at } \tfrac{1}{2} \text{ load}) = P_{\text{Cu}}(\tfrac{1}{2})^2 = 135(\tfrac{1}{4}) = 33.75 \text{ W}$$

In addition, the total kVA out at half-load is $\frac{1}{2}$ of 3 kVA, or 1.5 kVA.

$$\eta = \dfrac{1500 \times 0.8}{1500 \times 0.8 + 33.75 + 99 + 36 + 81} \times 100$$

$$= 82.77\%$$

8-7.2 Maximum Alternator Efficiency

As with dc machines (see Eq. 5-16), the peak efficiency occurs when the fixed losses equal the variable losses. In the alternator, the mechanical, core, and field losses are relatively constant. The copper loss, however, varies with load. Thus the maximum efficiency occurs at that fraction of rated load (k) where Eq. 8-18 is satisfied. The losses in this equation are all rated losses.

$$(k)^2 P_{Cu} = P_{core} + P_{mech} + P_{field} \qquad (8\text{-}18)$$

Example 8-9

At what percent of rated load does the alternator of Example 8-8 have maximum efficiency?

Solution

Substituting the losses from Example 8-8 into Eq. 8-18 and solving for k, we get

$$k^2 \times 135 \text{ W} = 99 \text{ W} + 81 \text{ W} + 36 \text{ W}$$

$$k^2 = \frac{216}{135} = 1.6$$

$$k = 1.26$$

Multiplying this by 100, we get 126% of rated load. In other words, peak efficiency for the alternator of Example 8-8 occurs when it is supplying 26% more than its rated load. Typically, an alternator's peak efficiency occurs around rated load or slightly less.

8-8   TYPICAL ALTERNATOR CHARACTERISTICS

There are several characteristics that can describe an alternator's behavior. All of them can be determined experimentally in the laboratory. Here we will examine them in a qualitative sense.

8-8.1 No-Load Saturation Curve

This characteristic is a plot of alternator line voltage versus field current at no load ($I_L = 0$). The alternator is run at rated speed. As can be seen in Figure 8-10, the voltage increases linearly until the flux begins to saturate. Rated voltage is typically obtained in the nonlinear region of the curve. This characteristic is also referred to as the no-load magnetization curve.

If the field current is varied while keeping the alternator at full load (rated line current being supplied), we can obtain the full-load characteristics also shown in Figure 8-10. Note that these curves differ depending on the power factor at which full load is supplied.

A couple of points should be made regarding the full-load curves. They all go through the same point ($V_L = 0$, field current $= I_f$). This happens because the terminals are short circuited to obtain zero line volts. Thus the only thing in the alternator circuit is its own impedance. Therefore, there is only one field current that can produce rated current with the output shorted. Note that this field excitation is the same as that measured in the short-circuit test (see Section 8-5.2).

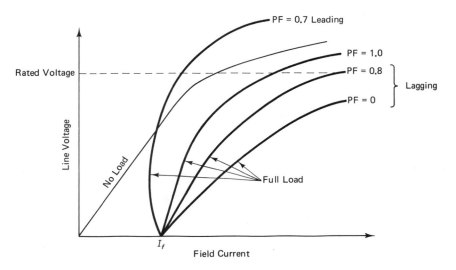

Figure 8-10 Typical alternator load characteristics.

8-8.2 Load Curves

Another characteristic useful in predicting an alternator's behavior is the set of load curves shown in Figure 8-11. A load curve is a plot of line voltage versus line current. Since it shows the variation of line voltage from no load to full load, it gives us a picture of the alternator's voltage regulation. Three curves are shown for three different load power factors.

Data for the curves in Figure 8-11 are obtained by first adjusting the field current to obtain rated line voltage at no load. The excitation is not changed after it is set at no load. The load is then increased (at each power factor condition) and the line voltage is recorded simultaneously.

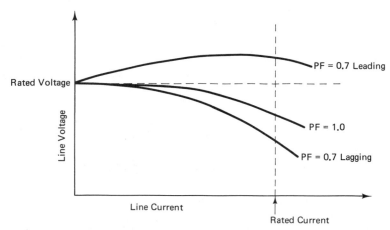

Figure 8-11 Typical alternator load curves showing the effect of power factor variation (no-load voltage common).

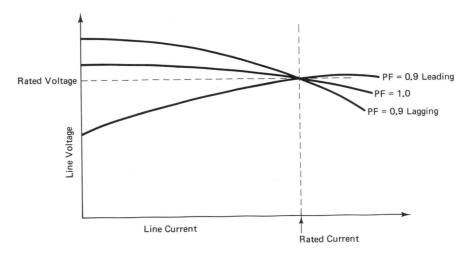

Figure 8-12 Typical alternator load curves with rated voltage and rated current as a common point.

A different set of load curves, shown in Figure 8-12, can be obtained with the following procedure. The field current is adjusted to get rated line current at rated voltage. The field excitation is not changed for the remainder of the test. Keeping the load power factor constant, the line current (load) is gradually reduced to zero, recording the line voltage simultaneously. Note that this is a more useful characteristic since we usually desire rated voltage at rated current. It also represents qualitatively the calculations done in Examples 8-6 and 8-7.

8-8.3 Additional Characteristics

Sometimes rather than plotting versus line current, parameters are graphed as a function of output power. The curves shown in Figure 8-13 are typical for alternators. The rated power point is usually indicated with a dashed line.

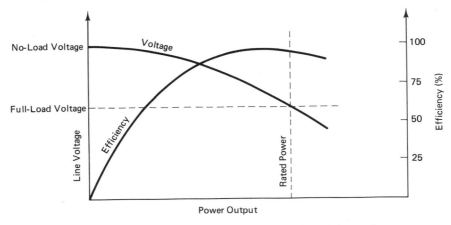

Figure 8-13 Efficiency and line voltage versus power output for an alternator.

SYMBOLS INTRODUCED IN CHAPTER 8

Symbol	Definition	Units: English and SI
E_p	Generated voltage for a single-phase stator winding	rms volts
E_{gL}	Generated line voltage	rms volts
z	Number of conductors in a single stator winding	—
f	Frequency of generated voltage	hertz
V_t	Line-to-line terminal voltage of a three-phase alternator	rms volts
r_p	Effective resistance of stator winding for one phase	ohms
x_p	Synchronous reactance of stator winding for one phase	ohms
Z_s	Synchronous impedance of stator winding for one phase	ohms
r_a	Resistance between any two terminals of a three-phase alternator with the third terminal open	ohms
r_p'	Calculated resistance of stator winding for one phase	ohms
V_{oc}	Phase voltage from open circuit test equal to measured value divided by $\sqrt{3}$	rms volts
I_{sc}	Phase current from short-circuit test measurement	amperes
I_f	Alternator field current	amperes
I_p	Rated phase current	amperes
I_L	Rated line current	amperes
V_{NL}	No-load phase voltage	rms volts
V_{FL}	Full-load phase voltage	rms volts
$\cos \theta$	Power factor of one phase of a balanced three-phase load	—
k	Fraction of rated load where maximum efficiency occurs	—

QUESTIONS

1. In what way is an alternator different from a dc generator? Why is this difference beneficial in the ac machine?

2. Define the following terms: exciter; synchronous reactance; synchronous impedance; rotor; stator.

3. When does maximum efficiency occur in an alternator?

4. Which are the fixed losses in an alternator? Which loss is variable?

5. What is the no-load saturation curve? By what other name is it called?

6. Where on the no-load saturation curve is rated voltage obtained?

7. What is an alternator load curve?

PROBLEMS

(English)

1. How fast must an alternator be driven so that it generates a voltage at 60 Hz if it has
(a) 4 poles?
(b) 6 poles?

(c) 12 poles?

(d) 36 poles?

2. An alternator generates a 400-Hz voltage when it is driven at 2670 rev/min. How many poles does it have?

3. A single-phase alternator generates 120 V at 60 Hz. If the stator has a total of 800 turns, find the needed flux per pole.

4. A single-phase alternator generates 28 V at 400 Hz. If the stator has 360 turns, find the needed flux per pole.

5. Repeat Problem 4 if the alternator is three-phase with a Y-connected stator. Note that 28 V is the generated line voltage.

6. How much voltage will an eight-pole alternator generate if it is driven at 900 rev/min? Its stator has 640 turns and the flux per pole is 60 kilolines.

7. An alternator whose stator has 240 turns has a flux per pole of 110 kilolines. It generates 60 Hz. Find the generated voltage.

(SI)

8. How fast must an alternator be driven so that it generates a voltage at 50 Hz if it has:

(a) 2 poles?

(b) 8 poles?

(c) 16 poles?

(d) 36 poles?

9. An alternator generates a 400-Hz voltage when it is driven at 420 rad/s. How many poles does it have?

10. A single-phase alternator generates 230 V at 50 Hz. If the stator has a total of 600 turns, find the needed flux per pole.

11. A single-phase alternator generates 28 V at 400 Hz. If the stator has 240 turns, find the needed flux per pole.

12. Repeat Problem 11 if the alternator is three-phase with a Y-connected stator. Note that 28 V is the generated line voltage.

13. How much voltage will a six-pole alternator generate if it is driven at 40π rad/s? Its stator has 600 turns and the flux per pole is 0.75×10^{-3} Wb.

14. An alternator whose stator has 400 turns has a flux per pole of 1.2×10^{-3} Wb. It generates 50 Hz. Find the voltage generated.

(English and SI)

15. A three-phase alternator is rated 230 V, 20 kVA. What is its full-load line current?

16. What is the rated line current of a three-phase alternator rated 208 V, 50 kVA?

17. A three-phase Y-connected alternator is rated 240 V, 10 kVA. From the test results below, determine:

(a) The resistance per phase

(b) The synchronous impedance per phase

(c) The synchronous reactance per phase

$$\text{Resistance test:} \quad V = 4\text{ V}, \quad I = 20\text{ A}$$

$$\text{Short-circuit test:} \quad I_L = \text{rated current}$$

$$I_f = 19\text{ A}$$

$$\text{Open-circuit test:} \quad I_f = 19 \text{ A}$$
$$V = 85 \text{ V}$$

18. A three-phase Δ-connected alternator is rated 230 V, 30 kVA. From the test results below, determine:
(a) The resistance per phase
(b) The synchronous impedance per phase
(c) The synchronous reactance per phase

$$\text{Resistance test:} \quad V = 3.6 \text{ V,} \quad I = 28 \text{ A}$$
$$\text{Short-circuit test:} \quad I_L = \text{rated current}$$
$$I_f = 23 \text{ A}$$
$$\text{Open-circuit test:} \quad I_f = 23 \text{ A}$$
$$V = 65 \text{ V}$$

19. A 600-V 50 kVA three-phase alternator is Y connected. Its resistance per phase is 0.2 Ω and the reactance per phase is 1.9 Ω. Find its voltage regulation if it supplies rated load at:
(a) PF = 1
(b) PF = 0.85 lagging
(c) PF = 0.85 leading
(d) What is the no-load line voltage in part (a)?

20. Repeat Problem 19, but this time assume that the alternator is Δ connected.

21. A three-phase 240-V 2-kVA Y-connected alternator has $r_p = 0.12$ Ω and $x_p = 0.9$ Ω. Find the generated voltage per phase when rated load is supplied at:
(a) PF = 1
(b) PF = 0.7 lagging

22. Repeat Problem 21, but this time assume that the alternator is Δ connected.

23. A 13,200-V 1000-kVA three-phase alternator is Δ connected. If $r_p = 0.16$ Ω and $x_p = 0.9$ Ω, find its voltage regulation if it supplies rated load at:
(a) PF = 1
(b) PF = 0.7 lagging
(c) PF = 0.90 leading

Table 8-2

Meter	Test 1	Test 2	Test 3
V_{dc} (V)	120	120	120
I_{dc} (A)	1.8	3.8	4.5
V_f (V)	0	8.4	2.5
I_f (A)	0	9.6	2.8
E_{gL} (V)	≈0	270	0
I_1 (A)	0	0	25
I_2 (A)	0	0	24
I_3 (A)	0	0	24.1

24. Repeat Problem 23, but this time assume that the alternator is Y connected.

25. A 240-V 10-kVA three-phase alternator is tested as shown in Figure 8-9 to determine its losses. The test results, all obtained at rated speed, are given in Table 8-2. If the dc motor has a constant efficiency of 80%, determine the alternator's efficiency at:
(a) Full load, PF = 1
(b) Full load, PF = 0.7 lagging
(c) Half load, PF = 0.7 lagging

26. At what percent of rated load is maximum efficiency obtained for the alternator in Problem 25?

27. A 660-V 50-kVA three-phase alternator is tested as shown in Figure 8-9 to determine its losses. The test results, all obtained at rated speed, are given in Table 8-3. If the dc motor has a constant efficiency of 85%, determine the alternator's efficiency at:
(a) Full load, PF = 1
(b) Full load, PF = 0.8 lagging
(c) Half load, PF = 1
(d) One-fourth load, PF = 0.7 lagging

Table 8-3

Meter	Test 1	Test 2	Test 3
V_{dc} (V)	230	230	230
I_{dc} (A)	10	19	25
V_f (V)	0	38	14
I_f (A)	0	11	4
E_{gL} (V)	≈0	730	0
I_1 (A)	0	0	42
I_2 (A)	0	0	44
I_3 (A)	0	0	45

28. At what percent of rated load is maximum efficiency obtained for the alternator in Problem 27?

29. The three-phase alternator whose load curve is shown in Figure 8-14 (curve a) is rated 230 V, 10 kVA. Find:
(a) The rated current
(b) The percent voltage regulation

30. The three-phase alternator whose load curve is shown in Figure 8-14 (curve b) is rated 210 V, 7.8 kVA. Find:
(a) The rated current
(b) The percent voltage regulation

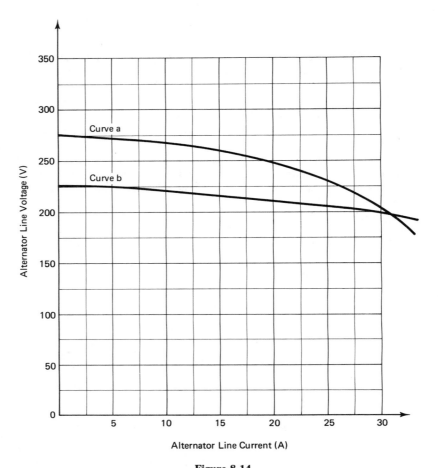

Figure 8-14

Chapter 9

The Three-Phase Induction Motor

The most widely used motor in factories and industries throughout the world is the induction motor. In particular, where very large machinery is to be operated, the three-phase induction motor is the workhorse. It is relatively simple to build, inexpensive, and highly reliable. Although Nikola Tesla is credited with inventing the induction motor, its principle of operation was actually demonstrated by Galileo Ferraris in 1885, a year before Tesla's discovery.

9-1   CONSTRUCTION

The motor has a frame or stationary part which is the **stator.** There are windings mounted in the stator in the same way as the alternator. The three-phase induction motor has a three-phase winding on the stator. Depending on how the stator is wound, it may have two, four, or any even number of poles (a four-pole motor would have two north and two south poles).

The rotating part (the shaft of the motor) is called the **rotor.** The most common rotor is the **squirrel-cage** type. It is called this because of its appearance. It looks like the exercise wheel used for hamsters and gerbils. It is made with copper rods embedded in slots in a laminated core. The ends of the rods are connected together with end rings as shown in Figure 9-1. The laminated rotor core is not shown. In actual operation the rods are the rotor conductors and the end rings short their ends together. Some rotors, called double-squirrel-cage rotors, have two sets of rods. One set made of brass is placed near the surface. It has a higher resistance than the inner set which is made of copper (see Figure 9-2). The inner set, however, being closer together, has a larger leakage reactance. Electrically, this appears in the rotor circuit as an inductive reactance. To understand why this double set produces improved characteristics, we

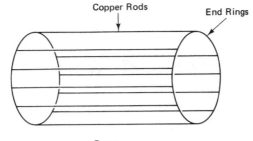

Figure 9-1 Simplified picture of a squirrel-cage rotor.

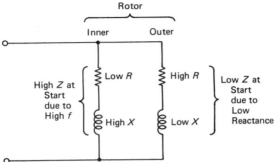

Figure 9-2 Equivalent circuit for a double squirrel-cage rotor.

must state the following, which will be explained in a later section of this chapter. The frequency of the rotor current is high on starting (the line frequency 50 or 60 Hz). When the motor builds up to a high speed, the frequency of the rotor current is low (approaching zero).

At starting the outer rods draw a large current because their total impedance is lower. In addition, having a high R and low X gives them a high power factor. The combination of high current and power factor produces a high starting torque.

At high speeds almost all the current goes through the low-resistance inner rods, whose reactance is negligible at low frequencies. This produces good high-speed running characteristics.

Another type is to have a **wound rotor** with an actual winding. By doing this an external resistance (called a **slip-ring rheostat**) can be connected with slip rings. In this way the rotor resistance can be varied to improve the motor's starting and running characteristics. A diagram of the wound rotor is shown in Figure 9-3. The winding must be of the same form as on the stator. In other words, it must be a three-phase winding with the same number of poles as on the stator. Examples of the two types of rotors are shown in Figures 9-4 and 9-5.

The air gap between rotor and stator is very important. Its presence (increasing the reluctance of the magnetic circuit) causes a large magnetizing current in the stator winding necessary to produce the flux. This lagging component of current causes the lagging power factor characteristic of induction motors. If the air gap is made too small, in an effort to improve the power factor, mechanical problems such as bearing wear and rotor–stator alignment are introduced. In addition, if the rotor and stator are not perfectly round, very small air gaps can cause motor noise and friction. The friction will increase the mechanical loss causing reduced efficiency and running

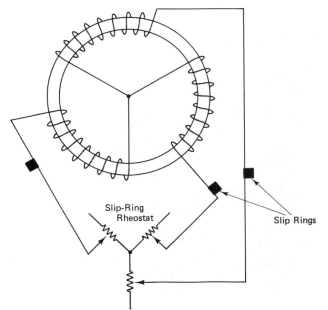

Figure 9-3 Simplified three-phase wound rotor with a slip-ring rheostat.

Slip-Ring Rheostat

Slip Rings

Figure 9-4 Squirrel-cage rotor for a three-phase induction motor. (Courtesy of Reliance Electric Company.)

speed. Very often the rotor conductors (slots) are not parallel to the shaft but are angled (**skewed**). This tends to eliminate noise and produce a smoother running motor.

It is important to note that the number of rotor slots should not equal the number of stator slots. In fact, they are very often designed so that their numbers are prime to each other (i.e., they do not have common factors). If they are not, several slots can line up at times, causing a pulsating flux. In some cases, if the slots are lined up, the rotor may even lock up on starting and not turn.

Slip Rings

Figure 9-5 Wound rotor for a three-phase induction motor. (Courtesy of Reliance Electric Company.)

9-2 *ROTATING FIELD CONCEPT*

Before we discuss the theory of operation for the induction motor (or even synchronous motor) a very basic concept, that of a rotating magnetic field, must be understood. To explain it we will examine the three-phase stator shown in Figure 9-6. It has been wound for two poles. The voltages applied to the stator are shown in Figure 9-7. The stator together with its magnetic field is shown at four different instances of time, which correspond to different instantaneous voltages as shown in Figure 9-7. The rotor has been left out of the picture so as not to confuse the discussion. We will assume a Y-connected stator (A', B', and C' connected). Furthermore, when $V_{A-A'}$ is positive, it means that winding end A is positive with respect to A'. This means that the current will be into the paper at the A end and out of the paper at the A' end of the winding. The times in Figure 9-7 were chosen to simplify the picture and to show a rotation of the magnetic field.

At time t_1, $V_{A-A'}$ and $V_{C-C'}$ are positive and $V_{B-B'}$ is a negative maximum. This produces the current directions shown in Figure 9-6a. Using the right hand with fingers curled in the direction of current indicates a resultant two-pole magnetic field in the direction shown.

A short time later, at t_2, $V_{A-A'}$ is a positive maximum with both $V_{B-B'}$ and $V_{C-C'}$ negative. This time corresponds to a displacement of 60 electrical degrees from time t_1 of the input voltage. The new current directions together with the resultant stator

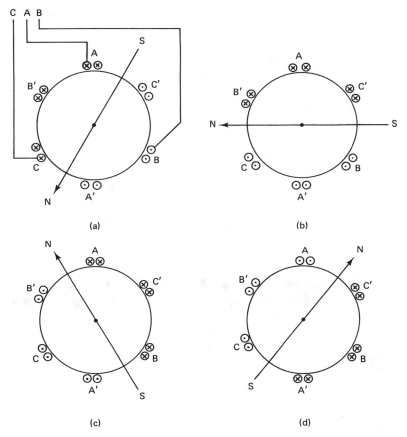

Figure 9-6 Stator field orientation: (a) at time t_1; (b) at time t_2; (c) at time t_3; (d) at time t_4.

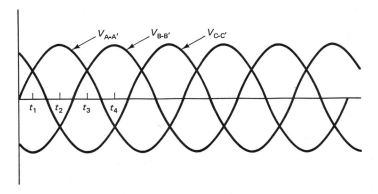

Figure 9-7 Three-phase stator voltages, phase sequence ABC.

field is shown in Figure 9-6b. Note that the resultant field has shifted (rotated) 60° clockwise from its position at t_1. Although it will not be shown here, if the stator was wired for four poles, the magnetic field would have rotated only 30° for a 60° electrical displacement.

At t_3, $V_{A-A'}$ and $V_{B-B'}$ are positive while $V_{C-C'}$ is now a negative maximum. This produces the currents and resultant field shown in Figure 9-6c. Note again a 60° clockwise rotation of the field.

Finally, at t_4, $V_{B-B'}$ is a positive maximum with $V_{A-A'}$ and $V_{C-C'}$ both negative. This is shown in Figure 9-6d.

The preceding discussion examined the field at four instants of time. This does not mean that the field jumps 60° at each instant. If we had examined thousands of points very close to each other, we would have seen that the field rotates continuously and uniformly in accordance with the frequency of the stator voltages.

9-2.1 Direction of the Rotating Field

The phase sequence in Figure 9-7 together with the winding arrangement on the stator of Figure 9-6 produced a clockwise rotation of the magnetic field. If the phase sequence were changed to ACB (rather than ABC) and we went through the analysis all over again, we would find that the rotation of the field reversed. It would now be turning counterclockwise. As far as the stator is concerned, this change in phase sequence can be accomplished simply by interchanging any two supply lines in Figure 9-6a (either A with B, C with A, or B with C).

9-2.2 Speed of the Rotating Field

The speed at which the magnetic field rotates is called **synchronous speed.** It can be calculated using an equation identical to Eqs. 8-2 (used for the alternator's generated frequency); however, now the subscript s is used for speed, indicating synchronous speed. It is important with induction motors to remember that synchronous speed refers to the speed of the **rotating stator field** and **not** the speed of the rotor or the rotor field. This is explained in more detail in Section 9-3.

(English)

$$S_s = \frac{120f}{\mathbf{P}} \tag{9-1a}$$

(SI)

$$\omega_s = \frac{4\pi f}{\mathbf{P}} \tag{9-1b}$$

In Eqs. 9-1, $\mathbf{P}$ is the number of stator poles, f is the frequency of the applied stator voltage in hertz, and S_s and ω_s represent the speed of the stator field (synchronous speed) in rev/min and rad/s, respectively.

Example 9-1 (English)

What is synchronous speed (in rev/min) for a three-phase four-pole 60-Hz induction motor?

Solution

Using Eq. 9-1a gives us

$$S_s = 120 \times \frac{60}{4}$$

$$= 1800 \text{ rev/min}$$

Example 9-2 (SI)

What is synchronous speed (in rad/s) for a three-phase four-pole 50-Hz induction motor?

Solution

Using Eq. 9-1b gives us

$$\omega_s = 4\pi \times \frac{50}{4}$$

$$= 157.1 \text{ rad/s}$$

9-2.3 Shape of the Rotating Field

Since the applied voltage in the stator is sinusoidal, the current in the stator winding will also be sinusoidal. In fact, the current magnitude in each stator conductor will vary sinusoidally as the three-phase input changes. The result, since the current produces the flux, will be a magnetic field whose magnitude resembles that of a sine wave. This net flux is actually the vector sum of three magnetic fields due to the three phase currents in the stator. The flux can be shown as the half-cycle of a sine wave rotating in a circular path. It is shown in Figure 9-8.

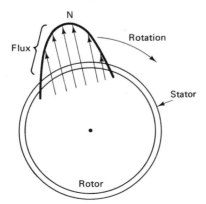

Figure 9-8 Diagram showing rotating flux with sinusoidal shape.

9-3 THEORY OF OPERATION

There are several techniques and analogies used to explain the operation of an induction motor. We will use the concept of the rotating field and a comparison of the induction motor with the transformer. In relating it to the transformer the stator winding and applied voltage will represent the primary circuit. The rotor winding or rods will represent the secondary circuit. The mechanical power delivered by the rotating rotor is analogous to the electrical power out of the transformer's secondary. There is one very basic difference in the transformer analogy. In the case of the transformer the frequency of the voltage or current in the secondary is the same as that in the primary. With the induction motor, the frequency of the rotor voltage or current will usually be different from that in the stator.

When a voltage is applied to the stator winding, a current will flow in it. This current is made up of two components. One of these is a magnetizing component. It will lag the applied voltage by 90°. Its purpose is to set up the magnetic field (rotating flux). The second component is in phase with the applied voltage and it supplies the real power (watts) to the motor. The picture is similar to that for the transformer (see Figure 7-14). When the rotor shaft has nothing connected to it, there is no mechanical power out of the motor. The only real power that must be supplied under this no load condition is for the mechanical, core, and I^2R losses in the motor. These losses are, however, very small and the angle θ_0 in Figure 7-14b is very close to 90°. This explains the very low power factor (cos 90° = 0) of the induction motor at no load. The condition when the rotor shaft is locked and held so that it cannot move is called a **blocked rotor.** This **standstill** condition will be referred to continuously and the subscript BR will be used to denote quantities when the rotor is blocked. This condition is also very similar to a transformer whose secondary has been short-circuited.

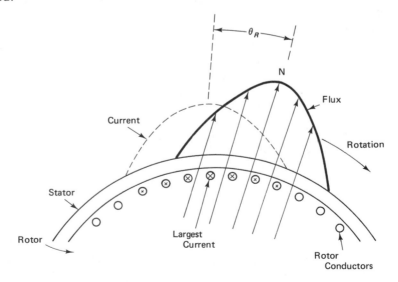

Figure 9-9 Diagram of flux and rotor current at standstill.

If we drew a picture of the rotor conductors and the rotating flux that cuts them, it would look like Figure 9-9. Applying the right-hand rule and noting that the direction of conductor with respect to flux is right to left (i.e., opposite to the flux motion), the current flow will be into the paper (indicated by X). The following should be noted about the rotor currents:

1. Current flows because the rotor conductors are shorted by end rings.
2. The frequency of the current at standstill is the same as the frequency of the applied stator voltage.
3. Since the flux is sinusoidally distributed, the magnitude of the voltage induced in each rotor conductor will vary sinusoidally in phase with the flux. However, since the conductors have both resistance and inductive reactance, the picture for current (dotted line) is shown lagging the flux (or voltage) by an angle θ_R (where θ_R is the rotor power factor angle).
4. The average magnitude of the rotor current and its frequency depend on the rate at which the flux cuts the conductors. Therefore, as the rotor builds up speed, the flux will cut the conductors at a slower rate, decreasing the rotor current and its frequency. In fact, if the rotor turned at synchronous speed (the speed of the rotating flux), the flux lines would not cut any conductors and the rotor current and frequency would both be zero.

Since there is a current flowing in the rotor conductors at standstill, there will be a force exerted on them (see Section 2-4.1). The force produces a torque in the same direction as the rotating flux. This can be verified by applying the left-hand rule (see Figure 2-14). The magnitude of the torque at any time is given by Eq. 9-2, which is very similar to the basic motor equation presented in Chapter 5 (Eq. 5-1). Now, however, we use the subscript R to denote a rotor quantity and since the flux and current are not exactly in phase, the expression is multiplied by the cosine of the rotor power factor angle.

$$T = K\phi I_R \cos \theta_R \qquad (9\text{-}2)$$

Since at starting the rotor current is proportional to flux and flux is proportional to the stator voltage, we can derive the following qualitative formula for the **starting torque** T_s of an induction motor.

$$T_s = K\phi(K_1\phi) \cos \theta_R$$
$$= K'V_L^2 \qquad (9\text{-}3)$$

Equation 9-3 merely states that the starting torque in an induction motor is proportional to the square of the applied stator line voltage (V_L). This starting torque accelerates the rotor depending on how much inertia is connected to the shaft (see Eq. 5-2). Note that once the rotor starts turning and builds up speed, Eq. 9-3 will no longer represent the torque developed. As the rotor turns faster and faster, fewer lines are cut by the rotor conductors, hence a lower voltage is induced in them. A lower voltage gives rise to a smaller current I_R, decreasing the torque given by Eq. 9-2.

Eventually, if the rotor speed keeps increasing, it will be turning at synchronous speed. When this happens no flux lines are cut by the conductors, the rotor voltage and current will then be zero, and the torque given by Eq. 9-2 will be zero. With zero torque the motor stops accelerating and now runs steady at synchronous speed. Wrong! If the motor was ideal and had no mechanical (friction and windage) losses, what we have just said would be true. However, this is a nonideal world we live in and no matter how good the motor is, there will always be mechanical losses present. As a result the motor never reaches synchronous speed. It attains a speed where just enough flux lines are cut to produce just enough torque to overcome the mechanical losses.

If a load is now placed on the shaft, the rotor will tend to slow down. As it slows down, more flux lines are cut until enough torque is developed to overcome the load placed on the shaft. The motor now runs under load at a slower speed than before the load was placed on the shaft.

It is important to note that the induction motor can never run at synchronous speed. For this reason it is called an **asynchronous** machine; that is, it runs at a speed other than the synchronous speed of its flux.

Furthermore, it should be pointed out that electrical power is transmitted to the rotor not by mechanical means (brushes or slip rings), but by **magnetic induction.** It is for this reason that it has been given the name "induction motor."

9-3.1 Motor Ratings

Induction motors are rated in terms of the voltage and frequency that should be applied to the stator. In addition, their rating includes the maximum mechanical power they can continuously supply (full load) and the speed they will run at when delivering this maximum or full-load power. Also included in the rating is the line current the motor will draw when it supplies full-load power. Sometimes the motor's rating may include its full-load power factor as well.

From the ratings much more information can be derived using basic equations and theory. This will be done later in this chapter in the numerical examples.

Example 9-3 (English)

A three-phase 208-V 60-Hz 2-hp induction motor develops a starting torque of 12 ft-lb when rated voltage is applied. Find:

(a) The starting torque when 240 V is applied

(b) The voltage needed to produce a starting torque of 18 ft-lb

Solution

From Eq. 9-3 the starting torque is proportional to the square of the line voltage.

(a) $T_{s1} = K' V_{L1}^2$ for the first voltage

$T_{s2} = K'V_{L2}^2$ for the second voltage

Dividing the first equation by the second, we have

$$\frac{T_{s1}}{T_{s2}} = \left(\frac{V_{L1}}{V_{L2}}\right)^2 \qquad (9\text{-}4a)$$

In this part the unknown quantity is T_{s2}. Solving for it, we get

$$T_{s2} = T_{s1}\left(\frac{V_{L2}}{V_{L1}}\right)^2$$

$$= 12 \text{ ft-lb}\left(\frac{240 \text{ V}}{208 \text{ V}}\right)^2$$

$$= 15.98 \approx 16 \text{ ft-lb}$$

(b) The equation derived above (9-4a) is rearranged to solve for V_{L2}.

$$V_{L2} = V_{L1}\sqrt{\frac{T_{s2}}{T_{s1}}}$$

$$= 208 \text{ V}\sqrt{\frac{18 \text{ ft-lb}}{12 \text{ ft-lb}}}$$

$$= 254.75 \text{ V}$$

Example 9-4 (SI)

A three-phase 440-V 50-Hz 2-kW induction motor develops a starting torque of 40 N-m when rated voltage is applied. Find:

(a) The starting torque when 460 V is applied

(b) The voltage needed to produce a starting torque of 60 N-m

Solution

From Eq. 9-3 the starting torque is proportional to the square of the line voltage.

(a) $T_{s1} = K'V_{L1}^2$ for the first voltage

$T_{s2} = K'V_{L2}^2$ for the second voltage

Dividing the first equation by the second gives us

$$\frac{T_{s1}}{T_{s2}} = \left(\frac{V_{L1}}{V_{L2}}\right)^2 \qquad (9\text{-}4b)$$

In this part the unknown quantity is T_{s2}. Solving for it, we get

$$T_{s2} = T_{s1}\left(\frac{V_{L2}}{V_{L1}}\right)^2$$

$$= 40 \text{ N-m}\left(\frac{460 \text{ V}}{440 \text{ V}}\right)^2$$

$$= 43.72 \text{ N-m}$$

(b) The equation derived above (9-4b) is rearranged to solve for V_{L2},

$$V_{L2} = V_{L1} \sqrt{\frac{T_{s2}}{T_{s1}}}$$

$$= 440 \text{ V} \sqrt{\frac{60 \text{ N-m}}{40 \text{ N-m}}}$$

$$= 538.9 \text{ V}$$

9-4   SPEED RELATIONSHIPS

The induction motor is basically a constant-speed device. This does not mean that its speed never changes. Its speed does change with variations in load; however, it will always be below synchronous speed. Unlike the dc motor, whose speed can be varied by changing the field, armature current, or terminal voltage, the induction motor changes speed only when the load it is driving changes. This, however, is not a way to vary a motor's speed. In general, the load is something that the user of the motor cannot control. The only way to actually control the induction motor's speed is to vary the synchronous speed of its field. Looking at Eqs. 9-1, we see that this can be done by changing the frequency of the applied voltage or the number of poles. In Chapter 13 we look at some special equipment and techniques that make this possible. In addition, by varying the rotor resistance in a wound rotor, the rotor current, hence torque, can be controlled. This will vary the speed somewhat; however, we will see later in this chapter that motor efficiency decreases due to the I^2R loss in the external resistance. It should also be noted that this technique is not possible with a squirrel-cage rotor.

The same variation of torque can be accomplished by line-voltage variation. Decreasing the voltage lowers the flux and thus the rotor current will also decrease. This technique is undesirable because it creates an unstable running condition where the motor is very sensitive to load changes. This is especially apparent in large motors.

9-4.1 Slip

SYN — SPEED

The **slip** of a motor is the difference between synchronous speed and rotor (shaft) speed. It is given by Eqs. 9-5 and has units of speed.

(English)

$$\text{slip} = \mathbf{s} = S_s - S \qquad (9\text{-}5a)$$

(SI)

$$\text{slip} = \mathbf{s} = \omega_s - \omega \qquad (9\text{-}5b)$$

The slip is frequently expressed as a percent of synchronous speed.

(English)

$$\text{percent slip} = \mathbf{s} = \frac{S_s - S}{S_s} \times 100 \qquad (9\text{-}6a)$$

$$\text{percent slip} = s = \frac{\omega_s - \omega}{\omega_s} \times 100 \qquad (9\text{-}6\text{b})$$

Example 9-5 (English)

A four-pole 208-V 2-hp 60-Hz three-phase induction motor has a no-load speed of 1790 rev/min and a full-load speed of 1650 rev/min. Find the slip in rev/min and percent for each case below.

(a) No load

(b) Full load

(c) Blocked rotor

Solution

Before any slip calculations can be made, we must first calculate synchronous speed using Eq. 9-1a.

$$S_s = 120 \times \frac{60 \text{ Hz}}{4 \text{ poles}} = 1800 \text{ rev/min}$$

(a) At no load using Eqs. 9-5a and 9-6a, we have

$$s = 1800 \text{ rev/min} - 1790 \text{ rev/min} = 10 \text{ rev/min}$$

$$s = (1800 - 1790) \times \frac{100}{1800} = 0.556\%$$

(b) At full load we use the same equations; however, $S = 1650$ rev/min.

$$s = 1800 \text{ rev/min} - 1650 \text{ rev/min} = 150 \text{ rev/min}$$

$$s = (1800 - 1650) \times \frac{100}{1800} = 8.33\%$$

(c) At standstill (blocked rotor), $S = 0$.

$$s = 1800 \text{ rev/min} - 0 = 1800 \text{ rev/min}$$

$$s = (1800 - 0) \times \frac{100}{1800} = 100\%$$

Example 9-6 (SI)

A four-pole 230-V 2-kW 50-Hz three-phase induction motor has a no-load speed of 155 rad/s and a full-load speed of 140 rad/s. Find the slip in rad/s and percent for each case below.

(a) No load

(b) Full load

(c) Blocked rotor

Solution

Before any slip calculations can be made, we must first calculate synchronous speed using Eq. 9-1b.

$$\omega_s = 4\pi \times \frac{50 \text{ Hz}}{4 \text{ poles}} = 157.14 \text{ rad/s}$$

(a) At no load using Eqs. 9-5b and 9-6b, we have

$$s = 157.14 \text{ rad/s} - 155 \text{ rad/s} = 2.14 \text{ rad/s}$$

$$s = (157.14 - 155) \times \frac{100}{157.14} = 1.36\%$$

(b) At full load we use the same equations. However, $\omega = 140$ rad/s.

$$s = 157.14 \text{ rad/s} - 140 \text{ rad/s} = 17.14 \text{ rad/s}$$

$$s = (157.14 - 140) \times \frac{100}{157.14} = 10.9\%$$

(c) At standstill (blocked rotor), $\omega = 0$.

$$s = 157.14 \text{ rad/s} - 0 = 157.14 \text{ rad/s}$$

$$s = (157.14 - 0) \times \frac{100}{157.14} = 100\%$$

From the preceding examples we can draw the following conclusions regarding slip:

1. Upon starting, the motor's slip is 100%. At times the slip will be used in equations. When this is done it is expressed as a decimal and the factor 100 in Eqs. 9-6 is left out. Therefore, as a decimal the slip at standstill is 1.
2. As the motor picks up speed the slip decreases to a very small number at the no-load condition. It is generally between 0 and 2%. As a decimal this is between 0 and 0.02.
3. As the load increases the motor slows, develops more torque, and the slip will increase as well. Thus we can see a relationship developing that will be derived shortly: torque is proportional to slip.

Example 9-7 (English)

A three-phase 60-Hz induction motor has a no-load speed of 595 rev/min and a full-load slip of 16%. How many poles does it have, and what is its full-load speed?

Solution

A little bit of trial and error with some common sense is needed to solve this problem. Knowing that the slip at no load is very small and the speed is slightly less than synchronous speed, we must find the synchronous speed, which is just a bit bigger than 595 rev/min.

First try eight poles in Eq. 9-1a,

$$S_s = 120 \times \frac{60}{8} = 900 \text{ rev/min}$$

This is too high!

Now try 10 poles,

$$S_s = 120 \times \frac{60}{10} = 720 \text{ rev/min}$$

Still too high!

Now try 12 poles

$$S_s = 120 \times \frac{60}{12} = 600 \text{ rev/min}$$

This is just right, 595 rev/min is a little less than 600 rev/min.

We can now conclude that the motor must have 12 poles and its synchronous speed is 600 rev/min. By substituting this into Eq. 9-6a, we can solve for the full-load speed.

$$16 = \frac{600 - S_{FL}}{600} \times 100$$

$$S_{FL} = 600 - 600 \times \frac{16}{100}$$

$$= 504 \text{ rev/min}$$

Example 9-8 (SI)

A three-phase 50-Hz induction motor has a no-load speed of 62.5 rad/s and a full-load slip of 20%. How many poles does it have, and what is its full load speed?

Solution

A little bit of trial and error with some common sense is needed to solve this problem. Knowing that the slip at no load is very small and the speed is slightly less than synchronous speed, we must find the synchronous speed that is just a bit bigger than 62.5 rad/s.

First try eight poles in Eq. 9-1b,

$$\omega_s = 4\pi \times \frac{50}{8} = 78.5 \text{ rad/s}$$

This is too high!

Now try 10 poles,

$$\omega_s = 4\pi \times \frac{50}{10} = 62.83 \text{ rad/s}$$

This is just right, 62.5 rad/s is a little less than 62.83 rad/s.

We can now conclude that the motor must have 10 poles and its synchronous speed is 62.83 rad/s. By substituting this into Eq. 9-6b, we can solve for the full-load speed.

$$20 = \frac{62.83 - \omega_{FL}}{62.83} \times 100$$

$$\omega_{FL} = 62.83 - 62.83 \times \frac{20}{100}$$

$$= 50.26 \text{ rad/s}$$

Since the solution of speed problems often involves the use of two equations (Eqs. 9-1 and 9-6), we can simplify things by combining the two. By substituting Eqs. 9-1 into Eqs. 9-6, we can eliminate the quantity "synchronous speed" from our calculations and come up with Eqs. 9-7 for rotor or shaft speed in terms of slip.

(English)

$$S = \frac{120f}{P}(1 - \mathbf{s}) \qquad\qquad (9\text{-}7a)$$

(SI)

$$\omega = \frac{4\pi f}{P}(1 - \mathbf{s}) \qquad\qquad (9\text{-}7b)$$

It is important to note that the slip in Eqs. 9-7 should be expressed as a decimal; that is, if the slip is 20% then in Eqs. 9-7, **s** equals 0.20.

Example 9-9 (English)

A four-pole 60-Hz three-phase induction motor has a full-load slip of 18%. What is its full-load speed?

Solution

The problem can be solved easily by substituting directly into Eq. 9-7a and noting that **s** = 0.18.

$$S_{FL} = \frac{120 \times 60}{4}(1 - 0.18) = 1476 \text{ rev/min}$$

Example 9-10 (SI)

A 12-pole 50-Hz three-phase induction motor has a full-load slip of 22%. What is its full-load speed?

Solution

The problem can be solved easily by substituting directly into Eq. 9-7b and noting that **s** = 0.22.

$$\omega_{FL} = \frac{4\pi \times 50}{12}(1 - 0.22) = 40.86 \text{ rad/s}$$

9-4.2 Speed Regulation

This quantity has the same meaning here as it does for any motor. It was covered in Section 5-4 for the dc motor and the equations are rewritten here for convenience.

(English)

$$\begin{matrix} \% \text{ speed} \\ \text{regulation} \end{matrix} = \frac{S_{NL} - S_{FL}}{S_{FL}} \times 100 \qquad (9\text{-}8a)$$

(SI)

$$\begin{matrix} \% \text{ speed} \\ \text{regulation} \end{matrix} = \frac{\omega_{NL} - \omega_{FL}}{\omega_{FL}} \times 100 \qquad (9\text{-}8b)$$

If we compare Eqs. 9-6 to 9-8, we can see that slip and speed regulation are calculated in a similar way. Both express a difference in speed as a percent, one of synchronous speed and the other of full-load speed. Although slip is used almost entirely as a measure of speed performance for induction motors, both quantities exist and can be calculated. They will be fairly close in magnitude to each other; however, they should not be confused.

Example 9-11 (English)

A six-pole 60-Hz three-phase induction motor has a no-load speed of 1190 rev/min and full-load speed of 980 rev/min. Find the percent slip at full load and the percent speed regulation.

Solution

First using Eq. 9-1a, find the synchronous speed.

$$S_s = 120 \times \frac{60}{6} = 1200 \text{ rev/min}$$

Using Eq. 9-6a, we have

$$s = (1200 - 980) \times \frac{100}{1200} = 18.33\%$$

Using Eq. 9-8a gives us

$$SR = (1190 - 980) \times \frac{100}{980} = 21.4\%$$

Example 9-12 (SI)

An eight-pole 50-Hz three-phase induction motor has a no-load speed of 77.8 rad/s and a full-load speed of 66 rad/s. Find the percent slip at full load and the percent speed regulation.

Solution

First using Eq. 9-1b, find the synchronous speed.

$$\omega_s = 4\pi \times \frac{50}{8} = 78.57 \text{ rad/s}$$

Using Eq. 9-6b, we have

$$s = (78.57 - 66) \times \frac{100}{78.57} = 16\%$$

Using Eq. 9-8b gives us

$$SR = (77.8 - 66) \times \frac{100}{66} = 17.9\%$$

Note that Eqs. 9-7 could also have been used in the preceding examples.

9-5   ANALYSIS OF ROTOR BEHAVIOR

We will now analyze the behavior of the rotor from the electrical relationships up to the torque output at any speed. There will be approximations made; however, the analysis presented is fairly accurate for both squirrel-cage and wound rotors. The subscript R when used will denote rotor quantities.

We have already pointed out (Section 9-3) that the frequency of the rotor current (f_R) is equal to the line frequency (f) at standstill. Also, as the rotor gains speed, f_R will decrease. This frequency relationship can be expressed mathematically by Eq. 9-9, where f and f_R are in hertz. In all of the equations that follow in this analysis, s is the slip as a decimal.

$$f_R = sf \tag{9-9}$$

At standstill the slip is 1 (100%) and f_R equals f (50 or 60 Hz). However, at a typical load condition of, say, 5% slip, the rotor current frequency is about 2.5 to 3 Hz, which is almost dc.

The inductance (L_R) of the rotor conductors is constant; however, since the reactance (X_R) is a function of frequency ($X_R = 2\pi f_R L_R$), it will vary with speed or slip. Thus, if we let X_{BR} be the reactance at standstill ($X_{BR} = 2\pi f L_R$), the reactance at any slip can be expressed as

$$X_R = sX_{BR} \tag{9-10}$$

Furthermore, if the rotor resistance is R_R, the total impedance at any slip would be

$$Z_R = \sqrt{R_R^2 + (sX_{BR})^2} \tag{9-11}$$

The voltage induced in the rotor E_R is a function of the stator voltage, the rate at which the rotor conductor cuts the flux and the turns ratio between stator and rotor.

In the squirrel-cage rotor the number of turns is a difficult quantity to define. Sometimes an effective turns ratio, which can be measured experimentally, is used. We will not, however, get involved in the transformation of voltage from stator to rotor. We will start out with the induced rotor voltage at standstill (E_{BR}), a constant, and derive our equations based on this quantity. Since E_R decreases as the rotor builds up speed, it is proportional to slip, which also decreases with an increase in speed.

$$E_R = sE_{BR} \tag{9-12}$$

The rotor current at any speed can be found by dividing Eq. 9-12 by Eq. 9-11 (Ohm's law).

$$I_R = \frac{sE_{BR}}{\sqrt{R_R^2 + (sX_{BR})^2}} \tag{9-13}$$

Typically, X_{BR} is about four or five times R_R. At full load the slip will be about 10%; hence the rotor impedance at this condition will be approximately R_R. In other words, $(sX_{BR})^2$ will be much smaller than $(R_R)^2$ and can therefore be neglected. Example 9-13 will illustrate this.

Example 9-13

What is the rotor impedance at a slip of 10% if $R_R = 0.08\ \Omega$ and $X_{BR} = 0.35\ \Omega$.

Solution

From Eq. 9-11,

$$Z_R = \sqrt{(0.08)^2 + (0.1 \times 0.35)^2} = 0.087$$

Note that Z_R is approximately equal to R_R; therefore, this substitution will be made in Eq. 9-13 to simplify our analysis.

$$I_R = \frac{sE_{BR}}{R_R} \tag{9-14}$$

Furthermore, since the rotor impedance at typical running conditions is basically resistive, we will assume that the rotor power factor ($\cos \theta_R$) is unity (1). Substituting this and Eq. 9-14 into our expression for torque, Eq. 9-2, we get

$$T = \frac{K\phi sE_{BR}(1)}{R_R} \tag{9-15}$$

Since the only variable in Eq. 9-15 is the slip, we can lump all the constant terms together to form a new constant K''.

$$K'' = \frac{K\phi E_{BR}}{R_R} \tag{9-16}$$

Note that the flux ϕ is, to a good approximation, dependent only on line voltage. The properties of the motor's magnetic circuit change only slightly during load variations.

Substitution of Eq. 9-16 into Eq. 9-15 gives us an approximate expression for torque developed in the rotor as a function of slip.

$$T = K''\mathbf{s} \tag{9-17}$$

It should be noted that T in Eq. 9-17 is the developed torque of the motor. The output torque is the developed torque minus the opposing torque due to the internal mechanical loss of the motor. However, since the mechanical loss is fairly constant, the output torque will vary with the developed torque. Thus the output torque will also vary with slip, as shown in Eq. 9-17. When the motor is running at constant speed (zero acceleration), the output torque is equal to the opposing load torque. Hence Eq. 9-17 can also be used to relate slip to the load, as in the following examples.

Example 9-14 (English)

A four-pole 60-Hz three-phase induction motor has a full-load speed of 1600 rev/min. How fast will it run at half load?

Solution

First find its full-load slip. This can be done easily using Eq. 9-7a.

$$1600 = \frac{120 \times 60}{4}(1 - \mathbf{s})$$

Solving for **s**,

$$\mathbf{s} = 1 - 0.89 = 0.11 \ (11\% \text{ slip})$$

From Eq. 9-17 slip is proportional to load; therefore, at half load it will be half of its full-load value. At half-load

$$\mathbf{s} = 0.11(\tfrac{1}{2}) = 0.055$$

The half-load speed can now be found by substituting this value of slip back into Eq. 9-7a and solving for S.

$$S = \frac{120 \times 60}{4}(1 - 0.055) = 1701 \text{ rev/min}$$

Example 9-15 (SI)

A two-pole 50-Hz three-phase induction motor has a full-load speed of 265 rad/s. How fast will it run at 75% of rated load?

Solution

First find its full-load slip. This can be done easily using Eq. 9-7b.

$$265 = \frac{4\pi \times 50}{2}(1 - \mathbf{s})$$

Solving for **s**,

$$\mathbf{s} = 1 - 0.84 = 0.16 \ (16\% \text{ slip})$$

From Eq. 9-17, slip is proportional to load; therefore, at 75% of full lo be three-fourths of its full-load value. At three-fourths load

$$s = 0.75 \times 0.16 = 0.12$$

The three-fourths load speed can now be found by substituting this value of s Eq. 9-7b and solving for ω.

$$\omega = \frac{4\pi \times 50}{2}(1 - 0.12) = 276.3 \text{ rad/s}$$

9-5.1 Torque–Power Relationships

In Chapter 5 we were introduced to Eqs. 5-3, which relate torque, speed, and power. These equations, together with the theory presented in this chapter, enable us to answer questions about the load on the motor simply by making a speed measurement and knowing its ratings.

Knowing its rated power and speed, we can calculate its rated (full-load) torque. By measuring its speed under load, we can calculate the slip, which in turn will tell us the load torque. The following examples will illustrate this.

Example 9-16 (English)

A 2-hp 208-V 60-Hz three-phase induction motor has a full-load speed of 1700 rev/min. It is running under light load at 1750 rev/min. Find:

(a) The rated torque

(b) The load torque

(c) The output power

Solution

First, we do not have to know how many poles the motor has. Common sense tells us that if its full-load speed is 1700 rev/min; its synchronous speed must be the next highest synchronous speed for a 60-Hz motor. This is 1800 rev/min and the motor must have four poles.

(a) Substituting the ratings into Eq. 5-3a, we can calculate the rated torque.

$$P = 2 \text{ hp} \times 746 \text{ W/hp} = 1492 \text{ W}$$

$$T = 7.04 \times \frac{1492 \text{ W}}{1700 \text{ rev/min}} \qquad T = 7.04 \frac{P}{SPEED}$$

$$= 6.2 \text{ ft-lb (rated torque)}$$

(b) From Eq. 9-6a,

$$\text{full-load slip} = (1800 - 1700) \times \frac{100}{1800}$$

$$s_{FL} = 5.56\% = 0.0556$$

Under light load,

$$s = (1800 - 1750) \times \frac{100}{1800}$$

$$= 2.78\% = 0.0278$$

Dividing the light-load slip by the full-load slip,

$$\frac{s}{s_{FL}} = \frac{0.0278}{0.0556} = 0.5 = \tfrac{1}{2}$$

This means that the motor is running at half load. It is therefore putting out one-half of rated torque. Since the load torque is equal to the output torque,

$$T_{load} = (\tfrac{1}{2})6.2 \text{ ft-lb} = 3.1 \text{ ft-lb}$$

(c) The output power can be calculated using the output torque and Eq. 5-3a.

$$P = \frac{TS}{7.04} = 3.1 \text{ ft-lb} \times \frac{1750 \text{ rev/min}}{7.04}$$

$$= 770.6 \text{ W}$$

$$= \frac{770.6 \text{ W}}{746 \text{ W/hp}} = 1.03 \text{ hp}$$

Example 9-17 (SI)

A 4-kW 230-V 50-Hz three-phase induction motor has a full-load speed of 275 rad/s. It is running under light load at 300 rad/s. Find:

(a) The rated torque

(b) The load torque

(c) The output power

Solution

First of all, we do not have to know how many poles the motor has. Common sense tells us that if its full-load speed is 275 rad/s, its synchronous speed must be the next highest synchronous speed for a 50-Hz motor. This is 314 rad/s and the motor must have two poles.

(a) Substituting the ratings into Eq. 5-3b, we can calculate the rated torque.

$$T = 1000 \times \frac{4 \text{ kW}}{275 \text{ rad/s}}$$

$$= 14.55 \text{ N-m}$$

(b) From Eq. 9-6b,

$$\text{full-load slip} = (314 - 275) \times \frac{100}{314}$$

$$s_{FL} = 12.4\% = 0.124$$

Under light load

$$s = (314 - 300) \times \frac{100}{314}$$

$$= 4.46\% = 0.0446$$

Dividing the light-load slip by the full-load slip, we have

$$\frac{s}{s_{FL}} = \frac{0.0446}{0.124} = 0.36$$

This means that the motor is running at 36% of full load. It is therefore putting out 36% of rated torque. Since the load torque is equal to the output torque,

$$T_{load} = 0.36 \times 14.55 \text{ N-m} = 5.238 \text{ N-m}$$

(c) The output power can be calculated using the output torque and Eq. 5-3b.

$$P = \frac{T\omega}{1000} = 5.238 \text{ N-m} \times \frac{300 \text{ rad/s}}{1000}$$

$$= 1.57 \text{ kW}$$

9-5.2 Torque–Slip Curve

The preceding examples were solved using an approximate relationship, Eq. 9-17. However, as the load increases, the motor slows down and the slip will increase. As this happens Eq. 9-17 becomes inaccurate and we must find an alternative approach. One way would be to use a **torque–slip curve.** This is a plot of output torque versus percent slip. It is similar to using the torque–speed curve (Figure 5-3) for a dc motor. A typical torque–slip curve is shown in Figure 9-10.

If we substituted Eq. 9-13 into Eq. 9-2 and noted that $\cos \theta_R = R_R/Z_R$, we could derive the following equation for torque as a function of slip:

$$T = \frac{(K\phi E_{BR} R_R)s}{R_R^2 + (sX_{BR})^2} \tag{9-18}$$

The torque–slip curve in Figure 9-10 is a plot of Eq. 9-18 as slip is varied. Upon starting, the slip is 100% and the motor develops its starting torque. It accelerates and settles at a speed (slip) in the normal operating range, which depends on the load. When the load increases to rated torque, the motor runs at rated slip. If the load increases further, causing the slip to increase, the output torque increases as well. Looking at Eq. 9-18, we see that as **s** increases, the term in the denominator begins to have a decreasing effect on T. The torque continues to increase with slip until the motor is putting out its **maximum torque.** This is also called the **pull-out torque** of the motor. When the load exceeds this torque, it is greater than the motor's maximum torque and the motor will gradually slow down until it stalls.

Mathematically, this maximum point on the curve occurs when the two terms in the denominator of Eq. 9-18 are equal.

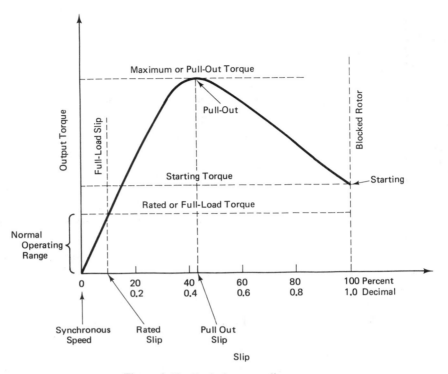

Figure 9-10 Typical torque–slip curve.

$$R_R^2 = (sX_{BR})^2$$

$$R_R = sX_{BR}$$

$$\text{pull-out slip} = s_{po} = \frac{R_R}{X_{BR}} \tag{9-19}$$

What is actually happening can be explained as follows. As slip increases, the reactive part (sX_{BR}) of the rotor impedance increases until it equals the resistive part (R_R). At that point the rotor power factor is 0.707. Further increases in slip (caused by an increase in load) decrease the power factor. This means that less real power (watts) is developed per ampere of rotor current than before. This reaction begins to have a snowball effect. Less power means increased slip, causing an even lower power factor and even less power. The breakdown process continues until the motor stalls.

Example 9-18 (English)

A two-pole 60-Hz 1-hp three-phase induction motor has the following per phase specifications: $R_R = 0.1\ \Omega$ and $X_{BR} = 0.4\ \Omega$. What is the slowest speed at which it can run continuously without stalling?

Solution

The slowest speed occurs just before pull-out. This corresponds to the maximum point on the torque–slip curve. From Eq. 9-19,

$$s_{po} = \frac{0.1 \ \Omega}{0.4 \ \Omega} = 0.25$$

Substituting this into Eq. 9-7a,

$$S = 120 \times 60 \times \frac{1 - 0.25}{2} = 2700 \ \text{rev/min}$$

Example 9-19 (SI)

A 10-pole 50-Hz 2-kW three-phase induction motor has the following per phase specifications: $R_R = 0.08 \ \Omega$ and $X_{BR} = 0.25 \ \Omega$. What is the slowest speed at which it can run continuously without stalling?

Solution

The slowest speed occurs just before pull-out. This corresponds to the maximum point on the torque–slip curve. From Eq. 9-19,

$$s_{po} = \frac{0.08 \ \Omega}{0.25 \ \Omega} = 0.32$$

Substituting this into Eq. 9-7b gives us

$$\omega = 4\pi \times 50 \times \frac{1 - 0.32}{10} = 42.7 \ \text{rad/s}$$

9-5.3 Wound Rotor

In the preceding discussion the only variable affecting the motor's running characteristic was the load. However, if the rotor is of the wound rotor type, its resistance per phase can be varied as shown in Figure 9-3. Not only will the rotor resistance affect the slip at pull-out, but it will affect the slip for any given torque. It can be shown from our approximate equations that slip is proportional to rotor resistance.

$$\frac{s_1}{s_2} = \frac{R_{R1}}{R_{R2}} \tag{9-20}$$

Thus increasing rotor resistance increases slip, which means for the same load we will now run at a slower speed. The increased rotor resistance will have no effect on the maximum or pullout torque of the motor. Figure 9-11 shows the effect of varying the rotor resistance on a motor's torque-slip curve. We see that we can vary the motor's speed (slip) by changing the rotor resistance. However, we should note the following: The speed can never exceed that obtained with a shorted rotor (R_R is a minimum). Too much rotor resistance (R_{R4} in Figure 9-11) will stall the motor before maximum torque can ever be reached, and increased rotor resistance will increase the starting torque of the motor.

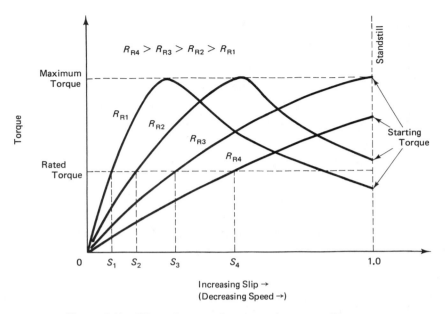

Figure 9-11 Effect of rotor resistance on the torque–slip curve.

Example 9-20 (English)

A six-pole 60-Hz wound rotor induction motor has the following specifications with the rotor short circuited.

$$= 0.05\ \Omega \qquad S_{FL} = 1050 \text{ rev/min}$$

$$_{BR} = 0.25\ \Omega$$

A slip-ring rheostat is used to add $0.15\ \Omega$ to each phase. Find:

(a) The shorted rotor pullout speed

(b) The new pull-out speed

(c) The new full-load speed

Solution

(a) From Eq. 9-19,

$$s_{po} = \frac{0.05\ \Omega}{0.25\ \Omega} = 0.2$$

From Eq. 9-7a,

$$S_{po} = 120 \times 60 \times \frac{1 - 0.2}{6} = 960 \text{ rev/min}$$

(b) The rotor resistance per phase is increased to $0.2\ \Omega$ $(0.05 + 0.15)$. From Eq. 9-19,

$$s_{po} = \frac{0.2\ \Omega}{0.25\ \Omega} = 0.8$$

From Eq. 9-7a,

$$S_{po} = 120 \times 60 \times \frac{1 - 0.8}{6} = 240 \text{ rev/min}$$

(c) Equation 9-20 is used to find the new full-load slip with the increased rotor resistance. We must first find the original full-load slip. From Eq. 9-1a,

$$S_s = 120 \times \frac{60}{6} = 1200 \text{ rev/min}$$

Thus the original slip using Eq. 9-5a is

$$s_1 = 1200 \text{ rev/min} - 1050 \text{ rev/min} = 150 \text{ rev/min}$$

Now using Eq. 9-20, we have

$$\frac{150 \text{ rev/min}}{s_2} = \frac{0.05 \ \Omega}{0.20 \ \Omega}$$

$$s_2 = 0.2 \times \frac{150}{0.05} = 600 \text{ rev/min}$$

Putting this back into Eq. 9-5a and solving for the new full-load speed, we get

$$600 \text{ rev/min} = 1200 \text{ rev/min} - S_{FL}$$

$$S_{FL} = 1200 - 600 = 600 \text{ rev/min}$$

Example 9-21 (SI)

A four-pole 50-Hz wound rotor induction motor has the following specifications with the rotor short-circuited:

$$R_R = 0.11 \ \Omega \qquad \omega_{FL} = 140 \text{ rad/sec}$$

$$X_{BR} = 0.5 \ \Omega$$

A slip-ring rheostat is used to add 0.2 Ω to each phase. Find:

(a) The shorted rotor pull out speed

(b) The new pull-out speed

(c) The new full-load speed

Solution

(a) From Eq. 9-19,

$$s_{po} = \frac{0.11 \ \Omega}{0.5 \ \Omega} = 0.22$$

From Eq. 9-7b,

$$\omega_{po} = 4\pi \times 50 \times \frac{1 - 0.22}{4} = 122.5 \text{ rad/s}$$

(b) The rotor resistance per phase is increased to 0.31 $\Omega(0.11 + 0.2)$. From Eq. 9-19,

$$s_{po} = \frac{0.31\ \Omega}{0.5\ \Omega} = 0.62$$

From Eq. 9-7b,

$$\omega_{po} = 4\pi \times 50 \times \frac{1 - 0.62}{4} = 59.7\ \text{rad/s}$$

(c) Equation 9-20 is used to find the new full-load slip with the increased rotor resistance. We must first find the original full-load slip. From Eq. 9-1b,

$$\omega_s = 4\pi \times \frac{50}{4} = 157\ \text{rad/s}$$

Thus the original slip using Eq. 9-5b is

$$s_1 = 157\ \text{rad/s} - 140\ \text{rad/s} = 17\ \text{rad/s}$$

Now using Eq. 9-20, we have

$$\frac{17\ \text{rad/s}}{s_2} = \frac{0.11\ \Omega}{0.31\ \Omega}$$

$$s_2 = 0.31 \times \frac{17}{0.11} = 47.9\ \text{rad/s}$$

Putting this back into Eq. 9-5b and solving for the new full-load speed, we get

$$47.9\ \text{rad/s} = 157\ \text{rad/s} - \omega_{FL}$$

$$\omega_{FL} = 157 - 47.9 = 109.1\ \text{rad/s}$$

9-6   EFFICIENCY

As with the preceding material on the induction motor, some reasonable approximations will be made in determining efficiency. The same approach will be used here as with the dc motor. Knowing the input power and the total losses at full load, the output can be calculated. The equations are the same as those presented in Section 5-5 and are listed here for convenience.

$$\eta(\%) = \frac{P_o}{P_i} \times 100 \tag{9-21}$$

$$P_o = P_i - \text{losses} \tag{9-22}$$

$$\eta(\%) = \frac{P_i - \text{losses}}{P_i} \times 100 \tag{9-23}$$

9-6.1 Description of Losses

The losses in Eqs. 9-22 and 9-23 represent the total internal losses of the motor. They fall into two basic categories. First, there are the constant losses, which are made up of the stator and rotor core losses (due to eddy currents and hysteresis) and the friction

and windage loss due to the rotating rotor. Second, there are the variable losses, the stator and rotor copper losses (I^2R), which depend on the load.

Rotor Core Loss. Under normal running conditions the rotor frequencies are very low; hence the hysteresis and eddy current losses will be very small. We will therefore assume that the core loss in the rotor is negligible.

Stator Core Loss. This loss is due to the main and leakage fluxes. As the load changes, these flux losses do change slightly, however, in opposite ways. It is therefore a good approximation to assume that this loss is the same whether at no load or full load.

Friction and Windage Loss. This loss is due to the mechanical rotation of the rotor. Since the speed does not change significantly from no load to full load, we will assume that this mechanical loss is constant.

Stator Copper Loss. This is the total I^2R loss in the stator winding. It is a function of load and will be calculated using a measured current and resistance.

Rotor Copper Loss. Since there are currents flowing in rotor conductors, there is an I^2R loss in the rotor. Since the rotor current and resistance cannot be measured directly in squirrel-cage motors, we will determine it indirectly from other measurements. To do this, we will have to know the slip **accurately** at full load. With today's sophisticated instruments, this can be done quite easily.

9-6.2 Description of Power Flow

Before we go on to the tests used to determine the losses, we will first examine the flow of power from input to output. This will give us a better understanding in our numerical calculations. As an aid, we will refer to the power flow diagram shown in Figure 9-12. When the stator winding is energized, electrical power will flow into the motor. Since it is three-phase power it is given by Eq. 9-24. In all of the equations that follow, power will be expressed in watts (or kW).

$$P = \sqrt{3}\, V_L I_L \cos\theta \tag{9-24}$$

When testing the motor, however, we do not know its power factor ($\cos\theta$); therefore, we cannot use the voltage and current measurement to determine the input power. The technique used to get P_i will be the two-wattmeter method described in Section 7-4.2.

$$P_i = W_1 + W_2 \tag{9-25}$$

In the stator there are two losses; the core loss (P_C), which depends on the applied voltage, and the copper loss (P_{CuS}), which depends on the line current (which in turn is a function of the load). If these two losses are subtracted from the input power, what is left is the power transferred across the air gap into the rotor (P_R).

$$P_R = P_i - P_C - P_{CuS} \tag{9-26}$$

In the rotor there is a negligible core loss and a copper loss that depends on the slip.

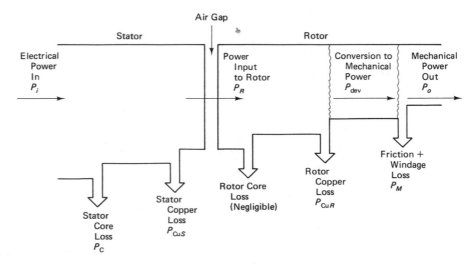

Figure 9-12 Power flow diagram for an induction motor.

As long as slip can be measured accurately, and it can, the rotor copper loss (P_{CuR}) is found using Eq. 9-27, where the slip is expressed as a decimal.

$$P_{\text{CuR}} = \text{s} \times P_R \tag{9-27}$$

After the rotor copper loss is subtracted from the rotor input, what is left is the mechanical power developed (P_{dev}) by the motor.

$$P_{\text{dev}} = P_R - P_{\text{CuR}} \tag{9-28}$$

Finally, the output power (P_o) is determined by subtracting the mechanical (friction and windage) losses (P_M) from the developed power.

$$P_o = P_{\text{dev}} - P_M \tag{9-29}$$

There is one minor problem encountered when we test the motor to find the losses. The total constant loss will be found; however, we will not know how much of it is stator core loss and how much is mechanical. We can assume that the constant loss is entirely a core loss and place it in the stator, or that it is entirely mechanical and place it in the rotor. Both are done in practice; however, in either case an error is introduced: that is, the rotor copper loss is calculated from the rotor power input. In the first case (constant loss entirely in stator) the rotor copper loss will be too small. In the second case (constant loss entirely in rotor) the rotor copper loss will be too big.

We will assume the second case for two reasons:

1. It will provide us with a conservative estimate for efficiency since the total losses will be slightly more than they really are.
2. Since it gives us a nonzero value for the mechanical loss (P_M), the output power will not equal the developed power, which is really the case (see Eq. 9-29).

As a result of this assumption, the stator core loss (P_C) in Eq. 9-26 will be zero in the examples that follow.

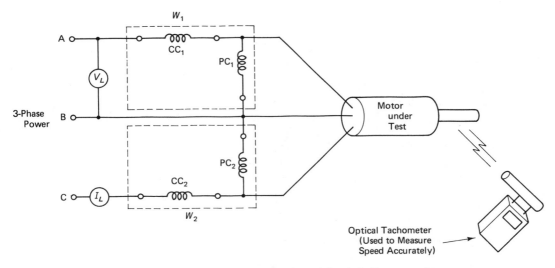

Figure 9-13 Sample arrangement for determining induction motor losses.

9-6.3 Tests to Determine Losses

There are two tests used to determine the losses of an induction motor. Before they are done, however, we must make a measurement to determine the effective resistance of the stator winding. It is similar to what was done for the alternator (see Section 8-5.1).

With the three stator lines disconnected from power, the resistance between any two stator lines is measured. We will call this resistance r_s. The measurement can be made either with an ohmmeter or from dc voltage and current measurements ($r_s = V/I$). For reasons beyond the scope of this book the effective stator resistance (r_e) is somewhat greater than this value. We will multiply it by 1.25.

$$r_e = 1.25 r_s \qquad (9\text{-}30)$$

It can be shown quite easily that the total copper loss in the stator winding in terms of the line current is given by Eq. 9-31.

$$P_{CuS} = \tfrac{3}{2} r_e I_L^2 \qquad (9\text{-}31)$$

It is interesting to note that Eq. 9-31 is valid whether the stator winding is in a delta or a wye configuration. In the tests that follow the arrangement shown in Figure 9-13 can be used.

No-Load Test. This test can be done simply by applying rated voltage to the stator without anything connected to the motor's shaft. Since we are at no load the speed is high, the slip is very small, and the rotor current is very small. Hence the rotor copper loss is negligible during this test. The input power is supplying the constant losses ($P_C + P_M$) and a small stator copper loss. If this copper loss is subtracted from the input power, what we are left with is the constant loss. By the assumption we have made, we will place this loss entirely in the rotor ($P_C = 0$).

$$P_{\cancel{C}}^{\;\;0} + P_M = W_1 + W_2 - P_{CuS} \tag{9-32}$$

In Eq. 9-32 W_1 and W_2 are the wattmeter readings and P_{CuS} is the small stator copper loss, which we will calculate with Eq. 9-31.

Rated-Load Test. The test setup in Figure 9-13 can also be used for this test; however, some means to load the motor mechanically must be devised. This can be done purely by mechanical means, that is, by use of a brake, or by having the motor drive a generator that is loaded electrically.

With the motor supplying rated load, all meters as well as the speed are recorded. The full-load point is determined when the stator draws rated line current. From the ammeter reading the rated stator copper loss (P_{CuS}) is calculated using Eq. 9-31. The tachometer reading allows us to calculate the slip. With these numbers, the wattmeter readings, and Eqs. 9-25, 9-26, and 9-27, we can determine the rotor copper loss (P_{CuR}) at rated load. Remember that P_C is zero. Equations 9-28 and 9-29 will give us the output power. Remember now that P_M is the constant loss determined in the no-load test.

In addition to the losses, we can find the motor's power factor at full load using Eq. 9-24. Note that we know V, I, and P_i from the rated-load test.

9-6.4 Calculation of Efficiency

The following examples will illustrate the calculation of efficiency from test results.

Example 9-22 (English)

A 5-hp 208-V 15-A four-pole 60-Hz three-phase induction motor is tested to determine its losses at full load. The test results are shown below.

$$\text{Stator resistance test: } r_s = 0.8 \ \Omega$$

$$\text{No-load test: } W_1 = 350 \text{ W}, \quad W_2 = -110 \text{ W}$$

$$V_L = 208 \text{ V}, \quad I_L = 4 \text{ A}$$

$$\text{Rated-load test: } W_1 = 2.6 \text{ kW}, \quad W_2 = 2.26 \text{ kW}$$

$$V_L = 208 \text{ V}, \quad I_L = 15 \text{ A}$$

$$S = 1710 \text{ rev/min}$$

From the test results, determine the motor's full load efficiency and power factor.

Solution

From the resistance test, calculate the effective stator resistance using Eq. 9-30.

$$r_e = 1.25 \times 0.8 \ \Omega = 1 \ \Omega$$

From the no-load test, calculate the small stator copper loss using Eq. 9-31.

$$P_{CuS} = \tfrac{3}{2} \times 1 \ \Omega \times (4 \text{ A})^2 = 24 \text{ W}$$

From Eq. 9-32,

$$P_C + P_M = 350 \text{ W} - 110 \text{ W} - 24 \text{ W} = 216 \text{ W}$$

Since we are placing the constant loss entirely in the rotor, $P_C = 0$ and $P_M = 216$ W. From the rated-load test, calculate the full-load stator copper loss using Eq. 9-31.

$$\text{rated } P_{\text{Cu}S} = \tfrac{3}{2} \times 1 \ \Omega \times (15 \text{ A}) = 337.5 \text{ W}$$

Calculate the rated power input using Eq. 9-25.

$$\text{rated } P_i = 2600 \text{ W} + 2260 \text{ W} = 4860 \text{ W}$$

Now calculate the power input to the rotor using Eq. 9-26.

$$\text{rated } P_R = 4860 \text{ W} - 0 - 337.5 \text{ W} = 4522.5 \text{ W}$$

The slip is calculated using Eq. 9-6a, the full-load speed of 1710 rev/min, and noting that synchronous speed for a four-pole 60-Hz motor is 1800 rev/min.

$$\text{rated } s = (1800 - 1710) \times \frac{100}{1800} = 5\%$$

or

$$s = 0.05 \text{ as a decimal}$$

From Eq. 9-27, calculate the rotor copper loss.

$$\text{rated } P_{\text{Cu}R} = 0.05 \times 4522.5 \text{ W} = 226.1 \text{ W}$$

Using Eqs. 9-28 and 9-29, we can now obtain the output power.

$$\text{rated } P_{\text{dev}} = 4522.5 \text{ W} - 226.1 \text{ W} = 4296.4 \text{ W}$$

and

$$\text{rated } P_o = 4296.4 \text{ W} - 216 \text{ W} = 4080.4 \text{ W}$$

or

$$P_o = \frac{4080.4 \text{ W}}{7\,6 \text{ W/hp}} = 5.47 \text{ hp}$$

The efficiency is calculated using Eq. 9-21.

$$\text{rated } \eta = \frac{4080.4 \text{ W}}{4860 \text{ W}} \times 100 = 83.96 \approx 84\%$$

The power factor can be calculated using the rated-load results and Eq. 9-24.

$$\text{rated PF} = \cos \theta = \frac{4860 \ W}{(\sqrt{3} \times 208 \text{ V} \times 15 \text{ A})}$$

$$\text{PF} = 0.899 \approx 0.9$$

Example 9-23 (SI)

A 3.5-kW 230-V 10-A four-pole 50-Hz three-phase induction motor is tested to determine its losses at full load. The test results are shown below.

Stator resistance test: $r_s = 0.9 \ \Omega$

No-load test: $W_1 = 290$ W,　$W_2 = -70$ W

$$V_L = 230 \text{ V},　I_L = 3.8 \text{ A}$$

Rated-load test: $W_1 = 2.3$ kW,　$W_2 = 2$ kW

$$V_L = 230 \text{ V},　I_L = 12 \text{ A}$$

$$\omega = 149 \text{ rad/s}$$

From the test results determine the motor's full-load efficiency and power factor.

Solution

From the resistance test, calculate the effective stator resistance using Eq. 9-30.

$$r_e = 1.25 \times 0.9 \ \Omega = 1.13 \ \Omega$$

From the no-load test calculate the small stator copper loss using Eq. 9-31.

$$P_{CuS} = \tfrac{3}{2} \times 1.13 \ \Omega \times (3.8 \text{ A})^2 = 24.5 \text{ W}$$

From Eq. 9-32,

$$P_C + P_M = 290 \text{ W} - 70 \text{ W} - 24.5 \text{ W} = 195.5 \text{ W}$$

Since we are placing the constant loss entirely in the rotor, $P_C = 0$ and $P_M = 195.5$ W. From the rated load test, calculate the full-load stator copper loss using Eq. 9-31.

$$\text{rated } P_{CuS} = \tfrac{3}{2} \times 1.13 \ \Omega \times (10 \text{ A})^2 = 244.1 \text{ W}$$

Calculate the input power using Eq. 9-25.

$$\text{rated } P_i = 2300 \text{ W} + 2000 \text{ W} = 4300 \text{ W}$$

Now calculate the power input to the rotor using Eq. 9-26.

$$\text{rated } P_R = 4300 \text{ W} - 0 - 244.1 \text{ W} = 4055.9 \text{ W}$$

The slip is calculated using Eq. 9-6b, the full-load speed of 149 rad/s and noting that synchronous speed for a four-pole 50-Hz motor is 157 rad/s.

$$\text{rated } s = (157 - 149) \times \frac{100}{157} = 5.1\%$$

or

$$s = 0.051 \text{ as a decimal}$$

From Eq. 9-27, calculate the rotor copper loss.

$$\text{rated } P_{CuR} = 0.051 \times 4055.9 = 206.85 \text{ W}$$

Using Eqs. 9-28 and 9-29, we can now obtain the output power.

$$\text{rated } P_{dev} = 4055.9 \text{ W} - 206.85 \text{ W} = 3849.1 \text{ W}$$

and

$$\text{rated } P_o = 3849.1 \text{ W} - 195.5 \text{ W} = 3653.6 \text{ W}$$

or

$$P_o = 3.65 \text{ kW}$$

The efficiency is calculated using Eq. 9-21.

$$\text{rated } \eta = \frac{3653.6 \text{ W}}{4300 \text{ W}} \times 100 = 84.96 \approx 85\%$$

The power factor can be calculated using the rated-load results and Eq. 9-24.

$$\text{rated PF} = \cos \theta = \frac{4300 \text{ W}}{\sqrt{3} \times 230 \text{ V} \times 12 \text{ A}}$$

$$\text{PF} = 0.899 \approx 0.9$$

From the preceding examples the efficiency of the motors at rated load was about 85%. This is typical with maximum efficiency occurring at slightly less than rated load (about 90% of full load).

The power factor of 0.9 is also typical. This increases from no load to full load, with the best (highest) power factor occurring slightly above rated load. For this reason it is desirable in factories to run induction motors as close to full load as possible. This is because the reactive power drawn from the utility grid limits the useful real power that can be used (see Example 7-24). In Section 10-6 we will see how a low lagging power factor can be changed (corrected) to a higher value.

9-7 TYPICAL CHARACTERISTICS

We have already examined one of the characteristic curves (the torque–slip curve, Figure 9-10) used to evaluate or predict a motor's performance. There are others that can be easily determined in the laboratory which we will look at here.

9-7.1 Characteristics versus Slip

Slip is an easy quantity to vary, measure, and observe. Therefore, performance parameters such as efficiency, power factor, and power output are frequently plotted as a function of it. Typical characteristics are shown in Figure 9-14 for a four-pole induction motor whose full-load slip is about 5%. Although determination of efficiency and power output is a little involved, the power factor curve can be easily obtained. An arrangement such as the one in Figure 9-13 is used. The motor is gradually loaded from no load to about 110% of rated load. At each setting all meters are recorded. The slip is calculated from the tachometer reading. Input power and power factor are calculated using Eqs. 9-25 and 9-24.

In examining Figure 9-14 note that the power factor has been multiplied by 100 (converted to percent). This enables us to use the same scale for both efficiency and power factor. Thus a power factor of 0.9 would appear on the graph as 90%.

It is interesting to note that the variation of efficiency between the half- and full-load points is insignificant. However, the power factor drops from 0.90 at full load to about 0.73 at half load. This drop in power factor is significant and its effects are severely felt when many motors such as these are operated at the same time in one factory.

Also, from Figure 9-14 we see that at no load the efficiency is 0% while the power factor is around 0.15, which is typical.

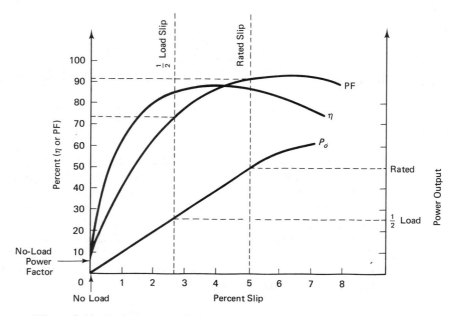

Figure 9-14 Typical curves of efficiency, power factor, and power output versus slip for a three-phase induction motor.

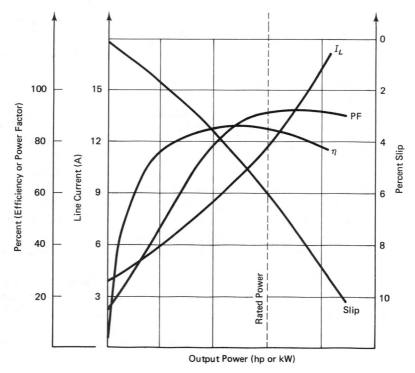

Figure 9-15 Typical picture of efficiency, power factor, slip, and line current versus output power.

9-7.2 Characteristics versus Power Output

Another common set of characteristics is shown in Figure 9-15. In this case efficiency, slip, power factor, and average line current are all plotted as a function of power output. Sometimes rotor speed is also plotted. Generally, a dashed line is drawn intersecting the abscissa at the rated power point.

As with Figure 9-14, the power factor has again been multiplied by 100. In this case a different percent axis is used for slip since its normal operating range is from 0 to 10%. If included, the speed can be plotted in either rev/min or rad/s. Its curve would be very similar to that plotted for slip.

9-7.3 Effect of the Wound Rotor on Characteristics

Often, the motor will be of the wound rotor type. In this way larger starting torques can be developed (see Figure 9-11). Increasing the resistance in the rotor circuit increases the starting torque. Once the motor comes up to speed, the resistance can be lowered. If it is not, the normal operating characteristics (those with a shorted rotor) will be adversely affected. Typical results are shown in Figure 9-16.

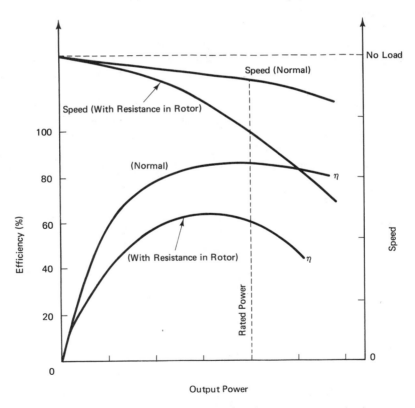

Figure 9-16 Typical effects of adding resistance to a rotor circuit.

9-8   STARTING TECHNIQUES

There are two basic ways that induction motors are started. The simplest is called **across-the-line starting.** In this technique, full (rated) voltage is applied to the motor at starting. This is done with switches or automatically with a pushbutton-controlled relay. However, there is a problem encountered with this method. A typical induction motor will draw a starting current from 5 to 10 times its full-load current when rated voltage is applied. Thus a 220-V motor with a full-load current of 10 A will take a starting current of 50 to 100 A when across-the-line starting is used. Induction motors are rugged and built to withstand this high surge of current. However, the lines supplying power to the motor may not be able to handle it. Consider a motor whose full-load current is 50 A. If started across the line, it could draw between 250 and 500 A instantaneously. When hit with this sudden excessive demand, the supply-line voltage might drop due to its imperfect regulation. This could cause all types of problems: from the momentary dimming of lights to the automatic disconnect of equipment due to undervoltage protection devices. With small motors this problem is not serious; however, with motors rated more than 5 hp (or 3.5 kW), something must be done to avoid it.

The second basic technique used for starting induction motors is called **reduced-voltage starting.** Since the induction motor at starting is a simple impedance (*RL* circuit), any reduction in the starting voltage will cause a proportionate reduction in starting current. A reduced starting voltage can be accomplished in a variety of ways.

In the laboratory the simplest way to reduce the starting voltage is to adjust it with a **three-phase variac.** This is nothing more than a hand-adjustable three-phase autotransformer.

In actual practice the techniques are more elaborate. One of these is called a **wye–delta starting connection,** which is shown in Figure 9-17. This can be used with a motor whose stator is normally wired in a delta connection. The stator windings are reconnected to switches as shown. The switches can be manual (triple-pole double-throw), or for automatic starters relays and solid-state switches (TRIAC). The TRIAC

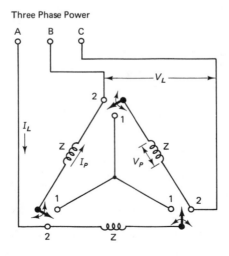

Figure 9-17 Wye-delta starting connection for a three-phase induction motor. Position 1, start; position 2, run.

eliminates the problem of contact corrosion that occurs in mechanical relays. The corrosion is due to oxidation caused by the switching of large currents in inductive loads.

Upon starting, the switches are in position 1. The stator is then in a wye connection. When the motor builds up speed (around 20% slip) the switches are thrown to position 2. The stator is now in a delta connection. The following analysis will show that the starting line current is reduced by a factor of 3. The starting current will be calculated in terms of the line voltage (V_L) and the impedance per phase (Z).

Δ connection:

$$I_p = \frac{V_L}{Z}$$

$$I_{L\Delta} = \sqrt{3}I_p = \frac{\sqrt{3}V_L}{Z}$$

Y connection:

$$V_p = \frac{V_L}{\sqrt{3}}$$

$$I_p = \frac{V_p}{Z} = \frac{V_L}{\sqrt{3}Z}$$

$$I_{LY} = I_p = \frac{V_L}{\sqrt{3}Z}$$

If we take the ratio of the starting current for the Y connection to that for the Δ connection, we get

$$\frac{I_{LY}}{I_{L\Delta}} = \frac{V_L/(\sqrt{3}Z)}{\sqrt{3}V_L/Z} = \frac{1}{\sqrt{3}\ \sqrt{3}} = \frac{1}{3}$$

or

$$I_{LY} = \tfrac{1}{3}I_{L\Delta}$$

Thus we can see that the starting line current for the Y connection will be one-third its value for the Δ connection.

Note that since the current is reduced, the torque will also be reduced. By starting out with a Y connection the voltage across a stator winding will be $V_L/\sqrt{3}$. From Eq. 9-3 we know that starting torque is proportional to the square of the line voltage. Thus the torque would be reduced by a factor of 3. The effect of this starting technique on the torque–slip characteristic is shown in Figure 9-18.

At start, we are at point 1 on the curve. As the motor builds up speed we get to point 2, where the switching to Δ takes place. There is a sudden increase in torque to point 3 and the motor accelerates to point 4 if rated load is applied. Note that if the switching did not take place, we would have stayed on the lower curve and ended up at point 5. The motor would end up running at a slower speed with a slip greater than rated.

Another way of starting with reduced voltage is to use a transformer whose secondary is an open delta, as shown in Figure 9-19. The secondary tap adjusts the starting voltage from one-fourth to one-half of its rated value.

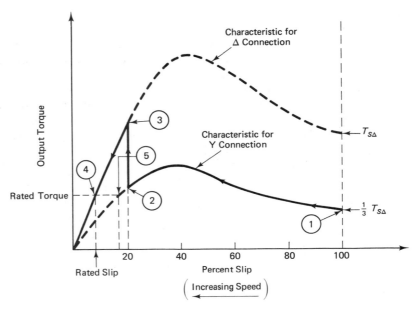

Figure 9-18 Torque–slip curve for the Y-Δ starting technique.

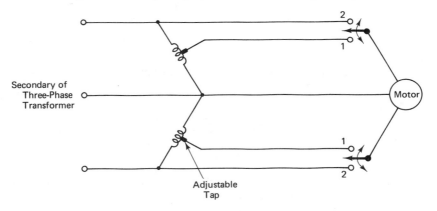

Figure 9-19 Typical wiring using an open delta starting compensator.

The motor is started with the switches (or relays) in position 1. Once the motor builds up speed, the switches are thrown to position 2 and the motor will now run at its rated conditions. A transformer used for this special application (reduced voltage starting) is called a **starting compensator.** It should be kept in mind that reduced voltage starting is not desirable in applications requiring high starting torques.

The problem of reduced starting torque is solved by using a wound rotor induction motor together with reduced voltage starting. As we have just seen, the reduced voltage lowers our starting current. If at starting a large resistance is added to the rotor using a slip-ring rheostat, the starting torque will be increased as shown in Figure 9-11. Once the motor builds up speed, the voltage is restored to its rated value and the rotor resistance is eliminated by setting the rheostat to zero (shorted rotor).

SYMBOLS INTRODUCED IN CHAPTER 9

Symbol	Definition	Units: English and SI
S_s	Synchronous speed in English system	rev/min
ω_s	Synchronous speed in SI	rad/s
I_R	Rotor current	amperes
I_L	Line current	amperes
ϕ	Stator field flux	lines or webers
θ_R	Rotor power factor angle	degrees
T	Torque developed in rotor	lb-ft or N-m
T_s	Starting torque	lb-ft or N-m
V_L	Applied stator line voltage	rms volts
s	Slip	rpm, rad/s, percent, or decimal
S	Speed of rotor in English system	rev/min
ω	Speed of rotor in SI	rad/s
SR	Speed regulation	percent
f	Frequency of applied stator voltage	hertz
f_R	Frequency of induced rotor current	hertz
L_R	Rotor inductance	henries
X_R	Rotor reactance	ohms
X_{BR}	Rotor reactance at standstill	ohms
R_R	Rotor resistance	ohms
Z_R	Rotor impedance	ohms
Z	Effective stator impedance per phase	ohms
E_R	Induced rotor voltage at any speed	volts
E_{BR}	Induced rotor voltage at standstill	volts
s_{po}	Slip at pull-out	decimal
W_1, W_2	Wattmeter readings	watts
P_R	Power transferred across air gap into rotor	watts
P_C	Stator core loss	watts
P_{CuS}	Stator copper loss	watts
P_{CuR}	Rotor copper loss	watts
P_M	Mechanical (friction and windage) loss of motor	watts
r_s	Resistance between any two stator lines	ohms
r_e	Effective stator resistance	ohms

QUESTIONS

1. Define the following terms: squirrel-cage rotor; double-squirrel-cage rotor; wound rotor; slip-ring rheostat.

2. With reference to rotor conductors, what is skewing?

3. What is the name given to the speed of the rotating stator field? Does this also refer to the motor shaft speed?

4. What is meant by the term "blocked rotor"?

5. Why is an induction motor an asynchronous machine?

6. What is slip? Is it the same as speed regulation?

7. When is the frequency of the rotor current equal to the line frequency? Can it ever be 0 Hz?

8. What is meant by the term "pull-out"?

9. What are the two basic techniques used to start induction motors?

10. Define the terms "three-phase variac" and "starting compensator."

11. For a given set of load conditions, what will be the effect of an increase in rotor resistance on the following quantities: rotor speed; pull-out speed; starting torque; pull-out torque of motor?

PROBLEMS

(English)

1. What is synchronous speed for a three-phase 60-Hz induction motor if it has:
(a) 2 poles?
(b) 4 poles?
(c) 8 poles?
(d) 12 poles?
(e) 32 poles?

2. A 1-hp 240-V 60-Hz three-phase induction motor runs at a speed of 1100 rev/min. How many poles does it have?

3. A three-phase 250-V 5-hp 60-Hz induction motor develops a starting torque of 30 ft-lb when rated voltage is applied. Find:
(a) The starting torque if the voltage drops to 230 V
(b) The line voltage needed to produce a starting torque of 35 ft-lb

4. If the motor in Problem 3 draws 12 A at a power factor of 0.80 lagging, what is its efficiency?

5. A 210-V 2-hp 60-Hz three-phase induction motor has a full-load speed of 550 rev/min. It develops a starting torque of 25 ft-lb. Find:
(a) The full-load output torque
(b) The starting torque developed if the line voltage becomes 230 V

6. If the motor in Problem 5 has a power factor of 0.90 lagging and is 82% efficient at full load, how much current should it draw?

7. An eight-pole 240-V 5-hp 60-Hz 790-rev/min three-phase induction motor has a no-load slip of 1%. Find:
(a) The full-load slip in percent
(b) The speed at which it will run if all the load is removed

8. A three-phase 60-Hz induction motor has a full-load speed of 400 rev/min. Find its rated slip;
(a) In rev/min
(b) As a decimal
(c) In percent

9. A three-phase 60-Hz induction motor is running at a speed of 1720 rev/min. An increase in load causes the slip to double. What is its new speed with the increased load?

10. A three-phase 50-Hz induction motor has a no-load speed of 1490 rev/min and a full-load slip of 7%. Find:

(a) The number of poles it has

(b) The full-load speed

11. A 12-pole 60-Hz three-phase induction motor has a full-load slip of 6%. Find its full-load speed.

12. An eight-pole 60-Hz three-phase induction motor has a no-load speed of 895 rev/min and a full-load speed of 830 rev/min. Find:

(a) The percent slip at full load

(b) The percent speed regulation

13. A four-pole 60-Hz three-phase induction motor has a no-load slip of 0.5% and a full-load slip of 4%. What is its percent speed regulation?

14. A two-pole 60-Hz three-phase induction motor has a no-load slip of 100 rev/min. If its speed regulation is 12%, what is its full-load slip in rev/min?

15. A 60-Hz three-phase induction motor has a rotor resistance of 0.1 Ω and blocked rotor reactance of 0.4 Ω. Find:

(a) The rotor impedance at standstill

(b) The rotor impedance at 10% slip

(c) The rotor impedance at 1% slip

16. For any 60-Hz three-phase induction motor, what is the frequency of:

(a) The stator current

(b) The rotor current at standstill

(c) The rotor current at 5% slip

(d) The rotor current at 0.5% slip

17. A 10-pole 60-Hz three-phase induction motor has a full-load speed of 660 rev/min. How fast will it run at:

(a) Half rated load

(b) One-fourth of rated load

18. A six-pole 60-Hz three-phase induction motor has a full-load speed of 1140 rev/min. If the load increases by 20%, what will be the new speed of the motor?

19. A four-pole 60-Hz three-phase induction motor is running at 1690 rev/min under full-load conditions. After some load is removed, the speed builds up to 1725 rev/min. At which percent of rated load is the motor now running?

20. A 208-V 5-hp 60-Hz three-phase induction motor has a full-load speed of 830 rev/min. It is running under a light-load condition at 860 rev/min. Find:

(a) The rated torque of the motor

(b) The load torque at the light-load condition

(c) The output power at the light-load condition

21. A four-pole 60-Hz 10-hp three-phase induction motor has the following rotor specifications; $R_R = 0.05$ Ω and $X_{BR} = 0.25$ Ω. Find:

(a) The percent slip at pull-out

(b) The slowest speed at which the motor will run continuously without stalling

22. A six-pole 60-Hz 5-hp three-phase induction motor has a full-load speed of 1120 rev/min. If $R_R = 0.07$ Ω and $X_{BR} = 0.3$ Ω, find:

(a) The full-load output torque

(b) The maximum torque of the motor (assume for this part that torque is proportional to slip all the way up to the pull-out point)

23. A four-pole 60-Hz three-phase wound rotor induction motor has the following specifications with the rotor shorted: $R_R = 0.08$ Ω, $X_{BR} = 0.35$ Ω, and a full-load speed of 1700 rev/min. A slip-ring rheostat is used to add 0.12 Ω to each phase. Find:

(a) The shorted rotor pull-out speed

(b) The pull-out speed with the added rotor resistance

(c) The full-load speed with the added rotor resistance

24. A 12-pole 60-Hz three-phase wound rotor induction motor has $R_R = 0.03\ \Omega$ and $X_{BR} = 0.2\ \Omega$ with the rotor shorted. Find:

(a) The pull-out speed with the rotor shorted

(b) The amount of resistance that must be added per phase with a slip-ring rheostat to reduce the pull-out speed to 350 rev/min.

25. A 10-hp 240-V 25-A eight-pole 60-Hz three-phase induction motor is tested to determine its losses and efficiency at full load. The results are shown below.

$$\text{Stator resistance test:} \quad r_s = 0.5\ \Omega$$

$$\text{No-load test:} \quad W_1 = 660\ \text{W}, \quad W_2 = -240\ \text{W}$$

$$V_L = 240\ \text{V}, \quad I_L = 10\ \text{A}$$

$$\text{Rated-load test:} \quad W_1 = 5.2\ \text{kW}, \quad W_2 = 4.9\ \text{kW}$$

$$V_L = 240\ \text{V}, \quad I_L = 28\ \text{A}$$

$$S = 860\ \text{rev/min}$$

Find:

(a) The full-load efficiency

(b) The full-load power factor

26. A 1-hp 210-V 4-A six-pole 60-Hz three-phase induction motor is tested to determine its losses and efficiency at full load. The results are shown below.

$$\text{Stator resistance test:} \quad r_s = 1.0\ \Omega$$

$$\text{No-load test:} \quad W_1 = 200\ \text{W}, \quad W_2 = -120\ \text{W}$$

$$V_L = 210\ \text{V}, \quad I_L = 0.5\ \text{A}$$

$$\text{Rated-load test:} \quad W_1 = 580\ \text{W}, \quad W_2 = 550\ \text{W}$$

$$V_L = 210\ \text{V}, \quad I_L = 3.5\ \text{A}$$

$$S = 1120\ \text{rev/min}$$

Find:

(a) The full-load efficiency

(b) The full-load power factor

(SI)

27. What is synchronous speed for a three-phase 50-Hz induction motor if it has:

(a) 2 poles?

(b) 4 poles?

(c) 6 poles?

(d) 16 poles?

(e) 36 poles?

28. A 1-kW 230-V 50-Hz three-phase induction motor runs at a speed of 95 rad/s. How many poles does it have?

29. A three-phase 230-V 5-kW 50-Hz induction motor develops a starting torque of 45 N-m when rated voltage is applied. Find:
(a) The starting torque if the voltage drops to 210 V
(b) The line voltage needed to produce a starting torque of 60 N-m

30. If the motor in Problem 29 draws 16.5 A at a power factor of 0.88 lagging, what is its efficiency?

31. A 230-V 1.5-kW 50-Hz three-phase induction motor has a full-load speed of 50 rad/s. It develops a starting torque of 40 N-m. Find:
(a) The full-load output torque
(b) The starting torque developed if the line voltage becomes 250 V

32. If the motor in Problem 31 has a power factor of 0.92 lagging and is 86% efficient at full load, how much current should it draw?

33. A six-pole 460-V 10-kW 50-Hz 98-rad/s three-phase induction motor has a no-load slip of 1%. Find:
(a) The full-load slip in percent
(b) The speed at which it will run if all the load is removed

34. A three-phase 50-Hz induction motor has a full-load speed of 35 rad/s. Find its rated slip:
(a) In rad/s
(b) As a decimal
(c) In percent

35. A three-phase 50-Hz induction motor is running at a speed of 152 rad/s. An increase in load causes the slip to double. What is its new speed with the increased load?

36. A three-phase 60-Hz induction motor has a no-load speed of 92 rad/s and a full-load slip of 5.5%. Find:
(a) The number of poles it has
(b) The full-load speed

37. A 36-pole 50-Hz three-phase induction motor has a full-load slip of 7.5%. Find its full-load speed.

38. A six-pole 50-Hz three-phase induction motor has a no-load speed of 103 rad/s and a full-load speed of 96 rad/s. Find:
(a) The percent slip at full load
(b) The percent speed regulation

39. A four-pole 50-Hz three-phase induction motor has a no-load slip of 0.7% and a full-load slip of 5.5%. What is its percent speed regulation?

40. A two-pole 50-Hz three-phase induction motor has a no-load slip of 3 rad/s. If its speed regulation is 10%, what is its full-load slip in rad/s?

41. A 50-Hz three-phase induction motor has a rotor resistance of 0.08 Ω and blocked rotor reactance of 0.35 Ω. Find:
(a) The rotor impedance at standstill
(b) The rotor impedance at 12% slip
(c) The rotor impedance at 0.5% slip

42. For any 50-Hz three-phase induction motor, what is the frequency of:
(a) The stator current
(b) The rotor current at standstill
(c) The rotor current at 4% slip
(d) The rotor current at 0.4% slip

43. A six-pole 50-Hz three-phase induction motor has a full-load speed of 100 rad/s. How fast will it run at:
(a) Half-rated load?
(b) One-fourth of rated load?

44. A four-pole 50-Hz three-phase induction motor has a full-load speed of 150 rad/s. If the load increases by 25%, what will be the new speed of the motor?

45. An eight-pole 50-Hz three-phase induction motor is running at 74 rad/s under full-load conditions. After some load is removed, the speed builds up to 76 rad/s. At what percent of rated load is the motor now running?

46. A 460-V 10-kW 50-Hz three-phase induction motor has a full-load speed of 48 rad/s. It is running under a light-load condition at 50 rad/s. Find:
(a) The rated torque of the motor
(b) The load torque at the light-load condition
(c) The output power at the light-load condition

47. A six-pole 50-Hz 7.5-kW three-phase induction motor has the following rotor specifications; $R_R = 0.07\ \Omega$ and $X_{BR} = 0.3\ \Omega$. Find:
(a) The percent slip at pull-out
(b) The slowest speed at which the motor will run continuously without stalling

48. A four-pole 50-Hz 10 kW three-phase induction motor has a full-load speed of 151 rad/s. If $R_R = 0.04\ \Omega$ and $X_{BR} = 0.2\ \Omega$, find:
(a) The full-load output torque
(b) The maximum torque of the motor (assume for this part that torque is proportional to slip all the way up to the pull-out point)

49. A six-pole 50-Hz three-phase wound rotor induction motor has the following specifications with the rotor shorted: $R_R = 0.1\ \Omega$, $X_{BR} = 0.5\ \Omega$, and full-load speed of 98 rad/s. A slip-ring rheostat is used to add 0.15 Ω to each phase. Find:
(a) The shorted rotor pull-out speed
(b) The pull-out speed with the added rotor resistance
(c) The full-load speed with the added rotor resistance

50. A four-pole 50-Hz three-phase wound rotor induction motor has $R_R = 0.1\ \Omega$ and $X_{BR} = 0.45\ \Omega$ with the rotor shorted. Find:
(a) The pull-out speed with the rotor shorted
(b) The amount of resistance that must be added per phase with a slip-ring rheostat to reduce the pull-out speed to 110 rad/s

51. A 5-kW 230-V 16-A four-pole 50-Hz three-phase induction motor is tested to determine its losses and efficiency at full load. The results are shown below.

$$\text{Stator resistance test:} \quad r_s = 1.1\ \Omega$$

No-load test:	$W_1 = 450$ W,	$W_2 = -150$ W
	$V_L = 230$ V,	$I_L = 5$ A
Rated-load test:	$W_1 = 3.2$ kW,	$W_2 = 2.5$ kW
	$V_L = 230$ V,	$I_L = 16$ A
	$\omega = 149.2$ rad/s	

Find:
(a) The full-load efficiency
(b) The full-load power factor

52. A 10-kW 550-V 14.5-A eight-pole 50-Hz three-phase induction motor is tested to determine its losses and efficiency. The results are shown below.

$$\text{Stator resistance test:} \quad r_s = 1.0 \ \Omega$$

$$\text{No-load test:} \quad W_1 = 920 \text{ W}, \quad W_2 = -260 \text{ W}$$

$$V_L = 550 \text{ V}, \quad I_L = 4.7 \text{ A}$$

$$\text{Rated-load test:} \quad W_1 = 6.3 \text{ kW}, \quad W_2 = 5.4 \text{ kW}$$

$$V_L = 550 \text{ V}, \quad I_L = 14.5 \text{ A}$$

$$\omega = 73.9 \text{ rad/s}$$

Find:
(a) The full-load efficiency
(b) The full-load power factor

Chapter 10

The Three-Phase Synchronous Motor

The basic theory of operation and construction of the synchronous motor is the same as that presented for the synchronous alternator in Chapter 8. There are a couple of minor differences. The rotors in synchronous motors are usually of the salient-pole type (see Figure 8-2). In addition, the windings are slightly different to provide a damping effect. Just as the dc shunt machine, a three-phase alternator can be used as a motor as well.

10-1   CONSTRUCTION

The **stator** of the motor is wound the same as the synchronous alternator and the induction motor. It must be wound for the same number of poles as the rotor. The rotor winding can be similar to that of an alternator. Sometimes, however, the rotor winding might be on a squirrel-cage rotor (see Figure 9-1). In this case there are two rotor circuits. We shall see in Section 10-3 that the function of the squirrel-cage rotor is to start the motor. Once it builds up speed, the other rotor winding will take effect and determine the motor's running characteristics. With today's advanced techniques, some of the smaller motors are made with a permanent-magnet rotor, thus eliminating the need for external excitation.

In some extremely rare cases the stator is not stationary but free to rotate. When this type of motor (called **supersynchronous**) is started, the stator actually builds up to a high speed. Then by gradually applying a brake to the stator, it will slow down. As it slows down the rotor starts to turn, gaining the speed that the stator is losing due to the braking action. Once the stator becomes stationary, the rotor attains its rated speed. This type of motor provides a higher-than-usual starting torque. However, it is very rare, and most high-torque applications use the induction motor.

When the stator winding is energized with a three-phase voltage, a stator field will be formed. Since the stator is wound in the same way as an induction motor, the stator field in a synchronous motor will also rotate at synchronous speed (see Section 9-2). The rotor is also energized so that it too has a magnetic field. We will refer to Figure 10-1 to illustrate the motor's behavior.

The rotor is initially at rest. The only thing moving is the stator field, which is shown rotating in a clockwise direction. As the stator north pole passes the rotor south pole, there will be a force of attraction between them. Thus the rotor will try to rotate in a clockwise direction also. However, before the rotor starts to move, the stator field will have rotated 180°. Its north pole will now be at the bottom of Figure 10-1. In this position it will repel the rotor (remember, like polarities repel) and try to rotate it counterclockwise. Thus we can see that during one revolution of the stator field, half the time it will try to turn clockwise and the other half it will try to turn counterclockwise. The net result is that the rotor will not turn at all. It is for this reason that synchronous motors do not develop a starting torque. Something additional is needed to start the motor. In Section 10-3 we will examine some of the starting techniques.

If, on the other hand, the rotor is brought up to a high speed by some other means, it will **lock** with the rotating stator field. Under these conditions there is a running torque developed. The rotor and stator will both rotate at synchronous speed in a direction determined by the stator field. Thus the direction of rotation can be reversed by interchanging any two of the supply lines connected to the stator windings (see Section 9-2.1).

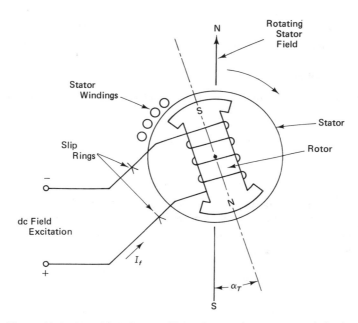

Figure 10-1 Simplified diagram illustrating synchronous motor behavior.

While the motor is running, the two rotating fields will not line up perfectly. The rotor pole will always lag behind the stator pole by some angle. This angle is called the **torque angle** and is shown as α_T in Figure 10-1. As the load on the motor's shaft increases, the torque angle will increase; however, the rotor will continue to turn at synchronous speed. This behavior continues until the torque angle is approximately 90°. At that point the motor is developing maximum torque. Any further increase in load will cause either of the following to occur:

1. If the increase in load is momentary or very little, the motor will **slip a pole.** In other words, the stator field will lose hold of the rotor and grab onto it again the next time around.
2. If the increase in load is big enough and not momentary, the motor will **lose synchronism** and stall.

In both cases above, a noticeable straining sound can be heard. It will not be derived here; however, this critical torque angle corresponding to a maximum torque is equal to the **synchronous impedance angle** (β) of the machine. If we refer to Section 8-5, we can see that the synchronous impedance (Z_s) is made up of resistance (r_p) and reactance (x_p). Thus Eq. 10-1 can be written

$$\beta = \arctan \frac{x_p}{r_p} \tag{10-1}$$

Since x_p is much bigger than r_p, β will be very close to 90°.

Example 10-1

The synchronous machine of Example 8-6 is run as a synchronous motor. What is the torque angle corresponding to a maximum developed torque?

Solution

From the information given in Example 8-6, $r_p = 0.12 \ \Omega$ and $x_p = 1.5 \ \Omega$. From Eq. 10-1,

$$\text{maximum } \alpha_T = \beta = \arctan \frac{1.5 \ \Omega}{0.12 \ \Omega}$$

$$= 85.43°$$

In light of the preceding discussion, it should be clear that since the two fields lock up, a synchronous motor must always run at synchronous speed, even with variations in load. It therefore has perfect (0%) speed regulation. It is not well suited in applications that require variable speeds.

10-2.1 Effect of Field Excitation

To help understand the effect of field excitation, we will refer to an equivalent circuit of one phase of the stator winding (Figure 10-2) and three vector diagrams (Figure

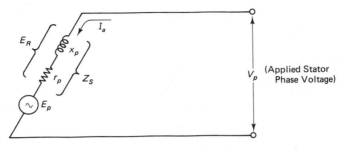

Figure 10-2 Diagram of one stator phase winding.

10-3). Note that Figure 10-2 is the same as that used for the synchronous alternator (Figure 8-5).

When the motor is running, the rotor field will be cutting the stator conductors. As a result, there will be a voltage (E_p) generated in the stator winding which opposes the applied phase voltage (V_p). This is very similar to the back EMF in a dc motor. If the rotor and stator fields were perfectly lined up ($\alpha_T = 0°$ in Figure 10-1), E_p and V_p would be 180° out of phase. However, since they are not, their phase difference will be less than 180° by an amount equal to α_T. The generated voltage, E_p, is the same

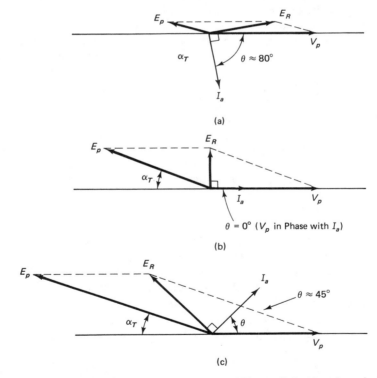

Figure 10-3 Vector diagrams for the circuit of Figure 10-2: (a) underexcited, lagging power factor; (b) normally excited, unity power factor; (c) overexcited, leading power factor.

as the one given by Eqs. 8-3 for the alternator. From those equations we can see that if the motor is running at constant speed (and it must), E_p will be a function of the rotor field flux (ϕ). In other words, the generated voltage (E_p) in Figure 10-2 is a function of the excitation (I_f) in Figure 10-1.

Referring to Figure 10-2, the resultant voltage (E_R) across the motor's impedance is given by the vector sum of V_p and E_p. Furthermore, the stator current (I_a) is, from Ohn's law, E_R divided by the impedance (Z_s). Since x_p is much greater than r_p, Z_s is basically inductive and I_a will lag E_R by about 90°.

The three diagrams in Figure 10-3 are all for the same load condition. If we neglect any change in the internal losses, the torque angle (α_T) is the same in each case. The field excitation in each case is different, thus producing a different E_p.

In Figure 10-3a the excitation is very small. Vector addition shows E_R almost in phase with V_p. The current I_a lags E_R by 90°; therefore, it will also lag V_p by a large angle θ (about 80°). Since the power factor is the cosine of the angle between V_p and I_a, it will be very small (PF = cos 80° = 0.17). Furthermore, it is a lagging power factor. This is called an **underexcited** condition.

If the excitation (hence E_p) is increased to a critical value, as shown in Figure 10-3b, the resultant voltage E_R will lead V_p by about 90°. Since I_a lags E_R by 90°, it will fall directly on top of V_p. Hence for this condition the angle between I_a and V_p is zero, and the power factor is unity (PF = cos 0° = 1.0). This is called a **normally excited** condition.

Finally, in Figure 10-3c we have an **overexcited** condition. The voltage E_p is so large that the resultant E_R leads V_p by more than 90°. Under this condition it can be seen that I_a actually leads the applied voltage V_p. For this overexcited case the power factor is therefore a leading one (PF = cos 45° = 0.707).

This analysis brings out one of the most important properties of a synchronous motor: By adjusting its excitation it can be made to operate at almost any power factor (lagging or leading) for a given load.

We should also note in Figure 10-3 that as the excitation increased, E_R went from a large value to a minimum (at unity power factor) and then increased again to a large value. This would cause a similar behavior with the armature current I_a.

10-2.2 Effect of Load

By further examination of Figure 10-3b we can make the following observations. If the load were to increase, α_T would increase as well. The resultant E_R would rotate toward V_p, causing I_a to rotate by the same angle. The power factor would no longer be unity. At this increased load, in order to get unity power factor again, E_p would have to be increased by increasing the excitation. This would rotate E_R back to where it is shown in Figure 10-3b.

Still referring to Figure 10-3b, there is one final note to be made. With the excitation constant (E_p does not change), an increase in load will increase α_T, rotating E_p clockwise. This will increase the resultant voltage E_R, causing a corresponding increase in I_a. We have just verified the basic motor equation (Eq. 5-1) presented in Chapter 5; that is, an increase in load (torque) is accompanied by an increase in stator ("armature") current I_a.

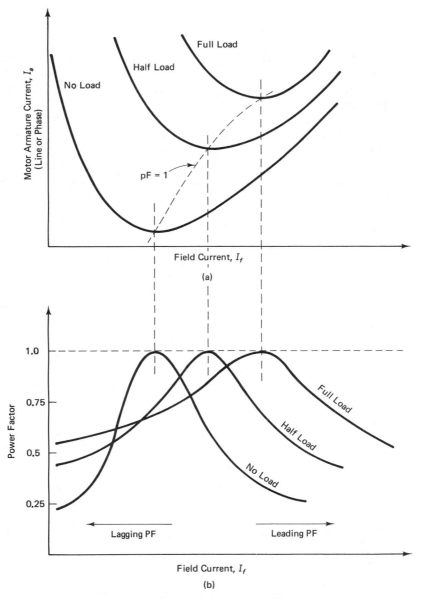

Figure 10-4 Synchronous motor V curves.

10-2.3 Synchronous Motor V Curves

The observations made in Sections 10-2.1 and 10-2.2 can easily be seen on a set of curves commonly plotted for a synchronous motor. One of these, Figure 10-4a, is a plot of stator current versus field excitation I_f (note that ϕ is proportional to I_f). It has the shape of a V. The other shown in Figure 10-4b is a plot of power factor versus field excitation. It has the shape of an upside-down V.

   It might be helpful now if we go back and read Sections 10-2.1 and 10-2.2, this time relating the observations made to Figure 10-4. Theoretically, the minimum line current at a given load and field excitation occurs when the power factor is unity. It is shown this way in Figure 10-4. However, practically speaking, the excitation for minimum line current might not give unity power factor. When the curves are obtained experimentally in the laboratory, this deviation of the maximum and minimum points might be due to a measurement error or the accuracy of a meter. However, this error is often caused by a phenomenon called **harmonic distortion.** This type of distortion occurs when the generated voltage E_p and/or the applied voltage V_p are not simple sine waves. Because of the way they are generated, E_p and V_p might contain small amplitudes of other sine waves. These other sine waves are called harmonics. Their frequencies are integer multiples of the fundamental frequency (50 or 60 Hz in our case).

Example 10-2

A 210-V 2-hp three-phase synchronous motor is tested in the laboratory with the arrangement shown in Figure 10-5. Data was taken at no load and at about one-half rated load (1 hp or 750 W). The meter readings are shown in Tables 10-1 and 10-2. The quantity I represents the average of the three line currents I_1, I_2, and I_3. Plot on one graph the no-load and half-load V curves for the motor.

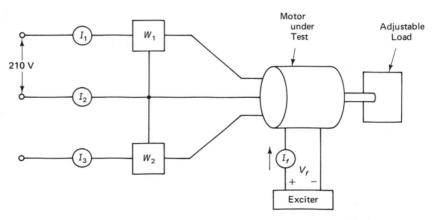

Figure 10-5 Test setup to determine a synchronous motor's V curves.

Table 10-1 No load

I_f (A)	I (A)	W_1 (W)	W_2 (W)	PF
3	5	680	−250	0.24
6	3.5	500	−100	0.31
7.5	2.6	400	0	0.42
9	1.8	300	60	0.55
11	1.05	190	190	0.995
12	1.25	150	230	0.84
13	1.55	100	300	0.71
15	2.4	−20	430	0.47
16	3	−100	500	0.37
17.5	4.4	−200	600	0.25

Table 10-2 Half load

I_f (A)	I (A)	W_1 (W)	W_2 (W)	PF
6.5	4.95	830	180	0.56
8.5	4.02	680	310	0.68
10	3.48	590	410	0.79
12	2.8	480	510	0.97
12.5	2.9	430	570	0.95
14	3.5	350	670	0.80
16	4.8	240	800	0.60
17.4	6.0	150	930	0.49

Solution

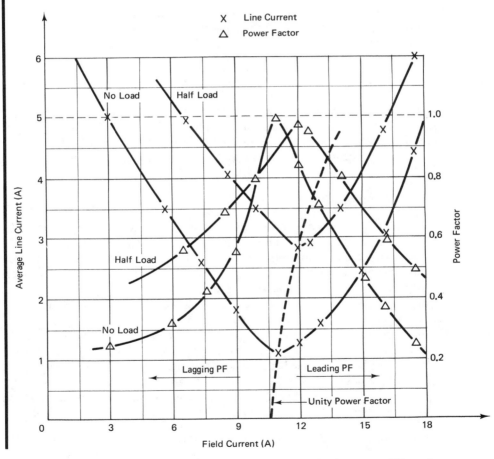

Figure 10-6 Experimentally determined V curves for the motor of Example 10-2.

The average line current versus field excitation can be plotted directly. The power factor, however, must be calculated using Eq. 7-51. The two-wattmeter method (Section 7-4.2) was used to determine total power input.

$$PF = \cos\theta = \frac{W_1 + W_2}{\sqrt{3}\,V_L I_L}$$

A sample calculation is shown below at no load and $I_f = 3$ A.

$$PF = \frac{680\text{ W} - 250\text{ W}}{\sqrt{3} \times 210\text{ V} \times 5\text{ A}} = 0.24$$

The remaining power factors were calculated in a similar fashion and are shown in Tables 10-1 and 10-2. The results are shown in Figure 10-6. It is interesting to note that when the power factor was unity (or close to it) in each case, the two-wattmeter readings were equal and positive. This is in accordance with the theory presented in Section 7-4.2.

A synchronous motor's V curves are very useful in selecting the appropriate field current. In this way the motor can be put to use with whatever power factor we desire. We will see in Section 10-6 how this can be applied in power factor correction.

10-3 SYNCHRONOUS MOTOR STARTING TECHNIQUES

It has already been pointed out that a synchronous motor does not develop a starting torque. Some other means must be used to start the motor and bring it up to a high speed. Once it is running close to synchronous speed and its field is excited, it will pull into synchronism. There are two basic techniques used to start synchronous motors.

10-3.1 Auxiliary Drive

In cases where the synchronous motor can be disconnected from the load, a small motor connected to the synchronous motor's shaft can be used to bring it up to speed. At this point the synchronous motor is excited, it pulls into synchronism, and power is disconnected from the small auxiliary motor. The load can then be connected to the synchronous motor. Since at starting the small motor only has to turn an unloaded large motor, its rating need only be about 10% of the synchronous motor's rating. In other words, a 1-hp (or 1-kW) motor would be used to start a 10-hp (or 10-kW) synchronous motor. The small motor is usually an induction or dc motor. If the former is the case, it will probably have two fewer poles than the synchronous motor. In this way its full-load speed will be close to the main motor's synchronous speed.

If a dc motor is used, its field can be adjusted to bring the main motor up to speed. An advantage of the dc motor is that it can be used as a **dynamic brake** to stop the main motor quickly (see Section 5-11).

10-3.2 Induction Start–Synchronous Run

This technique requires a special kind of synchronous motor. It is special in that its field winding is placed on a squirrel cage or wound rotor. In other words, it has two rotor windings. When three-phase power is applied to the stator (the field is not yet energized) the motor starts as an induction motor. It builds up speed to about 5 or 10% slip. At that point the field is energized and the motor, turning at a high speed, pulls into synchronism. When this occurs the rotor is turning at synchronous speed. The squirrel-cage winding will not be generating any currents and therefore is not affecting the synchronous motor's behavior. These additional rotor windings are called **amortisseur windings.** They serve a purpose in addition to starting the motor. Sometimes when the load changes frequently, the motor's speed is not steady as the torque angle oscillates (or **hunts**) back and forth trying to settle at its required value. This momentary change in speed creates a current, hence torque in the amortisseur windings. The momentary torque tends to stabilize (or damp) the oscillating torque angle. Therefore, these additional windings are sometimes called **damper windings.**

A special case of this type of motor is the **reluctance** (or **synchronous-induction**) motor. It is a synchronous motor that does not need dc field excitation. The rotor is basically a squirrel cage but it is not round; it has a salient pole appearance, as shown in Figure 10-7.

The motor starts like an induction motor. The rotor field is generated by the rotating stator flux. When it builds up to a high speed (low slip), the definite poles of the rotor lock up with the stator field. Obviously, both must have the same number of poles. Once it pulls into synchronism, the rotor maintains a small residual flux. It is therefore running like an underexcited synchronous motor. As such it will have a fixed power factor which is very low and lagging (see Figure 10-4b).

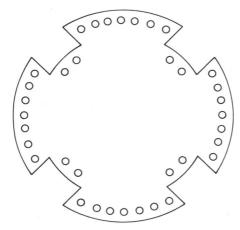

Figure 10-7 Squirrel-cage rotor with salient pole shape.

10-4 NUMERICAL ANALYSIS: POWER, EFFICIENCY, AND TORQUE

The synchronous motor will now be analyzed with some basic equations from preceding chapters. In some cases approximations will be made and noted. This will reduce the need to analyze complicated vector diagrams. For power factors in the normal operating range the approximations will introduce a small error (5% or less). The simplicity gained in the analysis, however, is well worth the slight loss of accuracy.

10-4.1 Power and Efficiency

Since we are dealing with a motor, the output power (a mechanical quantity) is difficult to measure. It will therefore be treated in the same way as the other motors; that is, the output power will be determined by subtracting the internal losses from the input; hence Eqs. 9-21 through 9-23 will be used. The losses will be described by examining the power flow diagram shown in Figure 10-8. Note that one loss not shown is the copper loss in the field winding. It is supplied by the ac source; however, the exciter must be included in efficiency calculations.

If the power factor is known, the input power can be calculated using Eq. 9-24; otherwise, the two-wattmeter method and Eq. 9-25 are used. If the stator copper loss (P_{Cu}) is subtracted from the input, we are left with the developed power (P_{dev}).

$$P_{Cu} = 3r_p I_a^2 \tag{10-2}$$

$$P_{dev} = P_i - P_{Cu} \tag{10-3}$$

Equation 10-2 is obvious by referring to Figure 10-2 and noting that there are three phases.

Although the core losses are distributed in both stator and rotor, we will combine them with the mechanical (friction and windage) losses of the rotor. We will call the total P_{MC}. If the mechanical and core losses are now subtracted from the developed power, we are left with the mechanical power out of the machine.

$$P_o = P_{dev} - P_{MC} \tag{10-4}$$

Substituting Eqs. 10-3 and 10-2 into Eq. 10-4, we obtain

$$P_o = P_i - P_{Cu} - P_{MC} \tag{10-5}$$

From Eq. 10-4 we can see that when P_o is zero (no-load run), the developed power is equal to the mechanical and core losses. Hence P_{MC} will be determined from a no-load run of the motor under test and we will assume that it is the same at full load.

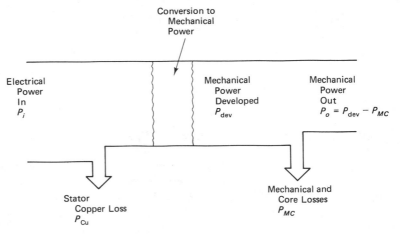

Figure 10-8 Power flow diagram for a synchronous motor.

The field loss (P_f) can be calculated from meter readings as shown in Figure 10-5 where R_f is the field resistance.

$$P_f = V_f I_f = R_f I_f^2 \tag{10-6}$$

Substituting the equations above into Eq. 9-23, we can develop an equation for the synchronous motor's efficiency:

$$\eta(\%) = \frac{P_i - P_{Cu} - P_{MC} - P_f}{P_i} \times 100 \tag{10-7}$$

Example 10-3

A 210-V 8-A three-phase Y-connected synchronous motor is tested as shown in Figure 10-5 to determine its efficiency. Its resistance and reactance per phase are 0.6 Ω and 6.3 Ω, respectively. The motor is underexcited at a constant field excitation of 10 A. The field resistance is 1.5 Ω. The test results are given below.

No-load test: $V_L = 210$ V, $I_L = 1.6$ A (average)

                 $W_1 = 300$ W, $W_2 = -30$ W

Full-load test: $V_L = 210$ V, $I_L = 8$ A (average)

                 $W_1 = 1400$ W, $W_2 = 1200$ W

Find:

(a) The no-load power factor

(b) The full-load power factor

(c) The full-load efficiency

Solution

(a) From the no-load results,

$$P_i = W_1 + W_2 = 300 \text{ W} - 30 \text{ W} = 270 \text{ W}$$

Rearranging Eq. 9-24, we have

$$PF = \cos \theta = \frac{270 \text{ W}}{\sqrt{3} \times 210 \text{ V} \times 1.6 \text{ A}}$$

no-load PF = 0.46 lagging (since it is underexcited)

(b) Repeating the above but using the full-load test data, we have

$$P_i = 1400 \text{ W} + 1200 \text{ W} = 2600 \text{ W}$$

$$PF = \frac{2600 \text{ W}}{\sqrt{3} \times 210 \text{ V} \times 8 \text{ A}}$$

$$= 0.89 \text{ lagging (since it is underexcited)}$$

(c) At no load the stator copper loss is, from Eq. 10-2,

$$P_{Cu} = 3 \times 0.6 \ \Omega \times 1.6 \text{ A}^2 = 4.6 \text{ W}$$

and from Eq. 10-3,

$$P_{dev} = 270 \text{ W} - 4.6 \text{ W} = 265.4 \text{ W}$$

Noting that at no load the developed power goes entirely to overcome mechanical and core losses, we get

$$P_{MC} = 265.4 \text{ W}$$

The stator copper losses are now calculated from the full-load data.

$$P_{Cu} = 3 \times 0.6 \, \Omega \times 8 \text{ A}^2 = 115.2 \text{ W}$$

The field loss is calculated using Eq. 10-6 and the given data.

$$P_f = R_f I_f^2 = 1.5 \, \Omega \times 10 \text{ A}^2 = 150 \text{ W}$$

The efficiency is now calculated using Eq. 10-7 and the power input from part (b).

$$\eta = \frac{2600 \text{ W} - 115.2 \text{ W} - 265.4 \text{ W} - 150 \text{ W}}{2600 \text{ W}} \times 100$$

$$= 79.6\%$$

10-4.2 Torque

In our torque calculations we will derive an approximate equation for the torque angle (α_T) of a synchronous motor. Looking at Figure 10-2, the resultant voltage E_r across the phase impedance is the vector sum of the $I_a r_p$ and $I_a x_p$ voltage drops. However, since x_p is much bigger than r_p, we will assume that E_R is given by the reactive voltage drop.

$$E_R = I_a x_p \qquad (10\text{-}8)$$

Referring to Figure 10-3, if we examine each case it should be clear the vertical component of E_R is given by Eq. 10-9.

$$\text{vertical component} = E_R \sin (90 - \theta) \qquad (10\text{-}9)$$

However, since the sin $(90 - \theta)$ is equal to cos θ (the power factor), we can rewrite Eq. 10-9.

$$\text{vertical component} = E_R \times \text{PF} \qquad (10\text{-}10)$$

Applying simple trigonometry to Figure 10-3, we have

$$\sin \alpha_T = \frac{\text{vertical component of } E_R}{E_p}$$

$$= \frac{E_R \times \text{PF}}{E_p} \qquad (10\text{-}11)$$

$$\alpha_T = \arcsin \frac{E_R \times \text{PF}}{E_p} \qquad (10\text{-}12)$$

Equation 10-12 is an approximate but useful equation. The generated phase voltage can be calculated using Eq. 10-13, where V_p is the applied stator phase voltage. Note the similarity of Eq. 10-13 to Eq. 8-10 for the alternator. Its derivation based on ac circuit analysis will not be shown.

$$E_p = \sqrt{(V_p \cos\theta - I_a r_p)^2 + (I_a x_p \pm V_p \sin\theta)^2} \tag{10-13}$$

$$(+ \text{ for leading PF}, \; - \text{ for lagging PF})$$

Example 10-4

For the motor in Example 10-3, find:

(a) The torque angle at no load

(b) The torque angle at full load

Solution

(a) First find the generated phase voltage using Eq. 10-13, the given data, and noting that since the motor is Y connected,

$$V_p = \frac{210 \text{ V}}{\sqrt{3}} = 121 \text{ V}$$

$$I_a = I_L$$

Also from Example 10-3, part (a),

$$PF = \cos\theta = 0.46$$

$$\theta = 62.6°$$

$$\sin\theta = 0.89$$

$$E_p = \sqrt{(121 \times 0.46 - 1.6 \times 0.6)^2 + (1.6 \times 6.3 - 121 \times 0.89)^2}$$

$$= 111.9 \text{ V}$$

From Eq. 10-8,

$$E_R = 1.6 \text{ A} \times 6.3 \text{ } \Omega = 10.1 \text{ V}$$

Now using Eq. 10-12, we have

$$\alpha_T = \arcsin\left(10.1 \times \frac{0.46}{111.9}\right)$$

$$= 2.38°$$

(b) In a similar way, but now using the full-load data,

$$PF = \cos\theta = 0.89$$

$$\theta = 27.13°$$

$$\sin\theta = 0.46$$

$$E_p = \sqrt{(121 \times 0.89 - 8 \times 0.6)^2 + (8 \times 6.3 - 121 \times 0.46)^2}$$

$$= 103 \text{ V}$$

From Eq. 10-8,

$$E_R = 8 \text{ A} \times 6.3 \ \Omega = 50.4 \text{ V}$$

Using Eq. 10-12, we have

$$\alpha_T = \arcsin\left(50.4 \times \frac{0.89}{103}\right)$$

$$= 25.82°$$

The developed and output torques can be calculated using the basic relationship presented in Chapter 5 (Eqs. 5-3). Since we are dealing with a synchronous motor, the speed in these equations is synchronous speed.

Example 10-5 (English)

The motor in Example 10-3 is a 60-Hz motor with four poles. Find its developed and rated torques at full load.

Solution

Equation 5-3a will be used. Note that the speed S is given by Eq. 9-1a.

$$S = S_s = 120 \times \frac{60 \text{ Hz}}{4 \text{ poles}} = 1800 \text{ rev/min}$$

From Example 10-3 the developed power at full load is (using Eq. 10-3)

$$P_{\text{dev}} = 2600 \text{ W} - 115.2 \text{ W} = 2484.8 \text{ W}$$

Substituting into Eq. 5-3a, we get

$$T_{\text{dev}} = 7.04 \times \frac{2484.8 \text{ W}}{1800 \text{ rev/min}}$$

$$= 9.7 \text{ ft-lb}$$

The output power is the developed power less the mechanical and core losses P_{MC} (see Figure 10-8).

$$P_o = 2484.8 \text{ W} - 265.4 \text{ W} = 2219.4 \text{ W}$$

Substituting this into Eq. 5-3a, we get

$$T_o = 7.04 \times \frac{2219.4 \text{ W}}{1800 \text{ rev/min}}$$

$$= 8.68 \text{ ft-lb}$$

Example 10-6 (SI)

The motor in Example 10-3 is a 50-Hz motor with two poles. Find its developed and rated torques at full load.

Solution

Equation 5-3b will be used. Note that the speed ω is given by Eq. 9-16.

$$\omega = \omega_s = 4\pi \times \frac{50 \text{ Hz}}{2 \text{ poles}} = 314 \text{ rad/s}$$

From Example 10-3 the developed power at full load is (using Eq. 10-3)

$$P_{\text{dev}} = 2600 \text{ W} - 115.2 \text{ W} = 2484.8 \text{ W} = 2.48 \text{ kW}$$

Substituting into Eq. 5-3b, we get

$$T_{\text{dev}} = 1000 \times \frac{2.48 \text{ kW}}{314 \text{ rad/s}}$$

$$= 7.9 \text{ N-m}$$

The output power is the developed power less the mechanical and core losses P_{MC} (see Figure 10-8).

$$P_o = 2484.8 \text{ W} - 265.4 \text{ W} = 2219.4 \text{ W} = 2.22 \text{ kW}$$

Substituting this into Eq. 5-3b, we get

$$T_o = 1000 \times \frac{2.22 \text{ kW}}{314 \text{ rad/s}}$$

$$= 7.1 \text{ N-m}$$

10-5   TYPICAL CHARACTERISTICS

As with the induction motor, quantities are frequently plotted versus power output. In the case of the synchronous motor, the characteristics are obtained with the field excitation held constant. Figure 10-9 is a typical set of curves for line current, efficiency, and power factor versus power output at constant excitation.

Two pictures for power factor are shown. One of them, labeled (a), was obtained by setting the excitation to get unity power factor at no load. The other, labeled (b), was obtained by setting the excitation to get unity power factor at full load. In both cases the unity power factor point can be found by varying the excitation until the line current is a minimum (see Figure 10-6). The dashed line of power factor curve (b) represents a leading power factor; the solid line represents a lagging power factor.

10-6   POWER FACTOR CORRECTION

The process of improving (raising) the power factor at which power is drawn from a source is called **power factor correction.** Before reading this section it would be helpful for us to review Sections 7-1 and 7-4.1. In particular we should read and understand Example 7-24 together with the observations made at the end of the example.

It was shown that the ideal conditions for both supplier and user of electrical energy occurred when the power factor was unity. In addition, we saw that as the power factor became smaller, both efficiency (Examples 7-19 and 8-8) and voltage regulation (Examples 7-16, 8-6, and 8-7) got worse.

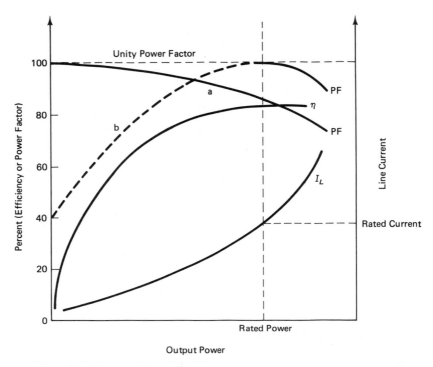

Figure 10-9 Typical characteristics versus output power for a synchronous motor. (*Note:* Excitation is constant.)

10-6.1 Causes of a Low Power Factor

In a typical industrial facility there are many devices that tend to lower the overall power factor.

1. Probably the biggest single contributor is the induction motor. When operated at light loads, its power factor is very low (see Figure 9-15)
2. A distribution transformer's power factor varies with load. When unloaded, its power factor is very low.
3. Electric arc welders typically have power factors of less than 70% (0.70).
4. Fluorescent and mercury vapor lamps have a typical power factor of 50% (0.50). Many of these lighting fixtures come with an internal device which corrects their power factor.

10-6.2 Disadvantages of a Low Power Factor

There are many disadvantages of low-power-factor operation. Aside from the loss of efficiency already mentioned, there are other indirect as well as direct (monthly) costs which are tied to a low power factor.

1. If a plant requires 700 kW and it operates at a 70% power factor, its distribution transformer must be rated 1000 kVA. This is much larger and more costly than one rated 700 kVA, which is the size needed if the power factor is unity.
2. To supply a given amount of power at a 70% power factor, the conductors would have to be twice as large (thick) than if the power factor were unity.
3. Many utilities charge a penalty if the power factor drops below a given number. In some cases the penalty increases with a decreasing power factor.
4. Sometimes a utility will charge a facility for kvar hours every month just as it would charge for energy used (kilowatt-hours).
5. A variation of the above would be a straight charge to the customer for the kVA used each month. This is done on occasion by utilities.

10-6.3 Ways to Improve the Power Factor

One of the obvious ways to improve the power factor would be to eliminate some or all of the causes mentioned in Section 10-6.1. However, it is frequently not practical and sometimes impossible to do this.

To understand how we could improve the power factor with some kind of device, we should understand what is happening electrically. We saw in Section 9-3 that there were two components of current delivered to an induction motor. One of these supplied the real power (kW) and was lost. The other was not lost, however, but rather was used to store energy in the motor's magnetic field (kvar). Since the current is alternating, the magnetic field undergoes cycles of building up and breaking down. When the field is breaking down, the reactive current flows out of the motor back to the supply. A crude picture of this behavior is shown in Figure 10-10a.

What is needed is some type of device that can be used as a temporary storage

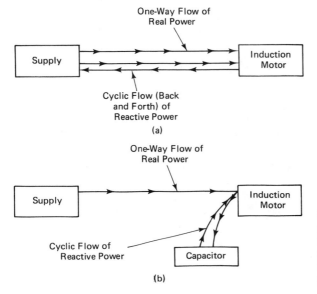

Figure 10-10 Simplified picture of power flow: (a) with supply acting as a temporary storage area for reactive power; (b) with capacitor acting as a temporary storage area for reactive power.

area for the reactive power when the motor's magnetic field breaks down. The ideal device for this is a capacitor, which also stores energy; however, it is stored in an electric field. By connecting a capacitor in parallel with the supply lines at the load, the cyclic flow of reactive power takes place between the motor and the capacitor. In this way the supply lines carry only the current supplying real power to the motor. The picture would now look as shown in Figure 10-10b. It should be noted that Figure 10-10b is for a unity-power-factor condition. For power factors other than unity the supply lines would carry some reactive power, however not as much as in Figure 10-10a. Sometimes the power factor is corrected to a value other than unity.

10-6.4 Capacitor Ratings

Capacitors manufactured for power factor correction are rated in kvar, operating voltage (rms), and frequency, and whether they are single-phase or three-phase delta. Typical values are from 1 to 60 kvar, 240 V, 480 V, and 600 V. More than one capacitor is often needed to obtain the required amount of power factor correction. For example, if 8 kvar of power factor correction is needed in a single-phase 240-V 60-Hz system, two capacitors would be used. One would be rated 5 kvar, 240 V, 60 Hz, and the other 3 kvar, 240 V, 60 Hz. Together they would provide a total of 8 kvar.

Sometimes a capacitor is used at a voltage and/or frequency lower than the rated values. When this is done the actual correction in terms of kvar will be less than the rated value.

The following equations are useful when dealing with power factor correction. Their derivation is left as an end-of-chapter problem for the student. It is a good but simple exercise in ac circuit analysis.

$$C = \frac{\text{rated var}}{2\pi f_r (V_r)^2} \qquad (10\text{-}14)$$

$$\text{actual kvar} = \frac{2\pi f C (V)^2}{1000} \qquad (10\text{-}15)$$

$$\text{rated kvar} = \text{actual kvar} \left(\frac{V_r}{V}\right)^2 \qquad (10\text{-}16)$$

$$\text{rated kvar} = \text{actual kvar} \frac{f_r}{f} \qquad (10\text{-}17)$$

In Eqs. 10-14 through 10-17, C is the capacitor's value in farads (F); f_r and V_r are the rated frequency (Hz) and voltage (rms volts), respectively; and f and V represent the actual frequency and voltage at which the unit is used.

Example 10-7

A capacitor rated 10 kvar, 240 V, and 60 Hz is to be used:

(a) At 120 V, 60 Hz

(b) At 240 V, 50 Hz

(c) At 230 V, 50 Hz

For each case find the actual kvar attained.

Solution

First we will use Eq. 10-14 to find the capacitance.

$$C = \frac{10000 \text{ var}}{2\pi \times 60 \text{ Hz} \times (240 \text{ V})^2} \approx 460 \ \mu\text{F}$$

The individual parts are solved by direct substitution into Eq. 10-15.

(a) $\quad$ kvar $= 2\pi \times 60 \times 460 \times 10^{-6} \times \dfrac{120^2}{1000}$

$\qquad = 2.5$ kvar

(b) $\quad$ kvar $= 2\pi \times 50 \times 460 \times 10^{-6} \times \dfrac{240^2}{1000}$

$\qquad = 8.32$ kvar

(c) $\quad$ kvar $= 2\pi \times 50 \times 460 \times 10^{-6} \times \dfrac{230^2}{1000}$

$\qquad = 7.64$ kvar

The following example will illustrate the use of a capacitor to correct the power factor of a load.

Example 10-8

A load draws 10 kVA at a lagging power factor of 0.8 from a single-phase 120-V 60-Hz supply line. Find the rating of the capacitor which when placed in parallel with the load will raise the overall power factor to unity.

Solution

The problem is best analyzed by constructing a power triangle (see Section 7-1, item 9) for the system. It is shown in Figure 10-11. Note that since the power factor is 0.8,

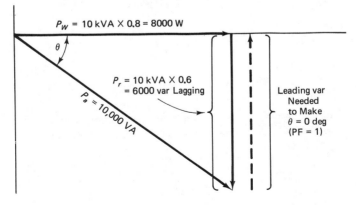

Figure 10-11 Power triangle for Exercise 10-8.

$$\cos \theta = 0.8$$

$$\theta = 37°$$

$$\sin \theta = 0.6$$

The reactive (P_r) and real (P_w) power components were calculated using Eqs. 7-10 and 7-11.

$$P_w = 10,000 \times 0.8 = 8000 \text{ W}$$

$$P_r = 10,000 \times 0.6 = 6000 \text{ var}$$

The dashed line in Figure 10-11 represents the amount of leading var that must be introduced to raise the overall power factor to unity, that is, to make $\theta = 0°$.

$$PF = \cos 0° = 1$$

Hence the capacitor must realize (attain) 6 kvar. If this rating (6 kvar, 120 V, 60 Hz) is available, the problem is solved. However, let us assume that the only capacitors available are rated 240 V. We must now work backwards using Eq. 10-16 to find the rating.

$$\text{rated kvar} = 6 \text{ kvar} \left(\frac{240 \text{ V}}{120 \text{ V}} \right)^2$$

$$= 24 \text{ kvar}$$

Hence two capacitors in parallel would be used. One would be rated 20 kvar and the other 4 kvar.

The procedure in a three-phase system is the same with one exception; we must be sure to use the three-phase equations introduced in Chapter 7 (i.e., Eqs. 7-49, 7-50, and 7-51) when they are applicable.

Example 10-9

Three-phase 230-V power is supplied to a factory which draws 60 kVA at a lagging power factor of 75%. How many leading kvar must be supplied by a three-phase capacitor bank to raise the overall power factor to:

(a) 0.9 lagging?

(b) 1.0?

(c) 0.9 leading?

Solution

In each case the kVA and kvar drain on the supply lines will change. The real and reactive power absorbed by the factory will remain the same and are calculated using Eqs. 7-11 and 7-10.

$$P_w = 60 \text{ kVA} \times 0.75 = 45 \text{ kW}$$

$$P_r = 60 \text{ kVA} \times \sin 41.4° = 39.7 \text{ kvar}$$

Note that initially

$$PF = \cos \theta = 0.75$$

and

$$\theta = 41.4°$$

For (b):

$$PF = 1 \quad \theta = 0°$$

For (a) and (c):

$$PF = 0.9 \quad \text{lagging or leading}$$
$$\cos \theta = 0.9$$
$$\theta = 25.8°$$

The power triangle for each case is shown in Figure 10-12.

(a) Referring to Figure 10-12a, at a power factor of 0.9 lagging, the lines must still supply Q kvar. Using simple trigonometry,

$$Q = 45 \text{ kW} \times \tan 25.8° = 21.75 \text{ kvar}$$

To do this the capacitor bank must supply R kvar, where

$$R = 39.7 \text{ kvar} - Q$$
$$= 39.7 \text{ kvar} - 21.75 \text{ kvar}$$
$$= 17.95 \text{ kvar}$$

This must be supplied by the capacitor bank.

(b) In this case since θ must be 0°, the capacitor bank must supply all 39.7 kvar.

(c) Here the capacitor bank must supply more than the 39.7 kvar needed by the factory. This is done in practice when the drain on the supply lines is expected to increase in the near future. When it does the excess kvar of the capacitors will supply the added reactive power drain of the factory. This will bring the overall power factor close to unity (100%).

Here

$$Q = 45 \text{ kW} \times \tan 25.8°$$
$$= 21.75 \text{ kvar (as before)}$$
$$R = 39.7 \text{ kvar} + Q$$
$$= 39.7 \text{ kvar} + 21.75 \text{ kvar}$$
$$= 61.45 \text{ kvar}$$

This must be supplied by the capacitor bank.

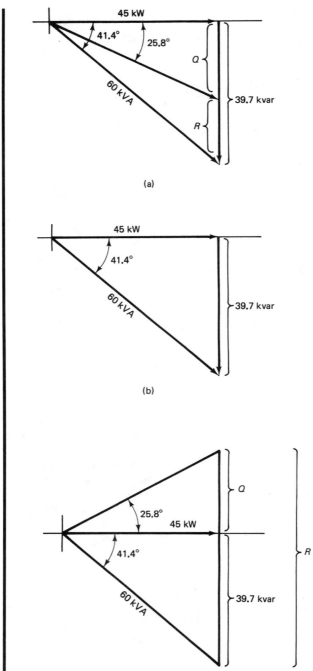

(a)

(b)

(c)

Figure 10-12 Power triangles for Example 10-9.

10-6.5 The Synchronous Condenser

In the preceding sections we saw that a device which draws leading reactive power, such as the capacitor, can be used for power factor correction. We also saw in Section 10-2.3 that by overexciting a synchronous motor it could be made to operate at a very low leading power factor. In other words, when placed across the supply lines it could be made to behave like a capacitor and thus be used for power factor correction. When this is done and it is operated at no load, it is called a **synchronous condenser.** Very often the motor will be run under load and by overexcitation can provide some power factor correction as well as driving a mechanical load.

Example 10-10

A small facility draws 25 kW at a lagging power factor of 0.85. What must be the kVA rating of a synchronous condenser (assume that PF = 0) which when placed across the lines will raise the overall power factor to unity?

Solution

First find the reactive power needed by the facility.

$$\text{total kVA} = \frac{25 \text{ kW}}{0.85} = 29.41 \text{ kVA}$$

$$\theta = \arccos 0.85 = 31.8°$$

$$\text{kvar} = 29.41 \text{ kVA} \times \sin 31.8° = 15.5 \text{ kvar}$$

Since all of this must be supplied by the synchronous condenser, it must have a rating of 15.5 kVA. Note that all of this 15.5 kVA is reactive since the PF is 0.

Example 10-11

A bank of three-phase induction motors operate off 210 V and draw 3 kVA total at a combined power factor of 0.80 lagging. The synchronous motor of Example 10-2 is connected across the lines. Find:

(a) The power factor if the synchronous motor is run at no load with a field current of 18 A (the V curves are shown in Figure 10-6)

(b) The field current necessary to raise the power factor to unity if the synchronous motor is run at half load

Solution

(a) From the no-load V curves, we obtain the following information at $I_f = 18$ A:

$$I_L = 5 \text{ A} \qquad \text{PF} = 0.2$$

V_L is given as 210 V in Example 10-2. For the synchronous motor,

$$P_a = \sqrt{3} \ V_L I_L = \sqrt{3} \times 210 \text{ V} \times 5 \text{ A}$$

$$= 1818.6 \text{ kVA}$$

Since PF $= 0.2 = \cos \theta$,

$$\theta = 78.46°$$

$$P_r = 1818.6 \text{ kVA} \times \sin 78.46°$$

$$= 1781.86 \text{ kvar (leading)}$$

$$P_w = 1818.6 \text{ kVA} \times 0.2 = 364 \text{ W (this goes for losses)}$$

For the induction motors,

$$\text{PF} = \cos \theta = 0.80$$

$$\theta = 36.87°$$

$$\sin \theta = 0.60$$

$$P_r = 3 \text{ kVA} \times \sin \theta = 1.8 \text{ kvar} = 1800 \text{ var}$$

$$P_w = 3 \text{ kVA} \times \cos \theta = 2.4 \text{ kW} = 2400 \text{ W}$$

The power triangle is shown in Figure 10-13a. The net reactive power is

$$Q = 1.8 \text{ kvar (lagging)} - 1.78 \text{ kvar (leading)}$$

$$= 0.02 \text{ kvar} = 20 \text{ var}$$

The net real power is

$$P_w = 2.4 \text{ kW} + 364 \text{ W} = 2.764 \text{ kW}$$

From Eq. 7-12,

$$P_a = \sqrt{(2764)^2 + 20^2} \approx 2764 \text{ VA}$$

and from Eq. 7-4,

$$\text{PF} = \frac{2764 \text{ W}}{2764 \text{ VA}} = 1.0$$

(b) The numbers for the induction motors are still the same.

$$P_w = 2.4 \text{ kW}$$

$$P_r = 1.8 \text{ kvar}$$

$$P_a = 3.0 \text{ kVA}$$

To raise the power factor to unity, the synchronous motor must supply the reactive power 1.8 kvar. The power triangle is shown in Figure 10-13b. From the half-load data for the synchronous motor (Table 10-2) we can compute the input power at any point. It should be fairly constant. At the high field-current points, it averages about 1050 W.

$$\text{At } I_f = 14: \quad P_i = W_1 + W_2 = 1020 \text{ W}$$

$$I_f = 16: \quad P_i = W_1 + W_2 = 1040 \text{ W}$$

$$I_f = 17.4: \quad P_i = W_1 + W_2 = 1080 \text{ W}$$

$$\text{average input } P_i \approx 1050 \text{ W}$$

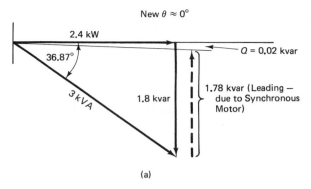

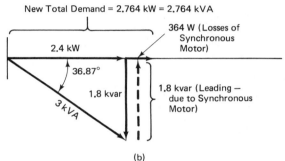

Figure 10-13 Power triangles for Example 10-11; (a) part (a); (b) part (b).

This makes sense since at half load the output is 750 W, which gives an efficiency of

$$\eta = \frac{750}{1050} \times 100 = 71\%$$

which is typical (see Figure 10-9). Knowing the real power (1050 W) and the needed reactive power (1800 var) of the synchronous motor, we can find the total kVA drawn using Eq. 7-12.

$$P_a = \sqrt{(1050)^2 + (1800)^2} = 2084 \text{ VA}$$

Substituting this into Eq. 7-49, we can find the line current.

$$I_L = \frac{2084 \text{ VA}}{\sqrt{3} \times 210 \text{ V}} = 5.73 \text{ A}$$

Using the half-load V curves (Figure 10-6), for a line current of 5.73 we need a field current of about 17.4 A. The power factor from the curves would be about 0.49 at that excitation.

Note that the power factor could also have been calculated from Eq. 7-51.

$$PF = \cos \theta = \frac{1050 \text{ W}}{\sqrt{3} \times 210 \text{ V} \times 5.73 \text{ A}}$$

$$= 0.50$$

Using this and the V curves would have given a field current of 17.3 A.

SYMBOLS INTRODUCED IN CHAPTER 10

Symbol	Definition	Units: English and SI
α_T	Torque angle	degrees
β	Synchronous impedance angle	degrees
x_p	Reactance per phase of stator winding	ohms
r_p	Resistance per phase of stator winding	ohms
Z_s	Impedance per phase	ohms
E_p	Generated stator phase voltage	rms volts
V_p	Applied stator phase voltage	rms volts
I_f	Field current	amperes
I_a	Stator phase current	amperes
E_R	Internal voltage drop across Z_s (this is similar to the armature voltage drop in a dc machine)	rms volts
θ	Angle between V_p and I_a	degrees
$\cos \theta$	Power factor of motor	—
W_1 and W_2	Wattmeter readings	watts
P_{Cu}	Total stator copper loss	watts
P_{MC}	Sum of mechanical and core losses	watts
V_f	Voltage applied to field winding	dc volts
R_f	Field winding resistance	ohms
C	Capacitor used for power factor correction	farads
f_r	Rated frequency of C	hertz
V_r	Rated voltage of C	rms volts
f	Actual frequency C is used at	hertz
V	Actual voltage C is used at	rms volts
Q	Reactive power delivered by supply lines	var
R	Reactive power supplied by capacitor bank	var

QUESTIONS

1. The induction motor is an asynchronous machine. Why is the synchronous motor called a synchronous machine?

2. What is meant by the expression "slip a pole"?

3. Define "torque angle" and "synchronous impedance angle."

4. As the load on a synchronous motor increases, what happens to the motor's torque angle and speed?

5. Explain the difference between underexcited, normally excited, and overexcited in terms of the motor's power factor.

6. What are V curves?

7. What problem in terms of starting torque is encountered when we try to start a synchronous motor?

8. Describe the two basic techniques used to start synchronous motors.

9. What are amortisseur windings? What is their purpose?

10. What are the major causes of low power factor? What are its disadvantages?

11. Why is a capacitor used for power factor correction?

12. What is a synchronous condenser?

PROBLEMS

(English and SI)

1. A synchronous motor has $r_p = 0.1 \ \Omega$ and $x_p = 1.2 \ \Omega$. What is the torque angle corresponding to maximum torque?

2. What is the largest torque angle a motor can have if $x_p = 2.8 \ \Omega$ and $r_p = 0.3 \ \Omega$?

3. A 230-V 1-kW ($1\frac{1}{3}$-hp) three-phase synchronous motor was tested as shown in Figure 10-5 (the line voltage, however, is 230 V). Tables 10-3 and 10-4 contain the no-load and full-load data. Plot on one graph the no-load and full-load V curves.

Table 10-3 No load

I_f (A)	Average I_L (A)	W_1 (W)	W_2 (W)
2	2.6	430	−180
4	1.7	310	−62
5	1.4	248	0
6	0.95	190	48
7.4	0.6	120	118
8	0.7	95	140
8.7	0.82	60	190
10	1.3	−20	265
10.8	1.6	−125	370

Table 10-4 Full load

I_f (A)	Average I_L (A)	W_1 (W)	W_2 (W)
2.4	7	1000	210
4.5	6	810	370
5.7	4.8	710	490
7	4.2	660	550
8.6	3.4	600	610
9.5	3.7	420	800
10.5	4.2	290	920
12	5.8	170	1050
13.4	6.9	65	1140

4. A 230-V 12-A Y-connected three-phase synchronous motor is tested as shown in Figure 10-5 (the line voltage, however, is 230 V). It has $r_p = 0.5 \ \Omega$, $x_p = 4.5 \ \Omega$, $R_f = 1.8 \ \Omega$, and the field current is kept constant at 11 A (underexcited). The test results are given below.

No-load test: $V_L = 230 \ \text{V},$ $I_L = 2 \ \text{A (average)}$

$W_1 = 500 \ \text{W},$ $W_2 = -210 \ \text{W}$

Full-load test: $V_L = 230 \ \text{V},$ $I_L = 12 \ \text{A (average)}$

$W_1 = 2250 \ \text{W},$ $W_2 = 2100 \ \text{W}$

Find:
(a) The no-load power factor
(b) The full-load power factor
(c) The full-load efficiency

5. A 460-V 10-A Δ-connected three-phase synchronous motor is tested as shown in Figure 10-5 (the line voltage, however, is 460 V). It has $r_p = 0.9$ Ω, $x_p = 7.5$ Ω, $R_f = 1.4$ Ω, and the field current is held constant at 18 A (overexcited). The test results are given below.

No-load test: $V_L = 460$ V, $I_L = 4.4$ A (average)

$W_1 = 920$ W, $W_2 = -340$ W

Full-load test: $V_L = 460$ V, $I_L = 10$ A (average)

$W_1 = 4.0$ kW, $W_2 = 3.5$ kW

Find:
(a) The no-load power factor
(b) The full-load power factor
(c) The full-load efficiency

6. For the motor in Problem 4, find:
(a) The no-load torque angle
(b) The full-load torque angle

7. For the motor in Problem 5, find:
(a) The no-load torque angle
(b) The full-load torque angle

8. A capacitor rated 5 kvar, 230 V, 50 Hz is to be used at different voltages and frequencies to correct a power factor. For each case below determine the actual kvar attained.
(a) 120 V, 60 Hz
(b) 210 V, 60 Hz
(c) 150 V, 50 Hz

9. A capacitor rated 20 kvar, 480 V, 60 Hz is to be used at different voltages and frequencies to correct a power factor. For each case below determine the actual kvar attained.
(a) 120 V, 60 Hz
(b) 230 V, 50 Hz
(c) 460 V, 50 Hz

10. A load draws 5 kVA at a lagging power factor of 0.75 from a single-phase 230-V 50-Hz supply line. Find the rating of the capacitor, in kvar, which when placed across the load will raise the power factor to:
(a) 0.9 lagging
(b) Unity
(c) 0.9 leading

11. Repeat Problem 10, but this time assume that the only available capacitors are rated 240 V and 60 Hz.

12. Derive Eq. 10-14.

13. Derive Eq. 10-15.

14. Derive Eqs. 10-16 and 10-17.

15. A load draws 15 kVA at a power factor of 0.85 lagging from a single-phase 120-V 60-Hz supply line. Capacitors available are rated 240 V and 60 Hz. It is desired to raise the power factor to unity. Find:
(a) The rating of the capacitor in kvar
(b) The capacitors value in farads

16. A factory draws 100 kVA at a power factor of 0.70 lagging from a 240-V 60-Hz three-phase supply. It is desired to raise the power factor to unity. Find the kvar rating of the three-phase capacitor bank needed if:

(a) The capacitors are rated 240 V and 60 Hz

(b) The capacitors are rated 600 V and 60 Hz

17. Repeat Problem 16, but this time it is desired to raise the power factor to 0.9 leading.

18. In Problem 16, if the kVA drain on the supply lines remains fixed at 100 kVA, what is the power supplied to the factory at:

(a) The power factor of 0.70 lagging?

(b) The corrected power factor of unity?

19. A factory has a total load of 500 kW at a lagging power factor of 0.67. The power is supplied by a 460-V 50-Hz three-phase supply line.

(a) Find the kVA drain on the lines.

(b) Find the kvar that must be supplied by a capacitor bank in order to raise the power factor to unity.

(c) Find the kVA drain on the lines at the corrected power factor of unity.

(d) If the only capacitors available are rated 600 V and 60 Hz, find their required rating.

20. The total power taken by a factory connected to a 240-V 60-Hz three-phase supply line is 350 kW at a lagging power factor of 78%. A capacitor bank rated 250 kvar, 240 V, 60 Hz is connected across the lines. Find the new power factor at which the lines will supply power to the factory.

21. Repeat Problem 20 if the capacitor bank was rated 250 kvar, 480 V, 60 Hz.

22. In Problem 20, what is the value of each capacitor (in farads) making up the three-phase bank?

23. A machine shop draws 38 kW at a lagging power factor of 75%. What must be the rating of a synchronous condenser (assume that PF = 0.2) which when connected to the lines and runs at no load will raise the overall power factor to unity?

24. A small production facility draws 2 kW at an 85% lagging power factor from a three-phase 210-V supply line. In an effort to improve the facility's power factor, the synchronous motor of Example 10-2 is used to drive a conveyor belt. To drive the belt, 750 W (about half the rated load of the motor) is required. To raise the facility's overall power factor to unity, find:

(a) The necessary field current

(b) The efficiency at which the motor will be operating

25. Referring to Problem 24, find:

(a) The initial kVA load on the lines

(b) The kVA load on the lines after power factor correction (*Hint:* The power consumed by the synchronous motor was not included in the original 2-kW load)

26. Repeat Problem 24, but this time it is desired to raise the overall power factor to 95% leading.

27. Referring to Problem 24, if the synchronous motor was accidentally excited with a field current of 9 A, find:

(a) The overall power factor

(b) The kVA load on the supply lines

(English)

28. How fast will a 240-V 16-pole 60-Hz three-phase synchronous motor turn?

29. If the motor in Problem 28 is powered by a three-phase 230-V 50-Hz supply line, how fast will it turn?

30. A 240-V 4-hp six-pole 60-Hz three-phase synchronous motor is run at full load. Find:

(a) The output torque of the motor

(b) The speed at which it would run if the load were cut by 70%

31. Repeat Problem 30, but the motor is now running at half load.

32. What is the percent speed regulation of the motor in Problem 30?

(SI)

33. How fast will a 230-V 24-pole 50-Hz three-phase synchronous motor turn?

34. If the motor in Problem 33 is powered by a three-phase 240-V 60-Hz supply line, how fast will it turn?

35. A 460-V 4.5-kW eight-pole 50-Hz three-phase synchronous motor is run at full load. Find:

(a) The output torque of the motor

(b) The speed at which it would run if the load were cut by 60%

36. Repeat Problem 35, but the motor is now running at half load.

37. What is the percent speed regulation of the motor in Problem 35?

Chapter 11

The Induction Generator

We have seen that synchronous and dc machines can be used as either generators or motors. In the same way an induction motor can be made to operate as an induction generator. When this is done it is also referred to as an **asynchronous generator** since the rotor does not turn at synchronous speed. In recent years the induction generator has received very little treatment in most books that cover ac machinery. The reason was simply that it found very little practical industrial use. The result is that many modern engineering and technical school graduates are not even aware of its existence.

The energy crisis of the mid-1970s sent many engineers and scientists to the drawing board. As a result of this energy research, the induction generator is now finding increased popularity in wind energy conversion systems (WECS). This chapter is devoted to the induction generator with the hope that it will once again be included in engineering and technical curricula.

11-1 THEORY OF OPERATION

To understand how and why the induction generator operates, it would be helpful to review Section 9-3. When an induction motor is running, we said that it drew two components of current from the supply lines. One provided the excitation (the rotating stator field) and the other provided real power (mechanical power output and internal losses). If the load is removed, the motor continues to draw the same excitation current; however, the real power current becomes very small. At this point if mechanical power is applied to the rotor shaft, causing it to gain speed and turn faster than synchronous speed, the following will be true:

1. The slip given by Eqs. 9-5 and 9-6 will be a negative number. This is permissible, and we should recall that nowhere in Chapter 9 did we say that the slip had to be positive. All of the equations introduced for the induction motor are valid for positive or negative slips.

2. Referring to Figure 9-9, the rotor conductors are now moving faster than the rotating flux. This means that the motion of rotor conductors relative to the flux will now be left to right. Hence the direction of rotor currents in Figure 9-9 will be opposite to what is shown. This can also be seen by inserting a negative slip into Eq. 9-13.

3. The reversal in rotor current will cause a reversal in the counter emf generated in the stator winding. This means that electric power will now flow out of the stator windings (generator action).

4. The internal losses of the machine will now be supplied by the prime mover turning the rotor shaft.

5. The machine continues to draw the exciting current from the supply lines. This supplies the rotating flux. Without this external excitation, generator operation could not exist. The induction generator is therefore **not self-excitating.**

6. The **frequency** of the generated stator voltage **depends on the speed of the rotating flux.** Hence it will be the same as the frequency of the applied stator voltage providing the excitation.

A simplified diagram showing power flow is shown in Figure 11-1.

Note that in order to operate, the generator must be connected to a source of reactive power (kvar). If this source is a three-phase supply line, the generator will draw reactive power from the lines and feed real power (kW) back into them.

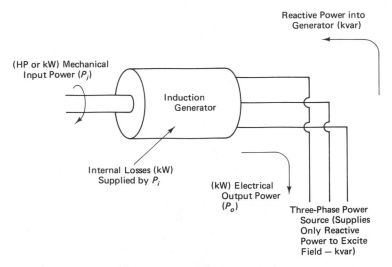

Figure 11-1 Diagram showing the flow of real and reactive power in an induction generator.

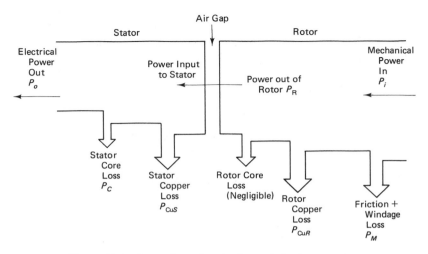

Figure 11-2 Power flow diagram for an induction generator.

A power flow diagram similar to Figure 9-12 for the induction motor is shown in Figure 11-2. It is important to note that when the slip is around zero (rotor turning at about synchronous speed), Figure 11-2 is not valid. At that intermediate condition when motor action stops and generator action begins there is actually input from both ends of the diagram. The internal losses are supplied in part by the mechanical input and the remainder by the electrical input. The actual crossover point could be determined by monitoring the power flow at the stator terminals with two wattmeters. The slip at which this power flow is zero ($W_1 + W_2 = 0$) would be the point at which motor action stops and generator action begins. It would actually be some small negative slip.

11-2   *EFFICIENCY*

Efficiency is determined in the same way as for the other generators (dc and synchronous). The input is determined by adding the internal losses to the electrical output.

The mechanical and core losses are determined by running a no-load test on the machine as a motor. They will be the same.

The stator copper loss (P_{CuS}) is calculated as before using the measured current, stator resistance, and Eq. 9-31.

The rotor copper loss is given by Eq. 9-27; however, now we must use Eq. 11-1 to find P_R (refer to Figure 11-2).

$$P_R = P_o + P_C + P_{CuS} \tag{11-1}$$

Note that now the output power (P_o) is given by Eq. 11-2.

$$P_o = W_1 + W_2 \tag{11-2}$$

If we treat the generator in the same way that we treated the motor, we will lump the stator core loss together with the friction and windage losses at the rotor. Hence the quantity P_C in Eq. 11-1 will be zero.

The output power is also given by Eq. 11-3, where $\cos \theta$ is the power factor of the induction generator.

$$P_o = \sqrt{3} \, V_L I_L \cos \theta \qquad (11\text{-}3)$$

The mechanical power input can now be found using Eq. 11-4.

$$P_i = P_o + P_{CuS} + P_{CuR} + P_M \qquad (11\text{-}4)$$

Note that in Eq. 11-4 P_M includes the stator core loss.

Example 11-1 (English)

The induction motor of Example 9-22 is run as a generator. The data for a load run is given below. Determine the efficiency and power factor. The mechanical and core losses (P_M) from the no-load test can be assumed constant at 216 W.

$$\text{Load run:} \quad W_1 = 2.1 \text{ kW}, \ W_2 = 1.4 \text{ kW}$$

$$V_L = 208 \text{ V}, \quad I_L = 11 \text{ A}$$

$$S = 1860 \text{ rev/min}$$

Solution

From Eq. 11-2,

$$P_o = 2100 \text{ W} + 1400 \text{ W} = 3500 \text{ W}$$

Using Eq. 11-3 gives us

$$\text{PF} = \cos \theta = \frac{3500 \text{ W}}{\sqrt{3} \times 208 \text{ V} \times 11 \text{ A}}$$

$$= 0.88 \text{ (or 88\%)}$$

From Example 9-22 the equivalent stator resistance r_e is 1 Ω. Using Eq. 9-31, we have

$$P_{CuS} = \tfrac{3}{2} (1 \ \Omega)(11 \text{ A})^2 = 181.5 \text{ W}$$

From Eq. 9-6a,

$$s = (1800 \text{ rev/min} - 1860 \text{ rev/min}) \times \frac{100}{1800 \text{ rev/min}}$$

$$= -3.33\%$$

Using Eq. 11-1, the power transferred across the air gap P_R is

$$P_R = 3500 \text{ W} + 0 + 181.5 \text{ W} = 3681.5 \text{ W}$$

and from Eq. 9-27,

$$P_{CuR} = 0.0333 \times 3681.5 \text{ W} = 122.7 \text{ W}$$

The input power can now be found using Eq. 11-4.

$$P_i = 3500 \text{ W} + 181.5 \text{ W} + 122.7 \text{ W} + 216 \text{ W}$$

$$= 4020.2 \text{ W}$$

Finally, from Eq. 9-21,

$$\eta = \frac{3500 \text{ W}}{4020.2 \text{ W}} \times 100 = 87\%$$

Example 11-2 (SI)

The induction motor of Example 9-23 is run as a generator. The data for a load run is given below. Determine its efficiency and power factor. The mechanical and core losses from the no-load test can be assumed constant at 195.5 W.

$$\text{Load run:} \quad W_1 = 1.7 \text{ kW} \quad\quad W_2 = 1.1 \text{ kW}$$

$$V_L = 230 \text{ V} \quad\quad I_L = 8 \text{ A}$$

$$\omega = 163 \text{ rad/s}$$

Solution

From Eq. 11-2, we have

$$P_o = 1700 \text{ W} + 1100 \text{ W} = 2800 \text{ W}$$

Using Eq. 11-3, we have

$$\text{PF} = \cos\theta = \frac{2800 \text{ W}}{\sqrt{3} \times 230 \text{ V} \times 8 \text{ A}}$$

$$= 0.88 \text{ (or } 88\%\text{)}$$

From Example 9-23 the equivalent stator resistance r_e is 1.13 Ω. Using Eq. 9-31 gives us

$$P_{\text{Cu}S} = \tfrac{3}{2}(1.13 \ \Omega)(8 \text{ A})^2 = 108.5 \text{ W}$$

From Eq. 9-6b,

$$s = (157 \text{ rad/s} - 163 \text{ rad/s}) \times \frac{100}{157 \text{ rad/s}}$$

$$= -3.82\%$$

Using Eq. 11-1, the power transferred across the air gap P_R is

$$P_R = 2800 \text{ W} + 0 + 108.5 \text{ W} = 2908.5 \text{ W}$$

and from Eq. 9-27,

$$P_{\text{Cu}R} = 0.0382 \times 2908.5 \text{ W} = 111.1 \text{ W}$$

The input power can now be found using Eq. 11-4.

$$P_i = 2800 \text{ W} + 108.5 \text{ W} + 111.1 \text{ W} + 195.5 \text{ W}$$

$$= 3215.1 \text{ W}$$

Finally, from Eq. 9-21,

$$\eta = \frac{2800 \text{ W}}{3215.1 \text{ W}} \times 100 = 87.1\%$$

11-3 TYPICAL CHARACTERISTICS

The characteristics of the induction generator are very similar to those of the induction motor with some minor differences. Figure 11-3 is a plot of power factor, efficiency, and mechanical power versus percent slip. It is plotted for both positive and negative slips; hence a comparison of the motor and generator action can be made easily.

The characteristics are somewhat symmetrical in their appearance. The dashed line represents the crossover point from motor to generator action. At that slip the prime mover is supplying all of the internal losses of the generator. Any further increase in speed will start generating electrical power (kW) out of the generator back into the supply lines. The only function of the supply lines at this point is to provide the magnetizing component of current. Note also that rated power is achieved at a lower slip for the generator than the motor. This is typically around −3 to −5% slip. Thus when the prime mover has a variable speed, that cannot be controlled, a speed

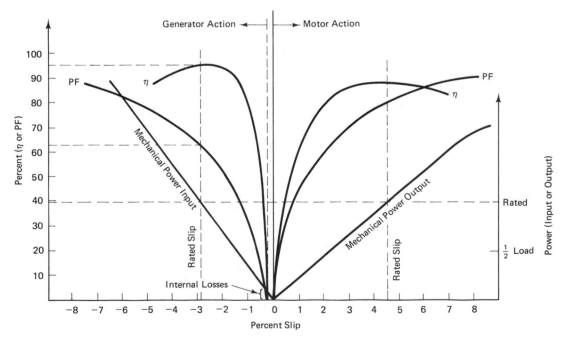

Figure 11-3 Typical curves of efficiency, power factor, and mechanical power for a three-phase induction motor and generator.

governor will be used to hold the generator's speed constant at its rated slip. If the generator slows down slightly, the power factor drops very quickly. This means that at light loads the generator will have a very poor (low) power factor which will degrade the overall system power factor. Thus to make the induction generator more suitable for use with existing power systems, some means of power factor correction should be used.

11-4 *POWER FACTOR CORRECTION OF THE INDUCTION GENERATOR*

When it is operated over various input (power and speed) conditions, the typical induction generator will have an average power factor of 50 to 65%. With the use of a capacitor bank (see Section 10-6.4) the average power factor can be increased fairly easily to 85%. This would be an acceptable figure from a utility's viewpoint if the induction generator was feeding power into the grid.

The use of capacitors brings up an interesting point. At any given power factor the capacitors can be selected large enough to supply all of the flux-producing kvar to the generator. If this is the case, the power lines can be disconnected and the generator will continue to operate and supply power to a load. In effect, the induction generator is now self-excited. As long as the capacitors are rated large enough to supply all of the needed kvar, this independent operation can exist.

It should be pointed out that in the absence of supply lines, the **frequency** of the **generated power** will **now** be a **function of the generator's speed.** This emphasizes more the need for a speed governor.

Example 11-3 (English)

A 2-hp 208-V three-phase four-pole 60-Hz induction motor is run at rated power as a generator. Its characteristics are given in Figure 11-3. Find the capacitor bank rating needed to run the generator without the use of external supply lines (self-excited). Also find the frequency of the generated voltage.

Solution

From Figure 11-3 we can obtain the following data for the generator at its rated power point;

$$s = -2.8\%$$

$$PF = 63\% \ (0.63)$$

$$\eta = 95\%$$

Using Eq. 9-21, we can find the output power.

$$P_o = 0.95 \times 2 \text{ hp} = 1.9 \text{ hp}$$

$$= 1.9 \text{ hp} \times \frac{746 \text{ W}}{\text{hp}} = 1417.4 \text{ W}$$

From Eq. 7-11,

$$P_a = \frac{1417.4 \text{ W}}{0.63} = 2249.84 \text{ VA}$$

Since

$$\cos \theta = 0.63$$

$$\theta = 50.95°$$

$$\sin \theta = 0.78$$

From Eq. 7-10,

$$P_r = 2249.84 \times 0.78 = 1747.2 \text{ var}$$

Hence the capacitor bank must provide a minimum of 1.747 kvar. Knowing the slip, we can find the speed using Eq. 9-6a. Note that synchronous speed is 1800 rev/min.

$$-2.8 = \frac{1800 - S}{1800} \times 100$$

Solving for the speed, we get

$$S = 1800 + 50.4 = 1850.4 \text{ rev/min}$$

Substituting this into Eq. 8-2a yields

$$f = 1850.4 \times \frac{4}{120} = 61.7 \text{ Hz}$$

This frequency when inserted into Eq. 10-17 will give the true capacitor bank rating.

Example 11-4 (SI)

A 5-kW 230-V three-phase six-pole 50-Hz induction motor is run at rated power as a generator. Its characteristics are given in Figure 11-3. Find the capacitor bank rating needed to run the generator without the use of external supply lines (self-excited). Also find the frequency of the generated voltage.

Solution

From Figure 11-3 we can obtain the following data for the generator at its rated power point;

$$s = -2.8\%$$

$$PF = 63\% \ (0.63)$$

$$\eta = 95\%$$

Using Eq. 9-21, we can find the output power.

$$P_o = 0.95 \times 5 \text{ kW} = 4.75 \text{ kW} = 4750 \text{ W}$$

From Eq. 7-11,

$$P_a = \frac{4750 \text{ W}}{0.63} = 7540 \text{ VA}$$

Since

$$\cos \theta = 0.63$$

$$\theta = 50.95°$$

$$\sin \theta = 0.78$$

From Eq. 7-10,

$$P_r = 7540 \times 0.78 = 5881 \text{ var}$$

Hence the capacitor bank must provide a minimum of 5.881 kvar. Knowing the slip, we can find the speed using Eq. 9-6b. Note that synchronous speed is 104.7 rad/s.

$$-2.8 = \frac{104.7 - \omega}{104.7} \times 100$$

Solving for the speed yields

$$\omega = 104.7 + 2.93 = 107.63 \text{ rad/s}$$

Substituting this into Eq. 8-2b, we have

$$f = 107.63 \times \frac{6}{4\pi} = 51.4 \text{ Hz}$$

This frequency when inserted into Eq. 10-17 will give the true capacitor bank rating.

In light of this ability to self-excite when capacitors are used for power factor correction, a potentially dangerous situation arises. Consider the case of a generator supplying power into a utility grid. Capacitors are not used and it is driven by either a windmill, gasoline engine, or water power. In the event of a grid power failure, the generator will lose its excitation and hence stop producing electrical power. A worker could safely repair the grid. If, however, capacitors are used for power factor correction the generator could self-excite and send power into the grid. An unknowing person doing repair work could be seriously injured. To safeguard against this danger, some sort of breaker should be used to disconnect the generator from the grid in the event of a grid failure.

11-5   APPLICATIONS OF THE INDUCTION GENERATOR

In the past the uses were quite limited. Some trains would use induction motors to power them. When the train went downhill, it would gain speed. When the speed was great enough (negative slip), the machine would run as a generator. Not only would this send power back into the train's power grid, but it would also provide some amount of dynamic braking (see Section 5-11).

Presently, there is an increased interest in alternate energy sources. As a result many manufacturers of wind energy systems are making use of the induction generator. We shall see in Chapter 15 that it greatly simplifies and reduces the overall cost of a wind energy system. It is particularly well suited where its generated power is to be used together with that of a local utility.

SYMBOLS INTRODUCED IN CHAPTER 11

Symbol	Definition	Units: English and SI
P_R	Power transferred from rotor to stator across air gap	watts
P_o	Electrical output power of generator	watts
P_M	Mechanical loss of generator	watts
P_{CuR}	Rotor copper loss	watts
P_{CuS}	Stator copper loss	watts
P_C	Stator core loss	watts
P_i	Mechanical power into generator	watts
W_1, W_2	Wattmeter readings	watts
V_L	Generated line voltage	rms volts
I_L	Generated line current	amperes
θ	Angle between V_L and I_L	degrees
$\cos \theta$	Power factor of generator	—
s	Slip	percent, decimal, rev/min, or rad/s
f	Frequency of generated voltage	hertz

QUESTIONS

1. Why is the induction generator an asynchronous machine?

2. If the shaft of an induction generator is turned at a high-enough speed, will it generate a voltage? If not, what else is needed?

3. What is the meaning of negative slip?

4. Does the frequency of the generated voltage depend on the rotor speed? Explain.

5. What is the potential danger in using capacitors to correct the power factor of an induction generator?

PROBLEMS

(English)

1. A 2-hp 210-V 6-A 60-Hz eight-pole three-phase induction motor is tested as a generator under load. The results are given below. Assume that the mechanical and core losses are constant at 100 W, and that its equivalent stator resistance is 1.4 Ω.

$$\text{Load test results:} \quad W_1 = 750 \text{ W}, \quad W_2 = 670 \text{ W}$$

$$V_L = 210 \text{ V}, \quad I_L = 5.8 \text{ A}$$

$$S = 925 \text{ rev/min}$$

Find:

(a) The generator's power factor

(b) The generator's efficiency

2. A 14-hp 240-V 35-A 60-Hz 12-pole three-phase induction motor is tested as a generator under load. The results are given below. Assume that the mechanical and core losses are constant at 600 W and that its equivalent stator resistance is 0.7 Ω.

Load test results: $W_1 = 5.2$ kW, $W_2 = 4.5$ kW

$V_L = 240$ V, $I_L = 33.5$ A

$S = 620$ rev/min

Find:
(a) The generator's power factor
(b) The generator's efficiency

3. A 35-hp 240-V 85-A 60-Hz 16-pole three-phase induction motor is tested as a generator under load. The results are given below. Assume that the mechanical and core losses are constant at 1600 W and that its equivalent stator resistance is 0.3 Ω.

Load test results: $W_1 = 12.4$ kW, $W_2 = 11.5$ kW

$V_L = 240$ V, $I_L = 83$ A

$S = 470$ rev/min

Find:
(a) The generator's power factor
(b) The generator's efficiency

4. A 5-hp 240-V eight-pole 60-Hz three-phase induction motor is run at rated power as an induction generator. Its characteristics are given in Figure 11-3. Find:
(a) The reactive power that must be supplied by a capacitor bank to self-excite the generator
(b) The frequency of the generated voltage when the generator is self-excited
(c) The required capacitor bank rating in kvar if those available are rated 480 V, 60 Hz

(SI)

5. A 5-kW 230-V 14-A six-pole 50-Hz three-phase induction motor is tested as a generator under load. The results are given below. Assume that the mechanical and core losses are constant at 280 W and that its equivalent stator resistance is 1.0 Ω.

Load test results: $W_1 = 2.2$ kW, $W_2 = 1.4$ kW

$V_L = 230$ V, $I_L = 12$ A

$\omega = 108$ rad/s

Find:
(a) The generator's power factor
(b) The generator's efficiency

6. A 15-kW 230-V 40-A eight-pole 50-Hz three-phase induction motor is tested as a generator under load. The results are given below. Assume that the mechanical and core losses are constant at 700 W and that its equivalent stator resistance is 0.6 Ω.

Load test results: $W_1 = 6$ kW, $W_2 = 4$ kW

$V_L = 230$ V, $I_L = 35$ A

$\omega = 81$ rad/s

Find:
(a) The generator's power factor
(b) The generator's efficiency

7. A 25-kW 460-V 34-A 16-pole 50-Hz three-phase induction motor is tested as a generator under load. The results are given below. Assume that the mechanical and core losses are constant at 1400 W and that its equivalent stator resistance is 0.4 Ω.

$$\text{Load test results:} \quad W_1 = 9.8 \text{ kW}, \quad W_2 = 6.5 \text{ kW}$$

$$V_L = 460 \text{ V}, \quad I_L = 30 \text{ A}$$

$$\omega = 41 \text{ rad/s}$$

Find:

(a) The generator's power factor

(b) The generator's efficiency

8. A 10-kW 460-V 12-pole 50-Hz three-phase induction motor is run at rated power as a generator. Its characteristics are given in Figure 11-3. Find:

(a) The reactive power that must be supplied by a capacitor bank to self-excite the generator

(b) The frequency of the generated voltage when the generator is self-excited

(c) The required capacitor bank rating in kvar if those available are rated 480 V, 60 Hz

Chapter 12

Single-Phase Motors

The typical home, office, or store has available only single-phase ac as a source of electrical energy. As a result the three-phase motors that we have discussed could not operate in these places. There is a tremendous need for a motor that can operate with single-phase ac as its power source. Among the applications are the following: refrigerators, freezers, washers, dryers, power tools, typewriters, copying machines, heating systems, water pumps, computer peripherals, and various small appliances.

There are three basic types of single-phase motors. First there is the single-phase induction motor. Its theory of operation is similar to that of the three-phase induction motor; hence it runs at a speed slightly lower than synchronous speed. Second, there is the single-phase synchronous motor. This is really an induction motor modified so that it pulls into synchronism and thus runs at synchronous speed. Finally, there is the series or universal motor. Its principle of operation is that of the dc motor; rotational torque is produced by a current-carrying conductor in a magnetic field.

12-1   SINGLE-PHASE INDUCTION MOTOR

"Single-phase induction motor" is a general term for an ac motor that runs at slightly less than synchronous speed. We will soon see, however, that this motor does not produce a starting torque. Some special techniques must be used to start the motor. Depending on the starting technique, different names are given to the motors.

Figure 12-1 is a picture of a single-phase squirrel-cage motor. When an ac voltage is applied to the single-phase stator winding, a pulsating magnetic field is developed. The field polarity will alternate (north and south) along the magnetic field axis. In other words, at one instant the field will be north at the top and south at the bottom. An instant later it will be south on top and north on the bottom. It will not

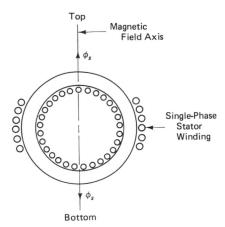

Top

Magnetic
Field Axis

ϕ_s

Single-Phase
Stator
Winding

ϕ_s

Bottom

Figure 12-1 Simplified picture of a single-phase induction motor.

rotate, however. As a result, the rotor conductors will not be cutting flux, neither voltage nor current will be induced in the rotor, and no torque will be developed to turn the rotor.

It should be clear that unless the field can be made to rotate, the motor will never start turning. The different types of induction motors that we will be discussing all develop a rotating magnetic field in one way or another.

Once a single-phase induction motor is turning, a running torque will be developed. This behavior can best be described by what is known as the **cross-field theory.** Another theory, called the double-revolving field theory, is used to explain this behavior. However, it requires a mathematical analysis, hence will not be discussed here.

Assuming that the rotor is turning, its bars (or conductors) will be cutting the stator field flux ϕ_s already described and shown in Figure 12-1. As a result, a voltage is induced in the rotor bars which is in phase with the stator flux. Since the rotor bars are shorted at their ends and are basically inductive, rotor currents that lag the induced rotor voltage by about 90° will flow. The rotor currents will set up a rotor flux ϕ_r in phase with them. Therefore, the rotor flux will lag the stator flux by about 90°; hence it is given the name **cross field.** Figure 12-2 shows the magnetic fields at different instances of time as the rotor makes one clockwise revolution. Keep in mind that since the applied voltage is a sine wave, the fluxes ϕ_r and ϕ_s will also be sinusoidal. Furthermore, due to the 90° phase difference, when ϕ_s is a maximum ϕ_r will be zero and when ϕ_r is a maximum ϕ_s will be zero.

If the rotor is turning at slightly less than synchronous speed, it will make one revolution (360 mechanical degrees) in about the time that the stator voltage takes for one electrical cycle (360 electrical degrees). At some instant of time (t_1) ϕ_s is a maximum and ϕ_r is zero. Ninety degrees (electrical and mechanical) later, at t_2, ϕ_s will be zero and ϕ_r will be at its maximum. As the rotor continues to turn and point X is now 180° from its initial position ϕ_s will be a negative maximum and ϕ_r will be zero. This corresponds to time t_3 in the figure. Finally, 90° later, at t_4, ϕ_s will again be zero and ϕ_r will be a negative maximum.

At time intervals between those shown in Figure 12-2, neither ϕ_s nor ϕ_r is at a maximum or zero but rather at some in-between value. In any event the vector sum

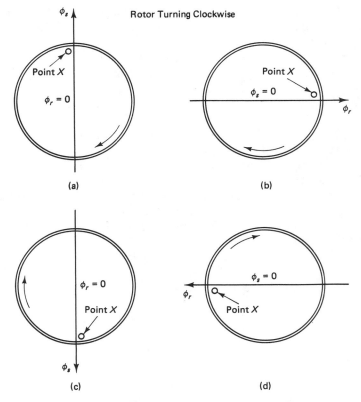

Figure 12-2 Field directions for one clockwise rotation of a rotor ($t_1 < t_2 < t_3 < t_4$): (a) at t_1; (b) at t_2; (c) at t_3; (d) at t_4.

of ϕ_s and ϕ_r produces a resultant flux ϕ' which is rotating at synchronous speed (note that the time it takes ϕ_s to complete one cycle depends on the frequency of the stator voltage).

We can see, then, that a resultant rotating field is established. As long as the rotor is turning slower than this field, its conductors will cut flux lines and a running torque will be maintained. At slow rotor speeds ϕ_r will be small compared to ϕ_s; hence the resultant field ϕ' will be elliptical in shape as shown in Figure 12-3a. As the rotor

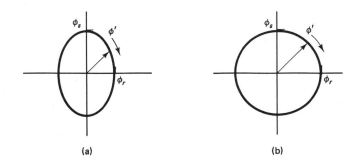

Figure 12-3 Diagrams showing resultant flux amplitude (ϕ') as it makes one clockwise rotation: (a) rotor speed less than synchronous speed; (b) rotor speed approximately equal to synchronous speed.

speed approaches synchronous speed, ϕ_r will approach ϕ_s in magnitude and the resultant flux will be uniform in magnitude (see Figure 12-3b).

Having established the existence of a rotating field, the theory presented in Chapter 9 for the three-phase induction motor can be applied to the single-phase induction motor. The only problem that remains is that of getting the motor started.

12-1.1 Split-Phase Motor

The split-phase motor uses two windings placed 90° apart on the stator. Electrically, they are connected in <u>parallel</u> with a single-phase voltage. A schematic diagram is shown in Figure 12-4a, with the vector diagram shown in Figure 12-4b. One of the windings, called the **main winding,** has many turns of a large wire. This provides for more inductance than resistance. Since the winding is basically inductive, the current through it (I_M) will lag the applied voltage (V_t) by a large angle (θ_1). The other winding, called the **auxiliary winding,** has fewer turns than the main winding and is of a thin wire. This provides for more resistance than inductance. Since this winding is basically resistive, the current through it (I_A) will lag V_t by a small angle (θ_2). The

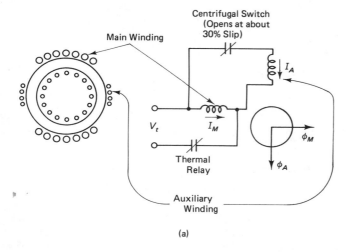

(a)

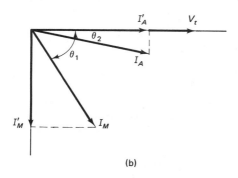

(b)

Figure 12-4 Split-phase induction motor: (a) schematic; (b) vector diagram.

auxiliary winding is also called the **starting winding** since its presence provides a starting torque.

From Figure 12-4b it is clear that I_A and I_M each has a component (I_A' and I_M') at right angles with each other. When this condition exists, the currents are said to be in **quadrature** with each other. In other words, I_A' leads I_M' by 90° or one-fourth of the electrical cycle. These currents will produce flux (ϕ_M and ϕ_A) which is in quadrature and displaced 90° in space. This is exactly the requirement for a rotating field. Since a rotating field is present, a starting torque will be developed. Usually, the angle between I_M and I_A ($\theta_1 - \theta_2$) is about 30°. This results in a rotating field whose magnitude varies significantly as it rotates in space and time. As a result of this nonuniform field, the starting torque will be nonuniform. Its average value, however, is sufficient to start the motor. Ideally, the angle $\theta_1 - \theta_2$ should be 90°. This is, however, physically impossible for this motor. Furthermore, the running characteristics of the motor determine the characteristics of the main winding; hence it cannot be made as inductive as we might think.

The starting winding will have a large I^2R loss and heat up quickly. To prevent it from burning out it is disconnected by a centrifugal switch when the motor reaches about 30% slip. Remember, this winding is only needed to start the motor. As the motor ages, the centrifugal switch contacts will pit and corrode. When this happens they may get stuck in the closed position and not open. To safeguard further against the winding burning up, a thermal relay is also used. If the motor draws the high starting current for more than 5 or 10 s, the relay will open up.

It is interesting to note at this point that when the motor is running (starting winding switched out) the only current it draws is I_M. Referring to Figure 12-4b, we can see that I_M will lag V_t by a large angle θ_1. The motor's power factor will therefore be quite small. This is a disadvantage in addition to its poor starting torque.

Typical sizes are from $\frac{1}{30}$ to $\frac{3}{4}$ hp (25 to 600 W). They normally run at around 4 to 6% full-load slip.

12-1.2 Shaded-Pole Motor

For very small size motors (down to $\frac{1}{2}$ W) the shaded-pole motor is used. It is very inefficient (5 to 20%); however, it is very reliable and does not need any kind of internal switching. Its operation can best be explained by examination of Figure 12-5. The motor shown has two poles. Around a portion of each of the poles is a **shading coil.** It can be one or more turns of wire (or a copper strap) forming a closed circuit. As such, a current will flow in it, which opposes the buildup of flux in that portion of the field pole. The net effect will be to cause a delay in the buildup of flux in that portion of the pole. Hence we will have two components of flux, ϕ_1 and ϕ_2, displaced in time and space. A rotating field will thus be set up and a starting torque developed.

For this motor, to reverse the direction of rotation, another set of shading coils are needed on the opposite edge of the field poles. In addition, switches are needed so that each set of coils can be opened and closed. Note that when a shading coil is open it will have no effect on the main field flux.

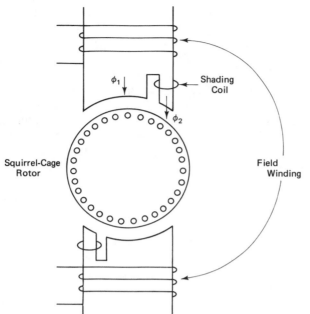

Figure 12-5 Diagram of a shaded-pole motor.

12-1.3 Capacitor-Start Motor

To develop a larger starting torque a capacitor is placed in series with the auxiliary winding of a split-phase motor. The wiring and vector diagrams are shown in Figure 12-6. The effect of the capacitor is to make I_A lead V_t by an angle θ_2. Note that for this motor the angle between I_A and I_M ($\theta_1 + \theta_2$) approaches 90°; thus the condition of quadrature is almost present for the actual winding currents. This angle is typically 80°. Compared to 30° for the split-phase motor, this is a tremendous improvement.

In general, the starting torque for a single-phase induction motor is given by

$$T_s = KI_AI_M \sin \theta_t \tag{12-1}$$

In this equation I_M and I_A are the main and auxiliary currents, respectively, and θ_t is the angle between them. The constant K depends on the windings, rotor resistance, and synchronous speed of the motor. Since sin 80° = 0.98 and sin 30° = 0.5, the starting torque (with everything else equal) would be about twice as much (0.98/0.5) for the capacitor-start motor.

The approximation of perfect quadrature between the two currents is not the only benefit. Since the capacitor makes I_A lead V_t, we do not have to worry about the inductance of the starting winding. It can therefore be made to have more turns and produce the same flux as the main winding. This would provide a uniform rotating magnetic field.

Furthermore, since the angle between I_A and I_M is larger, their vector sum (total motor current I_t) will be smaller. Thus each winding can be designed to draw more current. Thus for a given total motor current, the windings in the capacitor motor will have more current than those in a split-phase motor.

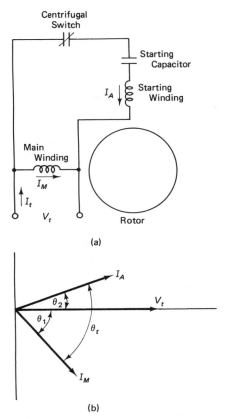

Figure 12-6 Capacitor-start motor: (a) wiring diagram; (b) vector diagram.

The net effect of the capacitor is to give its motor a starting torque of about four times its rated torque. The split-phase motor, on the other hand, produces a starting torque about one to two times its rated torque.

Once the capacitor motor has come up to speed and the starting winding has been disconnected, it will have the same running characteristics as the split-phase motor.

To reverse the direction of rotation of the capacitor-start (as well as the split-phase) motor, the connection of either winding would have to be reversed. Since the starting winding is disconnected at a high speed, this reversal can be accomplished only at standstill or at low speeds when the centrifugal switch is still closed.

The capacitor-start motor is made in sizes from $\frac{1}{4}$ to 10 hp (150 W to 7.5 W). The starting capacitor is the dry-type electrolytic made for ac use. Typical values are from 200 to 600 μF. The ideal value is that which produces quadrature. Values larger than this will only degrade the performance of the motor. Figure 12-7 shows a comparison of torque–slip curves for a split-phase and capacitor-start motor. It also shows the typical effect of the starting capacitor.

In the event of a capacitor failure the motor will not start and a noticeable hum can be heard. The motor can actually be started now (in either direction) by physically turning its shaft. Note that the single phase motor (Figure 12-7) does develop a torque at low speeds (high slip).

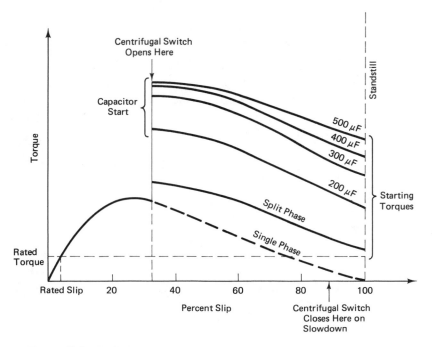

Figure 12-7 Typical torque–slip curves for split-phase and capacitor-start motors.

12-1.3.1 Permanent-Split Capacitor Motor.

A variation of the capacitor start motor is one in which the capacitor and auxiliary winding are not disconnected. The centrifugal switch in Figure 12-6a is eliminated and the auxiliary winding is left in all the time. This motor is called a **permanent-split capacitor** or **capacitor run motor.**

The angle θ_t in Figure 12-6b changes as the load and speed vary. Thus the ideal capacitor for running under load would not be the same as that needed for starting. Furthermore, the capacitors used for starting cannot be used for continuous operation. Since the capacitor used in this motor is in all the time, it must be of a different type, that is, one capable of operating continuously. The net result is that the motor has improved running characteristics; however, it does not provide a starting torque as large as that of the capacitor start motor.

Among the improvements are: higher efficiency and power factor at rated load, lower line current, and very quiet operation. The quiet operation is due to the fact that almost perfect quadrature can be obtained at rated load.

It should also be pointed out that the auxiliary winding must be designed so that it can operate continuously and not burn out. This makes the motor somewhat more costly.

12-1.3.2 Two-Value Capacitor Motor.

This motor (also called **capacitor start–capacitor run**) combines the good features of the capacitor start and permanent-split capacitor motors. This is accomplished by using two different capacitors as shown in Figure 12-8.

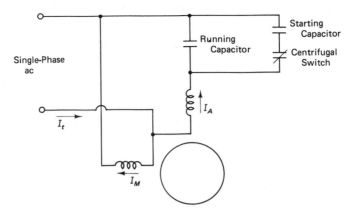

Figure 12-8 Diagram of a capacitor start–capacitor run (two value capacitor) motor.

When the motor starts the switch is closed, both capacitors are used, and the motor develops a high starting torque. As the motor comes up to speed, the switch opens, removing the starting capacitor. The motor now runs with the smaller running capacitor, which will provide optimum characteristics at rated load.

The starting capacitor is of the dry electrolytic type, typically 300 to 500 μF. The running capacitor is an oil-filled capacitor typically around 20 μF.

12-1.4 Repulsion-Start Motor

This type of motor was popular years ago but has since been replaced by the capacitor-start motor. The armature is similar to that of a dc motor having a commutator and brushes. The stator winding is similar to that on the induction motor. In this motor the brushes are shorted together and the brush rigging is movable. Since the brushes are shorted, an armature current will flow due to induction. If the brushes are mounted in line with the field axis or perpendicular to it, no torque will be developed. However, if they are shifted to an angle between these two positions, a starting torque will be developed and the motor will turn. As the motor builds up speed, a centrifugal device shorts out the entire commutator. The motor continues to operate, behaving very much like a squirrel-cage induction motor.

12-2 SERIES MOTOR

It was shown in Chapter 5 that a self-excited dc motor would continue to turn in a given direction even if the polarity of the applied voltage was reversed. This implies that dc motors could be operated just as well with an ac voltage applied. This is not entirely true. The shunt motor would hardly run at all for the following reasons. The field winding had many turns of a thin wire. This high resistance winding produced a strong flux due to the large number of turns. If an ac voltage was applied, the reactance of the field winding would result in a minimal field current. The flux would therefore be very weak. Furthermore, the inductance of the winding would make the

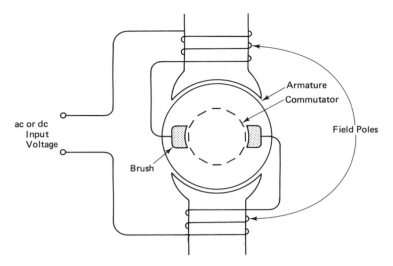

Figure 12-9 Simplified diagram of a universal motor.

flux lag the armature current by a considerable amount. The net effect of these conditions would be the development of an insufficient torque to start the motor.

In the case of the series motor these two problems do not exist. First, the field and armature currents are one and the same; therefore, the flux would always be in phase with the armature current. Second, the series-field winding has only a few turns of a heavy wire. Its reactance would therefore be negligible and the field current would be the same whether ac or dc was applied.

Some other modifications are made so that the series motor can operate equally well on ac or dc. When this is done it is called a **universal motor.**

To decrease the eddy current and hysteresis losses, which are larger with ac operation, the field pole is laminated. In addition, fewer turns are used for the field winding. A weaker flux will result in a lower core loss. To make up for the decreased flux, more armature turns are used. Furthermore, the field poles are made larger. The increased area decreases the flux density, which further decreases the core losses.

It is important to note that the universal motor does not depend on a rotating magnetic field for its operation. Torque development in this motor is due to the force on a current-carrying conductor in a magnetic field (see Section 2-4). Its appearance, except for the modifications mentioned above, is just like the dc series motor. A diagram is shown in Figure 12-9.

This motor has the same characteristics as the dc series motor. It provides relatively large torque for a small-size motor. It is ideal for many household appliances because it can run at extremely high speeds. Typical applications in the home are food blenders, electric tools, vacuum cleaners, and fans. In many cases the speed is reduced by the use of gears. The gear train also serves the purpose of a small load on the motor, which eliminates the danger of starting a series motor under no load (see Section 5-7.2). When there is a danger of running at too high a speed at no load, a speed governor will be used. This is typically a centrifugal switch which at some very high speed places a resistor in series with the armature. This reduces the armature voltage, which in turn lowers the speed.

12-3 SINGLE-PHASE SYNCHRONOUS MOTOR

The single-phase synchronous motor, as its name implies, runs at synchronous speed. It finds use where a constant speed is needed, such as in turntables and clocks. It is started in the same way as any of the single-phase induction motors and therefore has a rotating field. By having a modified rotor the motor pulls into synchronism and runs at synchronous speed.

12-3.1 Reluctance Motor

The reluctance motor is the single-phase version of the three-phase synchronous motor described in Section 10-3.2. Its rotor is similar to the one shown in Figure 10-7. Because the rotor is not cyclindrical there is a large loss of power (50 to 70%) when compared to a comparable-size induction motor. The motor does, however, provide a good starting torque. It is commonly used in timers, recorders, and turntables.

If the load it is driving increases significantly, the motor will slip out of synchronism. It will, however, continue to run with some slip just like an induction motor.

12-3.2 Hysteresis Motor

The hysteresis motor has a special rotor made of a special steel alloy. The material is such as to have a very large hysteresis (see Section 1-2.1). As the rotating flux passes the rotor, a starting torque develops due to the hysteresis in the steel. The rotor is further modified to have bars or cross-members running through it, as shown in Figure 12-10. The purpose of these cross-members is to provide a low-reluctance path which locks with the rotating field. This happens at very small slips and hence the motor runs at the speed of its revolving field, synchronous speed. To operate effectively the motor needs a uniform rotating field. Its stator is therefore of the shaded-pole or permanent-split capacitor type.

The hysteresis motor is very quiet and smooth running. Furthermore, it continues to run at synchronous speed when the load changes. The only effect is to change its torque angle. If the load increases sufficiently, increasing the torque angle beyond its maximum value, the motor will stall.

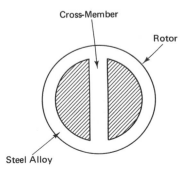

Figure 12-10 Typical hysteresis motor rotor.

12-4 CHARACTERISTICS

Typical characteristics of single-phase motors are presented in Figures 12-11 to 12-13. They are introduced without any kind of mathematical derivation. Their purpose is merely to compare the different motors described in this chapter.

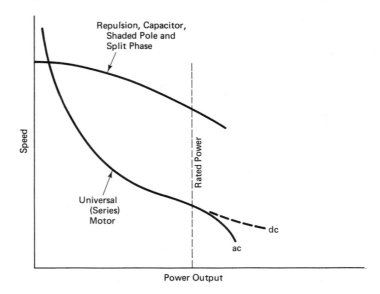

Figure 12-11   Speed versus power output for single-phase motors.

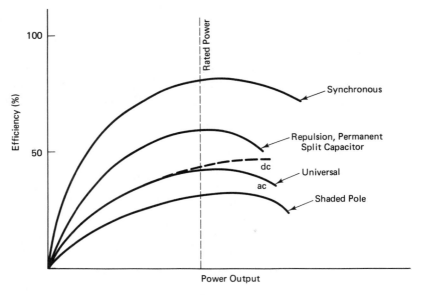

Figure 12-12   Efficiency versus power output for single-phase motors.

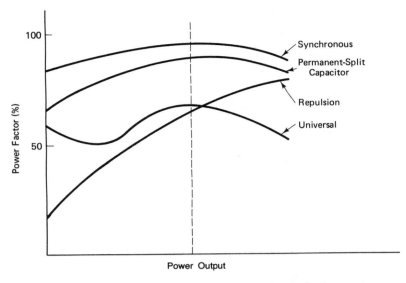

Figure 12-13 Power factor versus power output for single-phase motors.

Example 12-1 (English)

Given a 2-hp 220-V 60-Hz 1725-rev/min single-phase induction motor. It has an efficiency of 70% and operates at a power factor of 60% lagging when supplying rated load. How many poles does it have, and what line current does it draw?

Solution

Since its full-load speed is 1725 rev/min, its synchronous speed is obviously 1800 rev/min. Using Eq. 9-1a, we have

$$\mathbf{P} = \frac{120 \times 60 \text{ Hz}}{1800 \text{ rev/min}}$$

$$= 4 \text{ poles}$$

$$P_o = 2 \text{ hp} \times 746 \text{ W/hp} = 1492 \text{ W}$$

$$P_i = \frac{P_o}{\eta} = \frac{1492 \text{ W}}{0.70} = 2131.4 \text{ W}$$

$$\text{PF} = \cos \theta = 0.60$$

Using Eq. 7-11, we have

$$P_w = VI \cos \theta$$

$$I = \frac{P_w}{V \cos \theta}$$

$$= \frac{2131.4}{220 \text{ V} \times 0.60} = 16.1 \text{ A}$$

Example 12-2 (SI)

Given a 1-kW 230-V 50-Hz 300-rad/s single-phase induction motor. It has an efficiency of 65% and operates at a power factor of 55% lagging when supplying rated load. How many poles does it have, and what line current does it draw?

Solution

Since its full-load speed is 300 rad/s, its synchronous speed is obviously 314 rad/s (100π). Using Eq. 9-1b, we have

$$\mathbf{P} = 4\pi \times \frac{50 \text{ Hz}}{100\pi} = 2 \text{ poles}$$

$$P_o = 1 \text{ kW}$$

$$P_i = \frac{P_o}{\eta} = \frac{1000 \text{ W}}{0.65} = 1538.5 \text{ W}$$

$$\text{PF} = \cos\theta = 0.55$$

Using Eq. 7-11 gives us

$$P_w = VI \cos\theta$$

$$I = \frac{P_w}{V \cos\theta}$$

$$= \frac{1538.5 \text{ W}}{230 \text{ V} \times 0.55}$$

$$= 12.16 \text{ A}$$

Example 12-3 (English)

A 440-V 10-hp capacitor-start motor operates at a power factor of 85% lagging and draws 26 A at full load. What is its efficiency?

Solution

$$P_o = 10 \text{ hp} \times 746 \text{ W/hp} = 7460 \text{ W}$$

$$P_i = VI \cos\theta = 440 \text{ V} \times 26 \text{ A} \times 0.85$$

$$= 9724 \text{ W}$$

$$\eta = \frac{P_o}{P_i} \times 100$$

$$= \frac{7460}{9724} \times 100$$

$$= 77\%$$

Example 12-4 (SI)

A 230-V 1.5-kW capacitor-start motor operates at a power factor of 80% lagging and draws 12 A at full load. What is its efficiency?

Solution

$$P_o = 1.5 \text{ kW} = 1500 \text{ W}$$

$$P_i = VI \cos \theta = 230 \text{ V} \times 12 \text{ A} \times 0.80$$

$$= 2208 \text{ W}$$

$$\eta = \frac{P_o}{P_i} \times 100$$

$$= \frac{1500}{2208} \times 100$$

$$= 68\%$$

Example 12-5 (English)

A split-phase motor is rated 115 V, 7.2 A, 60 Hz, 1725 rev/min. Its efficiency is 60% and the power factor is 50% lagging at rated load. Find its rated power and torque.

Solution

$$P_i = VI \cos \theta$$

$$= 115 \text{ V} \times 7.2 \text{ A} \times 0.50$$

$$= 414 \text{ W}$$

$$\eta = \frac{P_o}{P_i} \times 100$$

$$P_o = \eta \times \frac{P_i}{100}$$

$$= 60\% \times 414 \text{ W}/100\%$$

$$= 248.4 \text{ W}$$

$$P_o = \frac{248.4 \text{ W}}{746 \text{ W/hp}} = 0.33 \text{ hp}$$

or

$$P_o = \tfrac{1}{3} \text{ hp (rated power)}$$

Rated torque can be found using Eq. 5-3a.

$$T = 7.04 \times \frac{248.4 \text{ W}}{1725 \text{ rev/min}}$$

$$= 1 \text{ ft-lb (rated torque)}$$

Example 12-6 (SI)

A split-phase motor is rated 230 V, 2.2 A, 50 Hz, and 152 rad/s. Its efficiency is 55% and the power factor is 45% lagging at rated load. Find its rated power and torque.

Solution

$$P_i = VI \cos \theta$$

$$= 230 \text{ V} \times 2.2 \text{ A} \times 0.45$$

$$= 227.7 \text{ W}$$

$$\eta = \frac{P_o}{P_i} \times 100$$

$$P_o = \eta \times \frac{P_i}{100}$$

$$= 55\% \times 227.7 \text{ W}/100\%$$

$$= 125.23 \text{ W} \approx 125 \text{ W (rated power)}$$

$$= 0.125 \text{ kW}$$

Rated torque can be found using Eq. 5-3b.

$$T = \frac{1000P}{\omega}$$

$$= 1000 \times \frac{0.125 \text{ kW}}{152 \text{ rad/s}}$$

$$= 0.82 \text{ N-m (rated torque)}$$

SYMBOLS INTRODUCED IN CHAPTER 12

Symbol	Definition	Units: English and SI
ϕ_s	Stator field flux due to stator current	lines or webers
ϕ_r	Rotor flux due to rotor current	lines or webers
ϕ'	Resultant flux: vector sum of ϕ_s and ϕ_r	lines or webers
V_t	Applied single-phase voltage	volts
I_M	Main stator winding current	amperes
I_A	Auxiliary stator winding current	amperes
θ_1	Angle between V_t and I_M	degrees
θ_2	Angle between V_t and I_A	degrees
I_A'	Component of I_A in phase with $V_t (= I_A \cos \theta_2)$	amperes
I_M'	Component of I_M in quadrature with $V_t (= I_M \sin \theta_1)$	amperes
T_s	Starting torque of single-phase motor	ft-lb or N-m
I_t	Total motor current: vector sum of I_A and I_M	amperes
θ_t	Angle between I_A and I_M in a capacitor-start motor	degrees

QUESTIONS

1. Why is there a need for single-phase ac motors?
2. What are the three basic types of single-phase motors?
3. What is meant by the term "cross field"?
4. Name four different single-phase induction motors?
5. Why does a split-phase motor have a very low power factor while running?
6. What is the purpose of the centrifugal switch in a split-phase motor? What will happen if the switch fails?
7. What is another name for the auxiliary winding in a split-phase motor?
8. What is a shading coil?
9. Why is a capacitor used in the capacitor-start motor?
10. Compare the split-phase with the capacitor-start motor as far as starting and running characteristics.
11. How could we reverse the direction of rotation of a split-phase or capacitor-start motor?
12. What is a permanent-split capacitor motor?
13. What is a two-value capacitor motor?
14. Does a motor exist that can work equally well with ac or dc voltages? If so, what is it called?
15. What kind of motor is used for clocks? Why is it used?
16. Which single-phase motor has the poorest speed regulation?
17. Which single-phase motor is more efficient at full load, the universal or synchronous?

PROBLEMS

(English)

1. A 3/4-hp 120-V 60-Hz split-phase motor has a power factor of 55% lagging and is 65% efficient at rated load. What line current should it draw at full load?
2. A 240-V 5-hp 60-Hz capacitor-start motor has a 60% lagging power factor and draws 38 A at full load. What is its rated efficiency?
3. A split-phase motor is rated 1/4 hp, 115 V, 4.5 A, 1140 rev/min, and is 65% efficient at full load. Find:
(a) Its full-load power factor
(b) Its rated torque
4. A 2-hp 240-V 60-Hz capacitor-start motor is 75% efficient and draws 12 A at full load. Find:
(a) Its full-load input power:
(b) Its rated power factor
5. A permanent-split capacitor motor is rated 1/2 hp, 115 V, 1110 rev/min, 6 A, and has a 60% lagging power factor. Find:
(a) Its full-load efficiency
(b) Its rated torque
6. A capacitor start–capacitor run motor is rated 3/4 hp, 115 V, 1725 rev/min, 7.8 A, and is 90% efficient at full load. What is its full-load power factor?

7. A 2-hp 115-V 60-Hz 3450-rev/min capacitor-start motor draws a full-load current of 26 A at a 60% lagging power factor. Find:
(a) Its rated torque
(b) Its rated efficiency

8. A split-phase motor rated 115 V, 1140 rev/min, 4 A, 60 Hz has a 55% lagging power factor and puts out 0.77 ft-lb at full load. Find:
(a) The rated output in horsepower
(b) Its full-load efficiency

(SI)

9. A 0.5-kW 230-V 50-Hz split-phase motor has a power factor of 50% lagging and is 60% efficient at rated load. What line current should it draw at full load?

10. A 460-V 3.5-kW 50-Hz capacitor-start motor has a 63% lagging power factor and draws 15 A at full load. What is its rated efficiency?

11. A split-phase motor is rated 230 V, 200 W, 2.5 A, 99.4 rad/s, and is 60% efficient at full load. Find:
(a) Its full-load power factor
(b) Its rated torque

12. A 1.5-kW 230-V 50-Hz capacitor-start motor is 78% efficient and draws 13 A at full load. Find:
(a) Its full-load input power
(b) Its rated power factor

13. A permanent-split capacitor motor is rated 300 W, 230 V, 96.8 rad/s, 2.6 A, and has a 60% lagging power factor. Find:
(a) Its full-load efficiency
(b) Its rated torque

14. A capacitor start–capacitor run motor is rated 600 W, 230 V, 150.4 rad/s, 3.9 A, and is 86% efficient at full load. What is its full-load power factor?

15. A 1.5-kW, 230-V, 50-Hz, 300.8-rad/s capacitor-start motor draws a full load current of 10 A at a 75% lagging power factor. Find:
(a) Its rated torque
(b) Its rated efficiency

16. A split-phase motor rated 230 V, 99.4 rad/s, 1.2 A, 50 Hz, has a 50% lagging power factor and puts out 0.6 N-m at full load. Find:
(a) The rated output in watts
(b) Its full-load efficiency

Chapter 13

Control of AC Machines

The ac motors we have discussed so far fall into three categories: those that run at the speed of their rotating field (synchronous), those that turn at a speed somewhat slower than their rotating field (asynchronous), and those that do not have a rotating field (universal or series motor). There are several techniques used to control the speed of these motors. What works well with one type of motor may not work well with another. The different methods of speed control and the motors they apply to will be covered here.

If a motor is of the synchronous type, it will run at a constant speed equal to its synchronous speed. Equations 9-1 tell us that this speed is strictly a function of the frequency of the applied stator voltage and the number of poles that the motor is wired for. Thus to vary a synchronous motor's speed we must change either f or P in Eqs. 9-1.

If the motor is of the asynchronous type, it will run at a speed given by Eqs. 9-7. Referring to these equations we see that the frequency and number of poles will also vary the speed of the asynchronous motor. In addition, we see that by changing the slip the motor's speed can be varied.

If the motor is a universal motor the techniques presented in Chapter 6 for the dc motor can all be used. These were field control, armature resistance control, and armature voltage control.

No matter what type of controller is used, it ultimately must vary one of the quantities mentioned above. There are several books available on motor control which cover the different schemes in depth. We will discuss them only in principle.

It should be pointed out that speed variation can be accomplished outside the motor by mechanical means such as gears and clutches.

13-1 POLE CHANGING

One of the ways to change the number of poles is to have two or more sets of independent windings on the stator. Alternatively, the various parts of one winding can be regrouped to form different numbers of poles. In either case switches are needed to accomplish the pole changing. Since the number of stator and rotor poles must be the same this is only practical on motors with a squirrel-cage rotor. In this type of rotor the poles are formed by magnetic induction and hence change automatically as the stator field changes its structure. To change the number of poles on a wound rotor or that of a synchronous motor would require several rotor windings and additional slip rings as well. It is therefore impractical to change the poles on these motors. When pole changing is done, the motor is called a **multispeed motor.** The speed selection is limited to a discrete set of speeds rather than a continuously variable speed selection. When pole changing is done, the voltage must very often be changed in order for the motor to operate properly. This is because a regrouping of the stator coils will change the flux strength, which in turn changes the required voltage.

Typically, if a four-pole motor is rated for 230 V, doubling the poles to eight will halve the rated voltage, making it 115 V. In addition to the voltage change, the power factor, power output, line current, and efficiency will drop, and the percent slip will increase when the number of poles increases.

13-2 SLIP CONTROL

Obviously, this type of control can only be used with asynchronous motors since they run with a slip. There are a few different techniques used. They all invariably change the running slip of the motor, hence its speed.

13-2.1 Line Voltage Control

By lowering the line voltage on an induction motor, the flux will decrease. From Eq. 9-15 it can be seen that the running torque will decrease as well. The slip must then increase (motor slow down) to make up for the loss in flux and maintain the required torque. This technique works well in single-phase induction motors, where the rotating flux is due to two independent oscillating fields (see Section 12-1). It also works in three-phase induction motors, however, not as well. When the stator voltage decreases, the motor torque decreases rapidly but the slip does not change much. As a result, the voltage must decrease considerably before the motor begins to slow down. When this finally occurs, the motor torque might be less than the maximum torque at the reduced voltage. If this happens, pull-out will occur and the motor will stall. Furthermore, if the motor does not stall a relatively spongy or unstable speed control can result due to the motor's long reaction time to line voltage changes. This technqiue works somewhat better when the load is viscous in nature, that is, decreases considerably with a reduction in speed.

13-2.2 Rotor Resistance Control

When the induction motor has a wound rotor, the speed can be controlled by varying the rotor resistance. We saw in Chapter 9 that this was done with a slip-ring rheostat. Figure 9-11 shows us that for a given torque (rated torque, for example) the slip increases as the rotor resistance increases. This technique works quite well for large loads. When the load is small, however, the variation in slip (hence speed) is very little in response to rotor resistance changes. The net effect is very poor speed regulation. In other words, if the resistance is adjusted for a speed at a light load, and the load now increases, there will be a big drop in speed. Furthermore, because of the increased rotor resistance the rotor efficiency, hence motor efficiency, drops considerably.

13-2.3 Rotor Voltage Control

In Section 13-2.2 we saw that increased rotor resistance decreased the motor speed. In effect, by increasing its resistance, the rotor was made to absorb (or dissipate) more of the input power. As a result, there is less power developed, hence a slower shaft speed.

It is logical to conclude, then, that if power were inserted into the rotor, the developed torque, hence speed, would increase. This is actually the case, and by adding or subtracting power to the rotor circuit, the speed can be increased or decreased. In fact, the speed can be increased so that the motor runs above synchronous speed. It is important to note that the voltage, and its frequency, introduced in the rotor circuit are critical. They must be in accordance with Eqs. 9-9 and 9-12. Thus for a motor to run at a speed 50% greater than synchronous speed, the following would be true. The slip would be -0.5; hence the voltage must be $E_{BR}/2$ and its frequency would have to be one-half the blocked rotor frequency (line frequency).

One of the common types of motors using this technique of speed control is called the **Schrage brush-shift motor.** It is also referred to in industry as a BTA motor.

13-3   FREQUENCY CONTROL

The final way in which speed can be controlled in motors where a rotating field exists is by varying the line frequency. The synchronous speed of the motor is directly proportional to the frequency of the applied stator voltage.

This is by far the most ideal way of speed control. The only problem is that of varying the frequency. In recent years, however, the developments in solid-state technology have led to practical ac motor speed controls utilizing the adjustable frequency technique. There is one additional problem that must be mentioned. The stator of an ac motor is analogous to that of a transformer. If we examine Eqs. 7-13, we can see that a reduction in frequency at constant voltage means that the flux must increase. An increase in flux means more core losses; hence the motor will overheat. Furthermore, increasing the frequency will lower the flux, which decreases the torque in accordance with Eq. 9-15. This problem is eliminated if the voltage is varied

proportionately with the frequency. Most adjustable-frequency drives do this automatically. If they do not, the motor would have to be overrated so that it could handle
the buildup of heat at low speeds. An alternative to the large motor would be a cooling
fan driven separately. Figure 13-1 shows the voltage-frequency variation for a 60-Hz
motor rated 230 V.

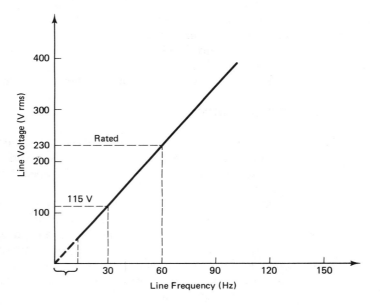

Figure 13-1 Variations of line voltage with frequency to control motor speed.

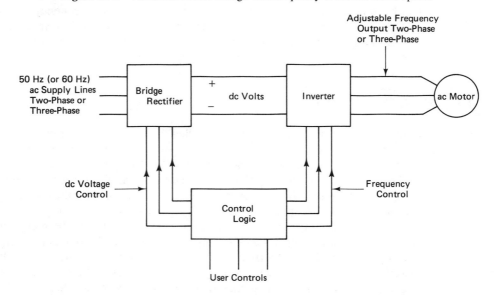

Figure 13-2 General block diagram of an adjustable frequencey drive.

The drive used for this motor would need a constant volt/cycle rating. In other words, if its output at 60 Hz were 230 V, at 30 Hz the output would drop to 115 V. Speed controllers are available which can vary the frequency from 6 to 100 Hz with ratings up to 75 hp (56 KW). A single drive unit can control several motors; however, you are limited to the control unit's rating.

Adjustable-frequency drives all do the same basic thing. A general block diagram of the process is shown in Figure 13-2. The actual components within the boxes differ since these can be done in different ways.

The purpose of the bridge rectifier is to convert the available ac power to dc. It is either a single-phase or a three-phase full-wave bridge. A picture of each is shown in Figure 13-3. Note that instead of using all diodes, half of each leg uses an SCR (see

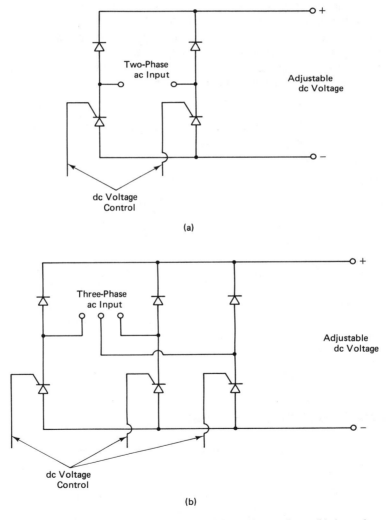

(a)

(b)

Figure 13-3 Adjustable output full-wave bridges: (a) two phase; (b) three phase.

Chapter 6). By varying the firing angle of the SCRs, the average output (dc voltage) is controlled in accordance with Figure 6-6f.

The inverter in Figure 13-2 converts the dc voltage to an ac voltage whose magnitude and frequency vary. Inverters are covered in detail in Section 13-6. The magnitude of the ac voltage out of the inverter is proportional to its dc input. The frequency of the output is a function of the signals coming from the control logic.

The user controls are for starting, stopping, and adjusting the speed of the motor. It should be noted that the output of the inverter used does not have to be sinusoidal in appearance. In fact, it is typically a series of dc levels that approximate a sine wave as shown in Figure 13-4. This does not make a difference to the motor and it will operate just as well with this type of signal.

Another scheme that basically does the same thing makes use of digital computer circuitry. Here the binary codes for the different levels approximating a sine wave are stored in a computer memory. The memory output (in binary) is then converted to an analog waveform (sine wave). By varying the rate at which the data comes out of the memory, the frequency of the sine wave can be controlled. This can be done with an adjustable clock or a computer program.

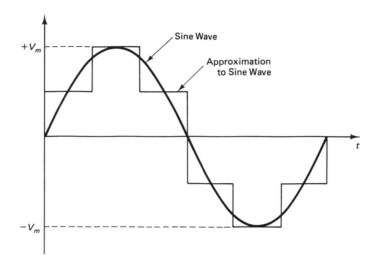

Figure 13-4 Typical inverter output of an adjustable frequency drive.

13-4 *UNIVERSAL MOTOR CONTROL*

Since the universal motor is in effect a dc series motor, its speed can be varied in any of the ways discussed in Chapter 6. These were field control, armature current, and armature voltage control.

Universal motors find much use where high-speed and/or large ranges of speed control are needed. Typical applications in the home are shavers, vacuum cleaners, blenders, sewing machines, and portable hand tools. The most common way of speed control is by variation of the armature voltage. This is most often done by changing a resistance in series with the armature. Continuous speed control, such as that in a sewing machine, can be obtained with a variable resistance (rheostat). Incremental speed control, such as that in a blender, can be obtained with pushbutton switches and different resistors. In any case tremendous ranges of speed control, from starting to 20,000 rev/min (2100 rad/s) and more can be obtained quite easily. Solid-state electronics (the DIAC, TRIAC, and SCR) are becoming increasingly popular. By using them for speed control as shown in Figures 6-13 and 6-17, the energy waste due to a series resistance is eliminated. As a direct result, better efficiency is obtained.

13-5 COMPUTER CONTROL

The computer control schemes covered in Section 6-2.3 will work quite well for the universal motor. Although both of these schemes make use of an ac voltage, they eventually drive the motor with a variable dc voltage. As we have seen, however, the universal motor will work equally well on either ac or dc. In this section we examine a technique that controls the speed of a single-phase induction motor. It will do this by varying the magnitude of the stator voltage.

Consider the configuration shown in Figure 13-5. The input is a pulse train that goes high (+5 V) a controlled time t_1 after the line voltage has a zero crossing. When this happens, the TRIAC driver and the TRIAC are both turned on. When the TRIAC is turned on, it is like a closed switch; thus the line voltage is applied to the stator. These devices were covered in Section 6-2.2.3 and 6-2.2.4. By varying t_1, the firing delay angle of the TRIAC is controlled. This in turn varies the rms value of the stator voltage. The *RC* snubber is included to eliminate the effect of transients on the TRIAC's operation. These transients are caused by the switching of an inductive load. Note the similarity between the load voltage in Figure 6-9 and the stator voltage in Figure 13-6. The only difference is that the former is a dc signal, whereas the latter is an ac signal. This is the case because a TRIAC is being used here rather than an SCR.

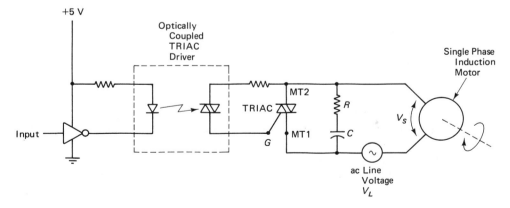

Figure 13-5 TRIAC control of stator voltage.

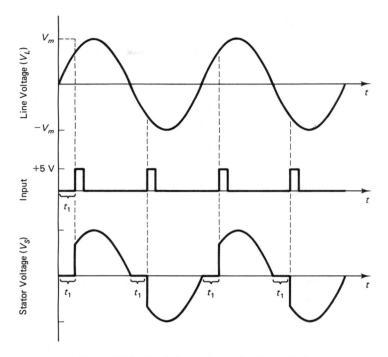

Figure 13-6 Typical waveforms for Figure 13-5.

The overall control scheme using a microprocessor is shown in Figure 13-7. The zero voltage switch was explained in Section 6-2.3 (see Figure 6-15). The TRIAC control circuit of Figure 13-5 is shown as a block to simplify the picture. This scheme is basically the same as that shown in Figure 6-23. It has been simplified in the following ways:

1. The command speed is entered manually with a potentiometer rather than digitally with software.
2. As a result of item 1, the digital-to-analog converter (DAC) is not needed.
3. As a result of item 2, the peripheral interface adapter (PIA) is not needed. Note that only three 1-bit input/output ports are required. These are ports A, B, and C in Figure 13-7.

The operation of the speed control scheme can be described as follows. The motor speed is sensed by a tachometer and compared to the speed command. As long as the comparator output is high (an indication that the motor is running too slow) the computer program will decrease the firing delay angle of the TRIAC. This will increase the rms voltage applied to the motor's stator winding. The result will be a higher motor speed. This process continues until the comparator output goes low. This is an indication that the motor's speed is greater than or equal to the commanded

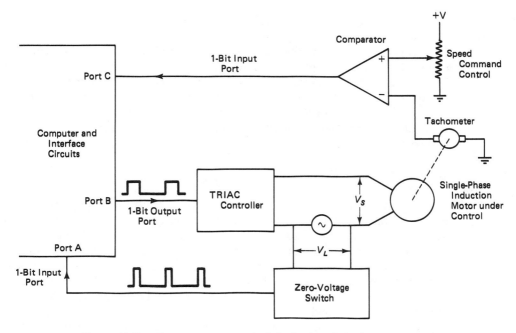

Figure 13-7 Microprocessor control of single-phase induction motor speed.

speed. At this point the firing delay angle is increased slightly lowering the rms stator voltage. In its steady-state condition the motor's speed will fluctuate slightly around the commanded speed.

The scheme is crude, but it will work. How well it works depends on the mechanical time constant of the system and how involved the computer program is. It could be made to work much better by including a proportional control of the motor's acceleration. By sensing the actual difference (error) between the tachometer output and the command voltage, the stator voltage could be varied proportionally. In this way the rms line voltage could be made relatively high on starting, causing the motor to accelerate quickly. As the motor approaches the commanded speed, the error decreases causing a proportionate decrease in stator voltage.

By increasing the amount of hardware and software, system operation can be optimized. The end result will be, as always, a more costly system.

13-6 INVERTERS

An **inverter** has already been defined as a device that converts dc power to ac power. Its function, then, is opposite to that of a rectifier. As a matter of fact, a circuit that looks just like a bridge rectifier is used by inverters. There are some differences, though. The inverter does not use diodes, but SCRs. As a result, some additional circuitry is required to fire the SCRs. This is done by applying a voltage to their gates at appropriate times (see Section 6-2.2.1).

Inverters are finding increased popularity in alternate energy systems. In many cases the alternate energy (sun, wind, water power, etc.) is converted to dc power. This power is then converted to ac power by an inverter and connected directly to a utility grid. When this is done by a utility company, its primary interest is to decrease the consumption of its power generated from fuel (oil, natural gas, etc.). This type of installation is called **supplemental.** On the other hand, if the installation is in a home, and some ac power is fed back to the utility for credit, it is called **cogeneration.**

13-6.1 Theory of Operation

An inverter's operation can be understood quite easily by examining Figure 13-8. Figure 13-8a shows a single-phase full-wave bridge rectifier. The circuit in Figure 13-8b looks very similar; however, it is made with SCRs. It is the circuit used in an inverter. By applying appropriate signals to the SCR gates, they can be turned on and off in the following way. Referring to Figure 13-8b, if SCRs, numbered 1 and 2, are fired at the same time, they will be on. At this time SCRs 3 and 4 are off and terminal X is connected to the positive dc while Y is connected to the negative dc. Conversely, if SCRs numbered 3 and 4 are fired at the same time, they will be on. During this time interval SCRs 1 and 2 are off and terminal X is now connected to the negative dc while

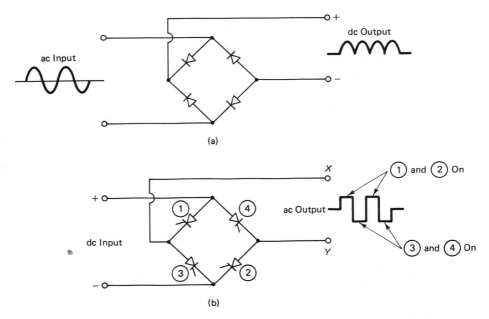

Figure 13-8 Comparison of bridge circuits: (a) single-phase full-wave bridge rectifier; (b) single-phase inverter circuit.

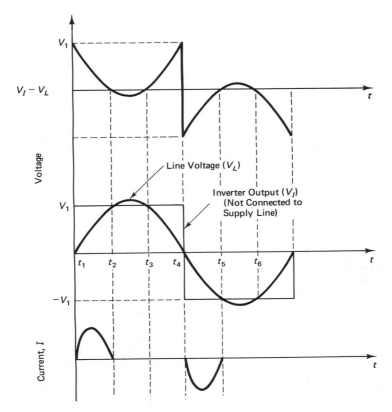

Figure 13-9 Typical voltage and current waveforms from an inverter connected to an ac supply line.

terminal Y is connected to the positive dc. Note that the dc polarity has been "inverted."

Although the output in Figure 13-8b is not a sine wave, it is alternating in nature. The rate at which the SCRs are turned on and off determines the frequency of the inverter output. To ensure that this frequency is the same as the line frequency, the sinusoidal line voltage is used to generate the SCR triggering pulses. When this is done, the inverter is said to be **line-commutated.*** The inverters used in adjustable-frequency ac motor drives must generate a whole range of frequencies. In addition, they are connected to a motor rather than an ac supply line. This makes them more complicated. Not only must they produce a voltage that more closely resembles a sine wave (Figure 13-4), but they must also generate the SCR gate triggering signals at different frequencies. Since they do this, they are said to be **self-commutated.**

Figure 13-9 shows a line-commutated inverter output waveform superimposed on the ac line voltage. To understand these waveforms, we must also refer to Figure

*Commutation is the process of switching the SCRs on and off.

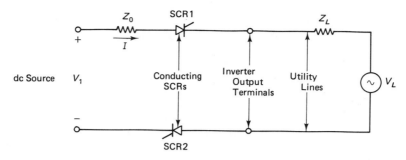

Figure 13-10 Equivalent circuit of a dc source connected through an inverter to an ac supply line during the time interval t_1 to t_4 (see Figure 13-9).

13-10. This is an equivalent circuit showing the source of dc voltage, its output impedance Z_o, and the conducting SCRs connected to the supply line with its impedance Z_L. For inverters to work properly the dc source impedance Z_o must be much larger than Z_L. In this way the voltage at the inverter output will always be sinusoidal. To see why this is so, we will solve for the voltage at the inverter output using superposition. Remember that Z_o is much larger than Z_L.

First, setting V_L to zero, the inverter output due to V_1 will be

$$V_I = V_1 \frac{Z_L}{Z_o + Z_L} \approx 0 \tag{13-1}$$

Since $Z_o \gg Z_L$, V_I will be close to zero. Second, setting V_1 to zero, the inverter output due to V_L will be

$$V_I = V_L \frac{Z_o}{Z_o + Z_L} \approx V_L \tag{13-2}$$

From the calculations above it should be clear that the inverter output will be an undistorted sine wave (the line voltage) as long as $Z_o \gg Z_L$. This assumption is a good one since the source impedance may be 1 or 2 Ω, whereas the line impedance will be very close to zero (note that $1 \gg 0$).

Why, then, should we go through all of this if the dc voltage V_1 is entirely across the source impedance Z_o? The answer is the current in the lines, hence power flow from dc source to ac supply.

Referring again to Figure 13-10, the quantity I is the current delivered by the dc source to the supply line. Since the inverter output is V_L (Eq. 13-2), the current is

$$I = \frac{V_I - V_L}{Z_o} \tag{13-3}$$

The voltage difference $V_I - V_L$ as well as the current I are shown in Figure 13-9. The following should be noted about the current:

1. It is alternating.
2. Although shown in phase with the line voltage, the current will actually lag it. This is due to many factors, one of which is the source impedance.

3. The current (hence power flow) is of a pulsating nature and not a continuous waveform.

4. The current is not sinusoidal. Note that the voltage causing it ($V_1 - V_L$) is not sinusoidal either.

The current is shown only for the time interval t_1–t_2. This is usually the case for line-commutated inverters. The reason for this is the following analysis. At time t_1 a pulse fires SCR1 and SCR2 turning them on. At time t_2 the voltage drop across the SCRs ($V_1 - V_L$) goes through zero, turning them off naturally. The same two SCRs could be fired with a pulse at time t_3. However, at the end of this interval, time t_4, the voltage across the SCRs, is large. Turning them off now, before SCR3 and SCR4 turn on, would require additional elaborate circuitry. Thus in order to keep the inverter relatively simple the SCRs are not fired at t_3 or t_6.

The dc voltage V_1 can be variable, and inverted power will flow into the lines. The larger it is, the more current will be delivered. It should not, however, be greater than the peak value of the ac line voltage. If it is, the natural commutation taking place at time t_2 would not occur.

It should also be pointed out that line-commutated inverters cannot operate in the event of a power-line failure. This is because they depend on the ac voltage for commutation. We can interpret this as both good and bad. It is an advantage over self-commutated inverters in the following way. Failed ac lines can have dc voltages present on them, which could be dangerous to people repairing the lines.

On the other hand, it can be considered a disadvantage. If the inverter is being used in a cogeneration installation and the dc source is a wind or solar generator, loss of the ac line voltage would mean loss of the inverted power as well. In any event, inverters should have relays that disconnect them from the lines and dc source in the event of failures.

13-6.2 Characteristics of Inverters

The efficiency of an inverter is extremely high. It is generally between 90 and 98%. This number, however, is based on the dc power into the inverter and not the total available dc power. In other words, a dc generator capable of delivering 100 W continuously may only average 40 W due to the limited time that the SCRs are conducting. The inverted power is difficult to calculate; hence inverter efficiency is determined from the dc input and the inverter losses. This was the same technique used for motors and thus the same equations can be used.

The efficiency is relatively constant as a function of input current. A typical curve for an inverter whose maximum input current is 50 A is shown in Figure 13-11. The power factor of an inverter is a lagging one and varies as a function of dc voltage and power output. As the dc voltage increases, the conduction time of the SCRs gets longer and the power factor increases proportionately. Figures 13-12 and 13-13 are typical characteristics.

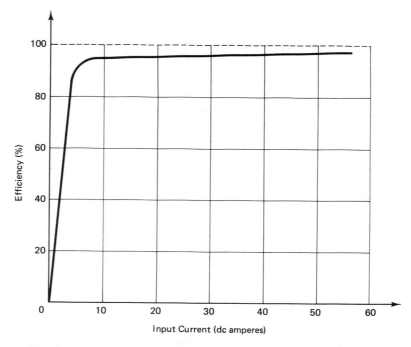

Figure 13-11 Typical curve of inverter efficiency versus dc input current.

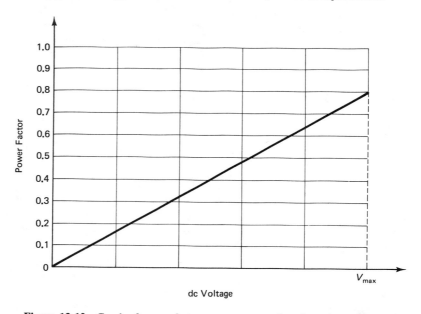

Figure 13-12 Graph of power factor versus average dc voltage for an inverter.

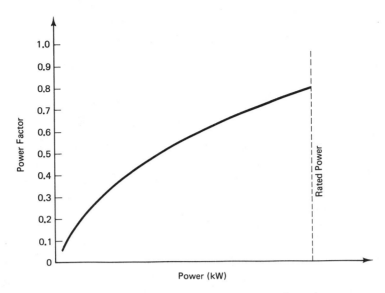

Figure 13-13 Graph of power factor versus power for an inverter.

SYMBOLS INTRODUCED IN CHAPTER 13

Symbol	Definition	Units: English and SI
V_1	Voltage from a dc source applied to an inverter	dc volts
V_f	Inverter output voltge	rms volts
Z_o	Output impedance of dc source	ohms
V_L	Supply line voltage	rms volts
I	Current delivered by dc source to ac supply lines	amperes
Z_L	Impedance seen looking into ac supply lines	ohms

QUESTIONS

1. What are the two ways that the speed of a synchronous motor can be varied?

2. Name three ways in which the speed of an asynchronous motor can be varied.

3. Name three ways in which the speed of a universal motor can be varied.

4. What is a "multispeed" motor?

5. If the number of poles on a motor is increased to lower the speed, what will be the effect on power factor, efficiency, slip, and power output?

6. Can line voltage control be used to vary the speed of a synchronous motor? An asynchronous motor? Explain.

7. What is a BTA motor?

8. If frequency variation is used to control an ac motor's speed, what else must be changed? Why?

9. What are the three basic parts of an adjustable-frequency drive?

10. What type of speed application is a universal motor well suited for? Name some typical applications.

11. What is a snubber circuit? What is its purpose?

12. Figure 13-7 shows a technique for controlling a single-phase induction motor's speed. In this scheme, what is the purpose of the tachometer? Of the zero-voltage switch?

13. What is an inverter?

14. With reference to inverters, what is the difference between a "supplemental" and "cogeneration" installation?

15. What is the difference between line-commutated and self-commutated inverters?

16. Does an inverter operate at a leading, lagging, or unity power factor?

17. Based on your knowledge of power factor, is it better to operate an inverter with a higher or a lower dc voltage?

18. Is there a limit as to how large the dc input to a line-commutated inverter can be? Explain.

Chapter 14

Solar Energy: Photovoltaics

The term "solar energy" refers to energy obtained from the sun. The energy travels from the sun to earth in the form of light. In this form it falls into the general category "electromagnetic energy." When the energy reaches earth, it can be captured by us and changed to other useful forms. The most common form that the sun's energy is converted to is heat. However, in recent years tremendous strides have been taken to efficiently convert light energy directly to electric energy. The term used to describe this conversion is **photovoltaic.**

14-1   *A BRIEF HISTORY*

With the increase in semiconductor technology came the invention of the **photovoltaic cell** (also called **solar cell**) in the early 1950s. Solar cell technology was given a big boost in the following years by the space program. Satellites orbiting the earth for months and years would need a continuous supply of electrical energy. The photovoltaic cell was the answer.

In the mid-1970s society saw the world's oil interests create havoc by controlling the supply of oil. This gave rise to new interest in alternate energy sources. One of these sources was, and still is, the sun. Today much work is being done to generate huge amounts of electrical power directly from the sun. The following are just a few of the many photovoltaic projects undertaken throughout the world:

1. The Solarex Corporation of Rockville, Maryland, has built a solar cell production facility which is powered entirely by a 200-kW photovoltaic system. The solar cells cover more than 3000 m^2 on the roof of the facility. It is not connected

to the local utility grid. It derives backup power from batteries that it charges during peak sunlight hours.

2. The Lockheed Corporation of California is building a foldable wing for a space shuttle. When fully extended the wing will generate more than 12 kW of power from solar cells mounted on it.

3. AEG-Telefunken of Frankfurt, Germany, is building a solar electric generator on the island of Pellworm in the West German North Sea. The solar cells cover 16,000 m² (two football fields) and will generate a maximum of 300 kW. The power will be transformed to 220 V and 380 V at 50 Hz using self-commutated inverters. It will be a cogeneration facility with excess power fed back to the local utility grid.

4. Solar cells manufactured by Siemens AG of Munich, Germany, will be assembled on the Greek island of Cythnos. More than 100 kW of electric power will be produced for use on the island, which has been totally dependent on imported diesel fuel.

5. On February 15, 1983, ARCO Solar of California dedicated a 1-MW photovoltaic power plant. It is located on 20 acres of desert land and can produce 3 million kilowatt-hours of electricity per year. The cells are mounted on panels that track the movement of the sun. The facility is unmanned, with all monitoring and tracking done by computer.

As a result of this increased research and development in photovoltaics, the price of solar cells has steadily decreased. A rule of thumb used to measure the cost of solar electricity is the cost per peak watt that can be obtained from photovoltaics. Since the sunlight is free, this is strictly a function of the solar cell's efficiency and the cost to manufacture the cell. In 1958 the cost was about $2000 per peak watt. By 1970 this had halved to about $1000 per peak watt. In 1972, a high-efficiency cell was invented and the cost dropped to $100 per peak watt. It steadily decreased and by 1980 was about $10 per peak watt. The U.S. Department of Energy has stated a goal for industry to lower the cost to $0.50 per peak watt by 1986.

14-2 THE SUN AND ITS ENERGY

It has been estimated that the sun is constantly emitting about 1.7×10^{23} kW of power. A very small part of this ($\approx 8.5 \times 10^{13}$ kW) reaches the earth. About 30% of this is lost and 70% ($\approx 6 \times 10^{13}$ kW) penetrates our atmosphere. What is really important, though, is the amount of power per unit area that we receive from the sun. The quantity used to describe this is **power density.** With the sun directly overhead on a clear day, the power density of sunlight is about 100 mW per square centimeter, which is equivalent to 1 kW per square meter. The power density of sunlight is also defined with a unit called the "SUN."

$$1 \text{ SUN} = 100 \text{ mW/cm}^2 = 1 \text{ kW/m}^2 \tag{14-1}$$

Thus on a cloudy day the power density of sunlight might be 0.3 SUN or 30 mW/cm².

Another quantity sometimes used to measure sunlight is **energy density.** This is just power density multiplied by time. The unit for it is the langley.

$$1 \text{ langley} = 11.62 \frac{\text{W-h}}{\text{m}^2} \tag{14-2}$$

The following example is introduced merely to help you distinguish between the quantities we have just introduced.

Example 14-1

If the strength of sunlight is 1 SUN for a period of 1 min, find the energy density in langleys.

Solution

$$1 \text{ SUN} = 1000 \text{ W/m}^2 = \text{power density}$$

$$\text{energy} = \text{power} \times \text{time}$$

$$\text{energy density} = 1000 \frac{\text{W}}{\text{m}^2} \times \frac{1 \text{ min}}{60 \text{ min/h}}$$

$$= 16.67 \text{ W-h/m}^2$$

From Eq. 14-2 there are 11.62 W-h/m² for every langley; thus

$$\text{energy density} = \frac{16.67}{11.62} = 1.434 \text{ langleys}$$

In other words, 1.434 langleys of energy are received by us from a bright overhead sun for every minute. It is important to note that if the sunlight had half of its given strength ($\frac{1}{2}$ SUN) we could still get 1.434 langleys, but it would take twice as long (2 min).

At a first glimpse the numbers in the preceding example are really meaningless to us. We can, however, relate the sun's power density to something we are all familiar with. If the roof of a home measuring 9 m × 12 m (about 30 ft × 40 ft) was covered with photovoltaic cells, it could provide all of the electrical energy needed by that home. Example 14-2 will illustrate this. First, however, we need some information that will be explained in detail in Section 14-4. We will assume that the efficiency of a solar cell is 10%. In other words, 10% of the sunlight power that falls on the cell is converted to electric power. This is a conservative figure for efficiency. Furthermore, we will assume that on the average the sun is at its peak (strength of 1 SUN) for 2.5 h every day. Remember this is an average; instantaneously, it is different from hour to hour and day to day. The sun is weaker in the morning and afternoon hours, on cloudy or overcast days, and there may be periods when the cells are in the shade.

Example 14-2

Calculate the average daily electrical energy converted by a rooftop covered with solar cells. The roof measures 9 m × 12 m.

Solution

The area of the roof is

$$9 \text{ m} \times 12 \text{ m} = 108 \text{ m}^2$$

Multiplying this by the power density of a full sun (Eq. 14-1) will give us the amount of power hitting the roof.

$$P = 1 \text{ kW/m}^2 \times 108 \text{ m}^2$$

$$= 108 \text{ kW}$$

If the cells are 10% efficient, the electrical power converted will be

$$P_{elec} = 0.1 \times 108 \text{ kW} = 10.8 \text{ kW}$$

Assuming that a peak sun exists for 2.5 h every day, the average daily energy is

$$10.8 \text{ kW} \times 2.5 \text{ h} = 27 \text{ kWh}$$

Note that this is more than enough to supply the daily needs of the average household in the United States.

It is interesting to note that the rooftop in Example 14-2 would cost more than $100,000 at 1980 prices (10.8 kW × $10 per watt). If, however, the price were to drop to $0.50 per peak watt, the cost would be about $5000 (10.8 kW × $0.50 per watt). This price is certainly attractive when we consider that you would never have to pay another electric bill.

There is actually more involved than what has been presented in Example 14-2. Before we can go into a solar generator in detail, we must understand the solar cell and how it works.

14-3   THE PHOTOVOLTAIC CELL

A solar cell is made of the semiconductor material silicon. This is the same material used for diodes and transistors. Silicon can be extracted from silicon dioxide, the chemical name for ordinary sand. It is extremely plentiful; about 25% of the earth's crust is silicon dioxide.

After pure silicon is obtained, it is melted down and doped with an "impurity." If the impurity is boron, the material formed is called P-type silicon. Doping with phosphorus will form N-type silicon. The molten material is then allowed to harden in a cylindrical shape (usually 3 or 4 in. in diameter).

The material is cut into thin wafers with a saw. The N-type slice is so thin that light will pass through it. The two types are bonded together forming a P-N junction, similar to a diode. When light falls on the N-type material, it penetrates it and enters the P-type material as shown in Figure 14-1.

In the area around the P-N junction, there are free electrons which had combined with holes in the P-type material, forming a depletion region at the junction. When

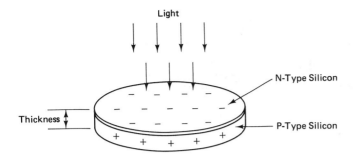

Figure 14-1 Picture of a solar cell.

the light enters the P-type material it imparts energy to the free electrons, causing them to jump up to the N-type material. As long as the light keeps penetrating the cell, the electrons will stay in the top of the cell, creating a difference in charge (voltage). In Figure 14-1 the thickness of the cell is exaggerated. It is actually about as thin as an eggshell. As a result, it should be handled with care. It can be broken very easily.

To make use of the voltage generated by a cell, wires must be connected to it. The bottom, P-type silicon, is generally coated with a good conductor (like solder). A wire is then connected to it and becomes the positive terminal of the solar cell. The N-type material presents a problem, for it is very difficult to connect a wire to the thin N-type layer. Furthermore, we would like to contact the entire surface. If this was done, however, the light could not enter the cell. As a compromise, many thin interconnected wires are bonded to the surface. Fewer slightly larger wires (called **subbuses**) connect the thin wires to a main bus. A wire can now be soldered to the main bus. This wire will be the negative terminal of the solar cell. The actual size and number of the wires on the surface is very critical to the cells efficiency. As more of the surface area is covered with wire, less light will enter the cell. On the other hand, if the wires are too thin and too few, their electrical resistance will be large and less current flow will result.

To reduce the loss of light energy due to reflection, the top surface of the cell is coated with an antireflective material. Diagrams showing typical cells are shown in Figure 14-2. The cell shown in Figure 14-2b is a special cell developed which can withstand a higher power density. The higher power density comes from a lens that focuses and concentrates the sunlight to a smaller area.

14-4   ELECTRICAL CHARACTERISTICS OF A CELL

The solar cell shown in Figure 14-2a will produce a voltage of about 0.57 V under open-circuit conditions when sunlight shines upon it. This means that there is nothing connected to the wires; hence the current is zero (no load). This voltage is the same regardless of how big the cell is. Note the similarity to a flashlight battery, which produces about 1.5 V independent of its size. However, a larger battery can provide

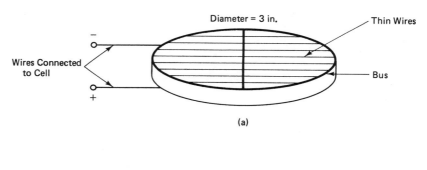

(a)

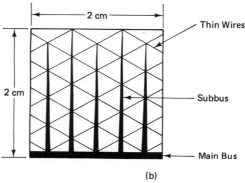

(b)

Figure 14-2 (a) Typical 3-in.-diameter cell; (b) high-efficiency concentrator cell.

more current at that voltage, hence more power. This is similar to a photovoltaic cell, where the current it can supply is directly proportional to the surface area of the cell. The difference between the two is that the solar cell can be considered a constant-current source as well as a voltage source. Unlike the battery or a dc generator, we cannot use equations to predict a photovoltaic cell's output. The solar cell is a nonlinear device; hence we are forced to use a characteristic curve. A typical one is shown in Figure 14-3. Note that when the current is zero (no load), the voltage (called the open-circuit voltage V_{oc}) is 0.57 V. This is point 1 on the curve. As the current increases due to a decreasing load resistance, the voltage stays relatively constant up to point 2 (530 mV, 620 mA). From point 2 to point 3 (the knee of the curve) the voltage drops significantly with an increase in current. At point 3, $V = 400$ mV and the current is approaching its maximum value. If the load resistance decreases to zero (a short circuit), the current stays constant at its short-circuit value ($I_{sc} = 800$ mA) while the voltage drops to zero (point 4 on the curve).

To operate at any particular point of the curve we must connect the necessary load resistance to the cell. The resistance can be found by dividing the voltage by the current at any point. Example 14-3 will illustrate the calculation of the necessary load.

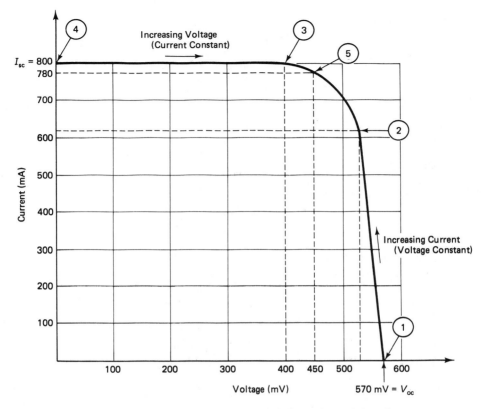

Figure 14-3 Typical *I–V* characteristic for a photovoltaic cell.

Example 14-3

The solar cell whose characteristic is shown in Figure 14-3 is connected to a resistive load as shown in Figure 14-4. Find the necessary value of R_L to obtain each of the operating points in Figure 14-3.

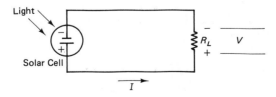

Figure 14-4 Circuit for Example 14-3.

Solution

Point 1: $V = 570$ mV, $I = 0$

$$R_L = \frac{V}{I} = \frac{570 \text{ mV}}{0 \text{ mA}} = \infty$$

$$= \text{open circuit}$$

Point 2: $V = 530$ mV, $I = 620$ mA

$$R_L = \frac{530 \text{ mV}}{620 \text{ mA}} = 0.85 \ \Omega$$

Point 3: $V = 400$ mV, $I = 800$ mA

$$R_L = \frac{400 \text{ mV}}{800 \text{ mA}} = 0.5 \ \Omega$$

Point 4: $V = 0$, $I = 800$ mA

$$R_L = \frac{0 \text{ mV}}{800 \text{ mA}} = 0 \ \Omega$$

$$= \text{short circuit}$$

Point 5: $V = 450$ mV, $I = 780$ mA

$$R_L = \frac{450 \text{ mV}}{780 \text{ mA}}$$

$$= 0.58 \ \Omega$$

As can be seen from Example 14-3, a simple cell will supply power (voltage and current) to loads having a small resistance (less than 1 Ω).

Our primary concern is to get power from the sun. We should therefore be calculating power output at various points along the characteristic curve. If there is a point or region where maximum power can be obtained, it should be selected as the operating point for the cell. The reason for this is simple. The power into the cell is a function of the cell area and the power density of the light. Once these are fixed, the peak efficiency occurs when the power output is a maximum. The voltage and current can then be changed electrically to their desired values.

Example 14-4

At each of the points in Example 14-3, calculate the electrical power obtained from the cell (power out).

Solution

Point 1: $P = VI$

$$= 570 \text{ mV} \times 0 \text{ mA} = 0 \text{ W}$$

Point 2: $P = VI$

$= 530 \text{ mV} \times 620 \text{ mA}$

$= 0.33 \text{ W}$

Point 3: $P = VI$

$= 400 \text{ mV} \times 800 \text{ mA}$

$= 0.32 \text{ W}$

Point 4: $P = VI$

$= 0 \text{ V} \times 800 \text{ mA} = 0 \text{ W}$

Point 5: $P = VI$

$= 450 \text{ mV} \times 780 \text{ mA}$

$= 0.35 \text{ W}$

Referring to Example 14-4, it should be clear that the extremes of the characteristic curve (points 1 and 4) are not good operating points. Good power output is obtained between points 2 and 3. In fact, maximum power out occurs somewhere around the center of the knee of the curve.

To see the variation of power output with operating point, we can plot it as a function of cell voltage. An approximate plot for the cell of Figure 14-3 is shown in Figure 14-5. Notice that the curve peaks at about 450 mV. Since this will most likely be the operating point selected, photovoltaic cells are specified by the minimum current they will supply at 0.45 V.

14-4.1 Effect of Sunlight on Cell Characteristics

In the preceding discussion on the solar cell's characteristic curve, there was no mention made of the sunlight's effect. It is extremely important, however, so we must consider it.

The characteristic shown in Figure 14-3 is representative of that particular cell under a bright noon time sun. If the power density of the sunlight were to decrease, the output of the cell would change accordingly. The decrease in power density can be attributed to any of the following reasons:

1. The sun is not shining directly on the cell due to the fact that it is just rising or setting (see Figure 14-6a).
2. The sun is not shining directly on the cell due to the fact that it is winter. In the northern hemisphere the sun has a southern exposure (see Figure 14-6b). *Note:* In the southern hemisphere the opposite condition exists. In the winter the sun has a northern exposure (see Figure 14-6c).
3. It is a cloudy or overcast day.

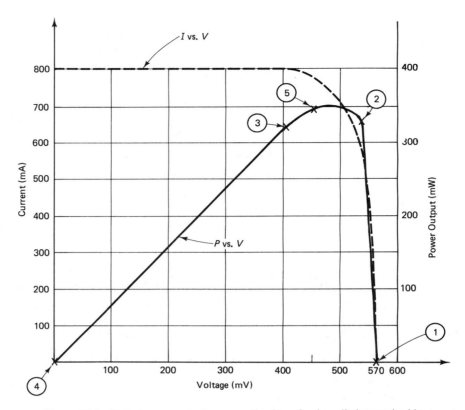

Figure 14-5 Plot of power output versus cell voltage for the cell characterized by Figure 14-3.

4. During certain hours of the day, there may be tall trees or structures that cast a shadow on the cell.

Latitude is the angular position of a point on the earth's surface north or south of the equator. The equator is defined as 0° latitude and divides the earth into two hemispheres. For every 69 miles that we move north or south from the equator, the latitude increases by 1°. The north pole has a latitude 90° north and the south pole is 90° south. Table 14-1 gives the approximate latitude of various locations throughout the world.

If we refer to Figure 14-6 and Table 14-1, we can understand the variation in the sun's position. At noon time the sun's position varies about 40° from its highest position in June to its lowest position in December. In March and September (equinox) the sun's center is directly over the equator. Its apparent position in the sky, then, is approximately equal to your latitude. Hence, if you live in New York City, the noon time sun will appear to be 40° south of vertical during the equinox. In June it will be 20° (40 − 20) south of vertical, and in December, 60° (40 + 20) south of vertical.

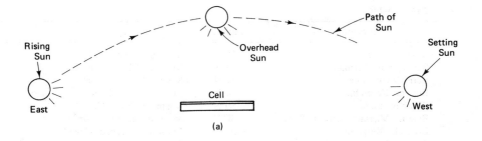

(a)

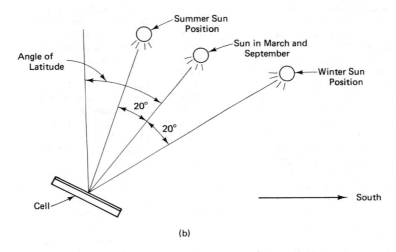

(b)

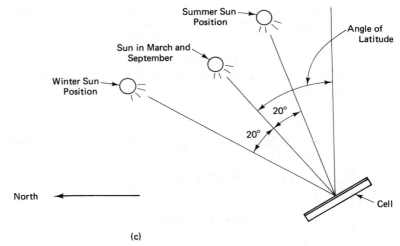

(c)

Figure 14-6 Variations in the sun's position: (a) from morning to night; (b) seasonal in the northern hemisphere; (c) seasonal in the southern hemisphere.

Table 14-1

Location	Latitude	Location	Latitude
Athens, Greece	38°N	Madrid, Spain	40°N
Berlin, East Germany	53°N	Manila, Philippine Islands	15°N
Bern, Switzerland	47°N	Mecca, Saudi Arabia	22°N
Bogota, Colombia	4°N	Mexico City, Mexico	20°N
Bombay, India	20°N	Miami, Florida, U.S.A.	26°N
Boston, Massachusetts, U.S.A.	42°N	Milwaukee, Wisconsin, U.S.A.	43°N
Brussels, Belgium	51°N	Montreal, Quebec, Canada	46°N
Buenos Aires, Argentina	35°S	Moscow, USSR	55°N
Cairo, Egypt	30°N	Munich, Germany	48°N
Capetown, South Africa	34°S	New York, New York, U.S.A.	40°N
Caracas, Venezuela	10°N	Norfolk, Virginia, U.S.A.	37°N
Detroit, Michigan, U.S.A.	42°N	Oslo, Norway	60°N
Djakarta, Java	6°S	Panama Canal, Panama	9°N
Edinburgh, Scotland	56°N	Paris, France	49°N
Edmonton, Alberta, Canada	54°N	Peking, China	40°N
Entebbe, Uganda	0°	Port-au-Prince, Haiti	18°N
Falkland Islands	52°S	Portland, Maine, U.S.A.	43°N
Guatemala, Guatemala	15°N	Quito, Ecuador	0°
Honolulu, Hawaii, U.S.A.	20°N	Rio de Janeiro, Brazil	23°S
Houston, Texas, U.S.A.	30°N	Rome, Italy	42°N
Jackson, Mississippi, U.S.A.	32°N	Saigon, South Vietnam	11°N
Johannesburg, South Africa	26°S	San Juan, Puerto Rico	19°N
Juneau, Alaska, U.S.A.	58°N	Singapore	2°N
Kansas City, Missouri, U.S.A.	39°N	Seattle, Washington, U.S.A.	47°N
Kiev, USSR	50°N	Sydney, Australia	35°S
Kingston, Jamaica	18°N	Tel Aviv, Israel	32°N
Las Vegas, Nevada, U.S.A.	36°N	Thule, Greenland	77°N
Lima, Peru	12°S	Tokyo, Japan	36°N
Little Rock, Arkansas, U.S.A.	34°N	Valparaiso, Chile	33°S
London, England	52°N	Vladivostok, USSR	43°N
Los Angeles, California, U.S.A.	34°N		

For maximum energy absorption from the sun, a solar cell should be tilted south (in the northern hemisphere) and north (in the southern hemisphere) by the angle of your latitude.

Example 14-5

At what angle should a solar cell be tilted to get the most energy from the sun if you are located in:

(a) Juneau, Alaska?

(b) Quito, Ecuador?

(c) Buenos Aires, Argentina?

Solution

Directly from Table 14-1, refer to Figure 14-7.

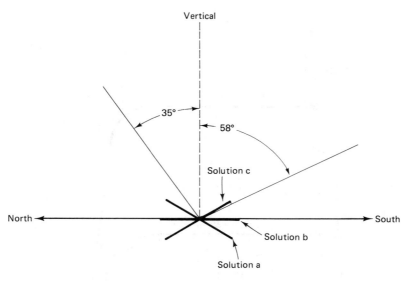

Figure 14-7 Solution to Exercise 14-5.

(a) 58° south of vertical

(b) Vertically upward

(c) 35° north of vertical

It is sometimes important to know how a solar cell will behave as the sun's strength varies. For this reason its characteristics will often include the effects of a variation in power density. A typical set of curves is shown in Figure 14-8. The following observation can be made with reference to this figure. The current output of a cell is directly proportional to the sunlight. Under a very weak sun the cell puts out very little current. On the other hand, the cell continues to put out a fairly high voltage. That is, under open-circuit conditions the voltage is relatively independent of the sunlight. What is more important, however, is the effect that the load has on the operating point as the power density varies.

Consider a fixed resistive load selected to make the cell operate at point a with full sunlight (1 SUN). Remember, the knee of the curve is where maximum power conversion occurs. The value of load R_L would be

$$R_L = \frac{0.45 \text{ V}}{0.58 \text{ A}} = 0.78 \text{ } \Omega$$

If now the power density were to drop to 0.5 SUN due to a cloud passing by, the operating point would change. To find it, we use the current output (300 mA) at the lower sun curve and the load resistance

$$V = R_L I = 0.78 \text{ } \Omega \times 0.3 \text{ A} = 0.23 \text{ V}$$

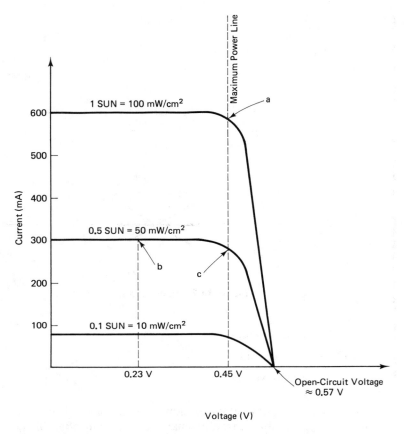

Figure 14-8 Variation of a solar cell's *I–V* curve with sunlight power density.

The new operating point would be that indicated at b. Note that we are no longer operating at the knee of the curve. This means that maximum power is not being converted. To do this the operating point would have to be at c. However, to get this operating point we need a different R_L.

$$\text{new } R_L = \frac{0.45 \text{ V}}{0.29 \text{ A}} = 1.55 \text{ }\Omega$$

The power at point c is one-half the power at point a.

$$P_a = 0.45 \text{ V} \times 0.58 \text{ A} = 0.26 \text{ W}$$

$$P_c = 0.45 \text{ V} \times 0.29 \text{ A} = 0.13 \text{ W}$$

The conclusion that we can draw from the preceding discussion is that the current output is directly proportional to the sunlight (power density). The power will not be proportional unless the load is changed as the intensity of light varies. In the case of a photovoltaic source, maximum power output is a function of more than one variable (the sunlight and the load). For this reason, inverters used for solar energy conversion systems have special tracking circuitry that constantly adjusts the loading on the solar cells while monitoring power output.

There is another problem caused by the variation in sunlight. It is rather difficult to predict how much power can be obtained from the sun from hour to hour or day to day. For this reason, average data on sunlight is used. To understand the average data we will consider the following illustrative example.

Example 14-6

In a city in the southwestern part of the United States, data was taken on a sunny day to determine how much energy was received from the sun. The sun rose at 7 A.M. and set at 7 P.M. The data collected is given in Table 14-2.

Table 14-2

Power Density (SUNS)	Time (h)	Energy Density (SUN-hours)
1.0	1	1.0
0.9	2	1.8
0.6	3	1.8
0.3	4	1.2
0.1	2	0.2
	Total energy density equals 6.0 SUN-hours	

If we wanted to predict the amount of energy that a cell could convert to electricity, we could do it in two ways.

One way would be to use a set of curves like the ones in Figure 14-8. For each power density of the sunlight in Table 14-2, we would compute the power at the knee of the curve. This assumes that the load is adjusted for maximum power. Multiplying each of the calculated powers by the time duration would give the energy for each time interval. Finally, adding each of these energies, we could find the total energy.

The other (much easier) way would be to note that the total energy density for the day is 6 SUN-hours. This is equivalent to having a full or peak sun (1 SUN) for 6 h. With this in mind we could calculate the power at point a in Figure 14-8 and find the total energy by multiplying that power by 6 h.

The second, easier way, described in Example 14-6 is in fact the technique that is used when predicting energy conversion. It is done with the help of charts that have been developed with data obtained by government agencies over a period of many years. They give the average number of peak SUN-hours per day over a 12-month (1-year) period. Figure 14-9 is an example of this type of chart for the United States. The figure tells us that someone living in Florida could expect an average of 4.7 peak SUN-hours per day. On the other hand, in Alaska one could expect an average of 3 peak SUN-hours per day.

The chart in Figure 14-9 is a yearly average. During parts of the year the number of peak SUN-hours would be greater. Naturally, then, during some periods the peak SUN-hours would be less. In any well-designed system it is best to do a worst-case analysis. With this in mind, a chart has been prepared for the 4-week period from

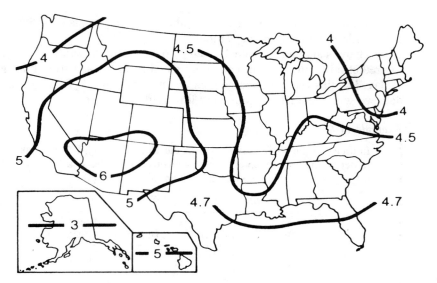

Figure 14-9 Yearly average peak sun-hours per day. (Courtesy of Solarex Ventures Group, Rockville, Md.)

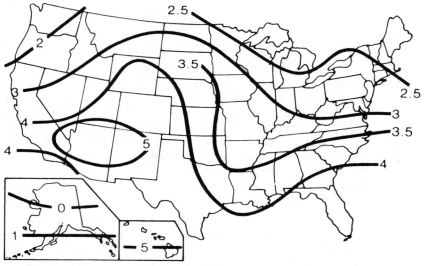

Figure 14-10 Peak sun-hours per day for a 4-week period from December 7 to January 4. (Courtesy of Solarex Ventures Group, Rockville, Md.)

December 7 to January 4. This is the period of minimum sunlight during the year for the northern hemisphere. The chart is shown in Figure 14-10.

It is important to note that both charts (Figures 14-9 and 14-10) are for a solar cell tilted at 45° toward the south. If the latitude of a particular location is different than 45°, tilting the cell at that angle (the latitude) would provide a slightly higher average than that indicated by the charts.

14-4.2 Effect of Temperature on Cell Characteristics

Solar cells are rated in terms of the minimum current they will supply at 0.45 V under a full sun at 25° Celsius (77° Fahrenheit). As the temperature of a cell changes, so will its output. As the temperature increases, the current will increase while the voltage will decrease (by about 2.1 mV/°C). The net effect will be a decrease in power output, hence cell efficiency. The opposite will happen as the temperature decreases; that is, cells operate more efficiently when they are cooler. For this reason commercially manufactured panels (groups of interconnected cells) have a metal underside (typically, aluminum) that acts as a heat sink. In areas of low latitude, cell temperatures can reach 80° Celsius (176° Fahrenheit) without a heat sink. Equations 14-3 and 14-4 can be used to solve for the new cell output at the different temperature. In these equations, E_R and I_R are the cell ratings in volts and mA at 25° Celsius. E_o and I_o will be the cell voltage and current at the new temperature T in degrees Celsius. Note that the current equation is also a function of the cell area A in square centimeters.

$$E_o = E_R - 0.0021(T - 25) \qquad (14\text{-}3)$$

$$I_o = I_R + 0.025(A)(T - 25) \qquad (14\text{-}4)$$

Example 14-7

A solar cell is rated 600 mA, 0.45 V, at 25°C. The cell area is 23 cm². While sitting in a full sun and supplying maximum power, the cell temperature rises to about 50°C. Find:

(a) The power output at 25°C

(b) The voltage, current, and power output at 50°C

Solution

(a) At 25°C the power output is just the product of the rated voltage and current.

$$P = 0.45 \text{ V} \times 600 \text{ mA} = 270 \text{ mW}$$

(b) To find the new voltage, Eq. 14-3 will be used.

$$E_o = 0.45 \text{ V} - 0.0021 \text{ V/deg } (50 - 25) \text{ deg}$$

$$= 0.40 \text{ V}$$

The new current is calculated using Eq. 14-4.

$$I_o = 600 \text{ mA} + 0.025 \frac{\text{mA}}{\text{deg-cm}^2} (23 \text{ cm}^2)(50 - 25) \text{ deg}$$

$$= 600 \text{ mA} + 14.4 \text{ mA} = 614.4 \text{ mA}$$

The new power output is

$$P' = 0.40 \text{ V} \times 614.4 \text{ mA}$$

$$= 246 \text{ mW}$$

The drop in power output is 24 mW (270 − 246). This represents a percentage drop of $24/270 \times 100 = 8.9\%$. Since this is significant, temperature effects should be taken into account when a photovoltaic system is designed. Power estimates should be increased by about 10% to allow for the loss due to increased cell temperature.

14-4.3 Solar Cell Efficiency

The efficiency of a photovoltaic cell is defined as the ratio of the electrical power output to the sunlight power it receives. A silicon solar cell has a **maximum theoretical efficiency** of about 25%. The cells manufactured today have rated efficiencies of 8 to 16%.

The efficiency of a cell depends on several factors. The most important are the number and thickness of the wires connected to the top of the cell (see Figure 14-2) and light reflected from the surface of the cell. In addition, when cells are connected together to form panels, the panel efficiency will be a function of the cell's shape.

Example 14-8

A circular cell has a diameter of 3 in. It has a rating at 25°C of 1200 mA and 0.45 V in a full sun. What is the cell's efficiency?

Solution

First find the cell area.

$$\text{radius } r = \frac{3 \text{ in.}}{2} = 1.5 \text{ in.}$$

$$r = 1.5 \text{ in.} \times 2.54 \text{ cm/in.} = 3.81 \text{ cm}$$

$$\mathbf{A} = \pi r^2 = \pi (3.81 \text{ cm})^2$$

$$= 45.6 \text{ cm}^2$$

$$1 \text{ SUN} = 100 \text{ mW/cm}^2$$

$$P_i = 100 \text{ mW/cm}^2 \times 45.6 \text{ cm}^2$$

$$= 4560 \text{ mW}$$

$$P_o = 0.45 \text{ V} \times 1200 \text{ mA} = 540 \text{ mW}$$

$$\text{efficiency} = \eta = \frac{P_o}{P_i} \times 100$$

$$= \frac{540 \text{ mW}}{4560 \text{ mW}} \times 100$$

$$= 11.84 \approx 12\%$$

14-4.4 Spectral Response of a Solar Cell

Up until this point we have only considered the source of energy as sunlight. The solar cell, however, responds well to all forms of visible light. This means that photovoltaic cells can operate indoors from florescent or incandescent lamps. The normalized

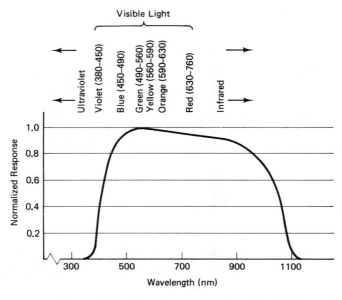

Figure 14-11 Spectral response of a solar cell.

spectral response of a solar cell is shown in Figure 14-11. The ordinate represents the response of the cell as a fraction of its maximum.

14-5 *INTERCONNECTION OF SOLAR CELLS*

Most practical applications require more than 800 mA at 0.45 V (typical solar cell output). Solar cells, however, can be treated just like batteries. By connecting them in series aiding, the net voltage can be increased, and by connecting them in parallel, the net current will increase. For example, if six identical cells each rated 1 A at 0.45 V in a full sun are connected as shown in Figure 14-12, the net output would be 2 A at 1.35 V.

Note that for each parallel path the current increases by 1 A and for each cell in series the voltage increases by 0.45 V. Solar cells are more flexible than batteries in that they can be broken into pieces to obtain odd ratings. The voltage output from a piece of a cell will still be the rated voltage; however, the current will be proportional to the area of the cell. Circular cells are commonly cut in halves and quarters to form semicircular and quadrant cells. When many cells are connected in series and parallel to form a permanent unit, the unit is called a **solar panel.**

It is important to mention that the current produced by a series connection of cells will be the minimum of all the cells. In other words, if a cell rated 1 A is connected in series with a quadrant of the same cell (rated 0.25 A), the current will be 0.25 A. Similarly, if one cell of a rooftop panel consisting of 100 cells in series is shaded by a small object (such as an antenna or vent pipe), the current output of the panel will be that of the shaded cell. This could easily cut the power output in half.

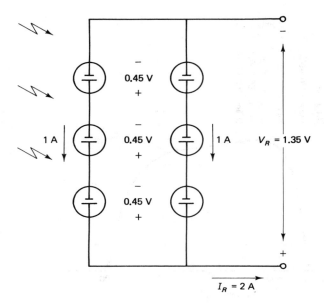

$V_R = 1.35$ V

$I_R = 2$ A

Figure 14-12 Interconnection of six solar cells to get a higher voltage and current rating.

Example 14-9

A student experimenting with solar cells has one 4-in.-diameter circular cell which is rated 2 A at 0.45 V. For a certain application, sunlight must supply power to a toy motor, which requires 1.8 V and draws 0.5 A. Show how the cell can be modified so that it supplies the necessary voltage and current.

Solution

If the cell is cut into quarters (quadrants), each piece will supply one-fourth of the rated current (0.5 A) at the rated voltage (0.45 V). By connecting the four quadrants in series as shown in Figure 14-13, the power requirement is met. Cells can be cut quite easily. First the under (metallic) side is scored with a razor. Then by lining the score up with the edge of a table, a little pressure on the extended half will break the cell along the score. To prevent an uneven cut, a flat object should be placed on the part of the cell that is on the table. Care should be taken because too much pressure could fracture the cell.

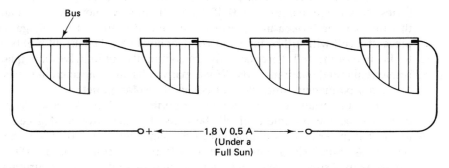

Figure 14-13 Cell connection for Example 14-9.

Solar panels are manufactured commercially in different sizes and ratings. The panels can be interconnected as well to form **solar arrays** or **modules.**

Example 14-10

A solar panel is rated 22 W at 10 V. How many of these panels are required to provide 11 A of current at 120 V? How should they be connected? What should be the value of the load resistance?

Solution

Since each panel produces 10 V, 12 panels are needed in series to get 120 V.

$$\frac{\text{total voltage}}{\text{volts/panel}} = \text{number of panels in series}$$

$$\frac{120 \text{ V}}{10 \text{ V/panel}} = 12 \text{ panels}$$

The current rating of each panel can be obtained from its power and voltage rating.

$$I = \frac{22 \text{ W}}{10 \text{ V}} = 2.2 \text{ A}$$

Each path (series connection of panels) will provide 2.2 A; therefore, five paths are needed.

$$\frac{\text{total curent}}{\text{current/path}} = \text{number of paths}$$

$$\frac{11 \text{ A}}{2.2 \text{ A/path}} = \text{five paths}$$

Thus five parallel paths each having 12 panels in series are needed to form an array having the power requirement. The panels can be placed in any arrangement as long as they are electrically connected properly. A total of 60 panels (5 × 12) are needed. One possible configuration is shown in Figure 14-14. To find the load resistance necessary to obtain the ratings, simply apply Ohm's law.

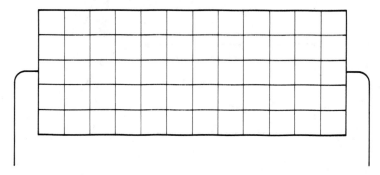

Figure 14-14   Array made with 60 panels to meet the power requirement in Example 14-10 (120 V and 11 A).

$$R_L = \frac{120 \text{ V}}{11 \text{ A}} = 10.91 \ \Omega$$

This means that the array should be connected to something having an input resistance of 10.9 Ω in order to get maximum power conversion in a full sun. As the lighting changes, the input resistance would have to be changed in order to maintain maximum power conversion. Note that the total power will be 120 V $\times$ 11 A = 1.32 kW. This can also be calculated by

$$60 \text{ panels} \times \frac{22 \text{ W}}{\text{panel}} = 1.32 \text{ kW}$$

Panels can be described with characteristic curves just as solar cells can. The characteristics for the panel in Example 14-10 are shown in Figure 14-15. The dashed curve represents the change in the characteristic due to a panel temperature of 70°C.

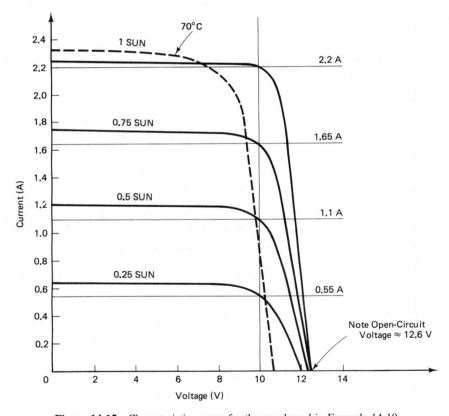

Figure 14-15 Characteristic curves for the panel used in Example 14-10.

14-6 OVERALL SYSTEM CONFIGURATION

Aside from experimental and small-scale applications, a solar generator will have two basic configurations. The first of these, shown in Figure 14-16a, is a stand-alone system. In this case there is no utility power available. During peak sun hours, the solar array must supply all of the ac power needs and keep the storage batteries fully charged. The batteries must have the capability of storing enough energy to supply the power needed when the sun goes down.

The second basic system is one in which the solar generator is connected to the utility grid. In this case, shown in Figure 14-16b, storage batteries are not needed. The dc power from the solar array gets inverted to ac power. During peak sun hours the inverter output supplies the ac power needs. If these needs are low, power will be fed back into the utility grid for credit. During nighttime hours the ac power needs are supplied by the utility grid.

In Figure 14-16, circuit breakers and meters have been left out to simplify the picture. The intent is merely to illustrate the idea rather than all of the parts required.

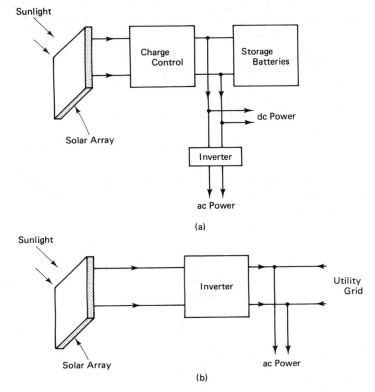

Figure 14-16   Two basic solar generator configurations: (a) stand alone system; (b) supplemental or cogeneration system.

It is important to note that although the solar array's efficiency is the dominant one, the efficiencies of the other components (batteries, inverter, and any additional circuitry) should be taken into account.

Regardless of the system configuration, one of two fundamental questions must be asked.

1. Given a specific location and electrical energy need, how big must the solar array be?

2. Given a specific location and size of a solar array, how much electrical energy can the array supply?

The first of the questions is answered by working backwards from the output (energy need) to find the output required from the array. Knowing this and the location, the size of the array and its rating can be determined. The second question is answered in a straightforward manner. Once the location and array size and rating are known, the energy output can be calculated. In the following examples we will assume the following efficiencies, which are typical; inverter 95%, batteries 80%. Furthermore, we will assume a 10% loss of solar cell power due to an increased operating temperature.

Example 14-11

A homeowner living in Los Angeles, California, wishes to place a photovoltaic array on his south-facing roof. The roof is unshaded and measures 15 by 30 ft (about 4.5 by 9 m). The power derived is to be converted to ac and used in his home. Excess power will be fed back into the local utility grid. Analyze the system configuration and calculate how much energy the homeowner can expect to realize from the installation.

Solution

The system will look like the one shown in Figure 14-16b. In addition, an ac kilowatt-hour meter and circuit breaker will be placed between the inverter and the utility grid. The purpose of the meter is to measure the energy fed by the homeowner's system back into the grid. He will be paid for this energy by the utility company. The panel chosen is rated 16.1 V, 2.3 A, 37 W, and is 10% efficient at 25°C in a full sun. It measures 1 by 4 ft. At 47°C the ratings become 14.6 V, 2.33 A, and 34 W. If we allow about 3 ft at the edges of the roof for a work area, 54 panels can be placed on the roof quite easily. They could be set end to end with six in a row (6 × 4 = 24 ft) and nine rows total (9 × 1 = 9 ft) (see Figure 14-17). The peak power output would be

$$54 \text{ panels} \times \frac{37 \text{ W}}{\text{panel}} = 1998 \text{ W}$$

A 2-kW single-phase inverter will be used. Its maximum input current is specified as 20 A and the input voltage can vary from 50 to 100 V dc. The output is 120 V ac.

With these specifications in mind, the six panels in each row will be connected in series. This will give a range of voltage (from 25 to 47°C) of

$$6 \times 16.1 = 96.6 \text{ V} \ (25°C)$$

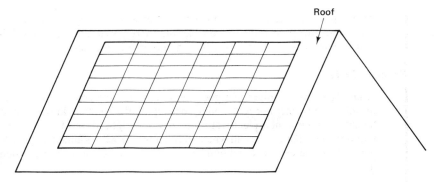

Roof

Figure 14-17 Solar panel arrangement on a rooftop for Example 14-11.

to
$$6 \times 14.6 = 87.6 \text{ V } (47°C)$$

which is within the inverter range. The nine rows will be connected in parallel. Thus the peak current from the solar array (at 47°C) will be

$$9 \times 2.33 \text{ A} = 20.97 \text{ A}$$

which is 5% over the inverter limit. This can most likely be handled by the inverter. From Table 14-1 the panel should be tilted 34° toward the south. This is probably very close to the angle of the roof itself. From Figure 14-9 someone in Los Angeles can expect an average of 5 peak sun-hours per day. Multiplying this by the peak power output at 47°C will give us the average daily energy supplied by the solar array.

$$\text{At } 47°C: \quad P = 87.6 \text{ V} \times 21 \text{ A} = 1.84 \text{ kW peak}$$

$$\text{Average daily energy} = 1.84 \text{ kW} \times 5 \text{ h} = 9.2 \text{ kWh}$$

If the inverter efficiency is 95%, the average energy produced by the system per day will be

$$9.2 \text{ kWh} \times 0.95 = 8.74 \text{ kWh}$$

Multiplying this by 30 days gives us

$$8.74 \text{ kWh} \times 30 \text{ days} = 262.2 \text{ kWh/month}$$

or

$$8.74 \text{ kWh} \times 365 \text{ days} = 3190 \text{ kWh/year}$$

At 10 cents per kilowatt-hour the homeowner will save about $320 per year in electric bills. The solar array would cost (using a 1980 price of $10 per peak watt)

$$1998 \text{ W} \times \$10 \text{ per watt} = \$19,980$$

It would take about 60 years for the homeowner to recover his initial investment (actually longer if we consider the interest lost on his $20,000). If, however, the cost for the array drops to $.50 per peak watt (predicted for 1986), the cost would be

$$1998 \times \$0.50 = \$999$$

At this price he could recover his investment within 4 years. The investment becomes even more attractive when we consider the energy investment credit and the fact that the cost of commercial electricity will certainly rise in the future.

Example 14-12

A family owns a camper and goes camping in remote areas. The owner would like to install a solar generator with battery backup on the camper. In this way the family will be able to supply their electrical needs day and night. Table 14-3 is a list of the energy requirements. For the area in which they camp, we will assume five peak SUN-hours per day. All of the electrical equipment requires 12 V dc. Specify the system and analyze it.

Table 14-3

Appliance	Current (A)	Time (h)	Battery drain (A-h)	Power (W)	Energy (W-h)
Refrigerator	2	20	40	24	480
Miscellaneous (radio, television, lighting, toaster, etc.)	5	4	20	60	240
Total			60	84	720

Solution

We will assume that power is always drawn directly from the batteries; therefore, they must be rated for 60 A-h. Assuming a battery efficiency of 80%, the solar panels must supply the batteries with $60/0.80 = 75$ A-h of charge every day. Since the peak sun exists on the average for 5 h, the panels must supply 15 A of charging current every hour of peak sun.

$$I = \frac{75 \text{ A-h}}{5 \text{ h}} = 15 \text{ A}$$

We will use a 44 W panel rated 14.6 V, 3 A. If five panels are connected in parallel, they will supply the required current of 15 A. The voltage is just right to supply a steady charge to the batteries.

To design the system properly, we should provide sufficient battery storage to allow for extended cloudy weather. If three car batteries rated 80 A-h were used for storage, an adequate supply of energy would be available. The system would look as shown in Figure 14-18. The diode is used to prevent current flow from the batteries back to the solar panels during darkness. Not only would it drain the batteries, but it would damage the solar cells as well. The diode should be rated in excess of 15 A; about 20 A would be just right. The resistor may or may not be necessary. Its function is to limit the battery charging current. Since the peak power of the panels is

$$5 \text{ panels} \times \frac{44 \text{ W}}{\text{panel}} = 220 \text{ W}$$

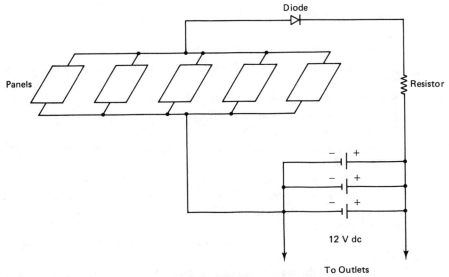

Figure 14-18

At $10 per peak watt (1980 prices), the cost of the panels alone would be

$$220 \text{ W} \times \$10 \text{ per watt} = \$2200$$

However, if the price were to drop to $0.50 per peak watt, the cost would be $110 for the panels. This is below the cost of the batteries and would be well worth the expense. It is important to note that the panels should be mounted on a frame that could be adjusted. In this way the panels could be tilted toward the sun for maximum power conversion.

SYMBOLS INTRODUCED IN CHAPTER 14

Symbol	Definition	Units: English and SI
1 SUN	Power density of an overhead sun on a clear day	100 mW/cm^2
1 langley	Unit of the sunlight's energy density	$11.62 \dfrac{\text{W-h}}{\text{m}^2}$
V_{oc}	Open-circuit voltage of a solar cell	mV
I_{sc}	Short-circuit current of a solar cell	mA
E_R	Rated voltage of a solar cell at 25°C	volts
I_R	Rated current of a solar cell at 25°C	amperes
E_o	Solar cell voltage at any temperature	volts
I_o	Solar cell current at any temperature	amperes
A	Solar cell area	cm^2
T	Solar cell temperature	degrees Celsius

QUESTIONS

1. What is a photovoltaic cell? By what other name is it called?
2. Define the terms "power density" and "energy density." How are they related to each other?
3. What is the meaning of the term "SUN" with respect to power density?
4. What are solar cells made from?
5. How thick is a solar cell?
6. When a solar cell is used to convert sunlight, which terminal (top or bottom) is the positive voltage end?
7. Why can't we use equations to predict a solar cell's output? What must we use?
8. At which point or region of a solar cell's I–V characteristic is maximum power obtained?
9. With respect to the earth's surface, what is the meaning of latitude? Where is 0°, 90° north, and 90° south latitude?
10. To obtain a maximum amount of energy from the sun, at what angle should a cell be tilted?
11. How does the current output of a solar cell vary with the sun's power density? With the cell's area?
12. How does the voltage output of a solar cell vary with the sun's power density? With the cell's area?
13. What is the meaning of the term "peak SUN-hours"?
14. Does the voltage and current output of a cell vary with cell temperature? If so, how?
15. What is the purpose of the metal underside on a commercially manufactured solar panel?
16. What is the maximum theoretical efficiency of a solar cell?
17. What is the difference between a solar panel and a solar array?

PROBLEMS

(English and SI)

1. If the strength of sunlight is 1 SUN for a period of 4 h, find:
(a) The energy density in W-h/m^2
(b) The energy density in langleys
(c) The amount of energy that strikes an area of 5 m^2
2. How long must sunlight whose strength is 0.7 SUN last in order to absorb 10 langleys?
3. How long must sunlight whose strength is 1 SUN last in order for an area of 70 m^2 to absorb 50 kWh?
4. What is the average daily electrical energy converted by a rooftop array? The roof measures 10 m × 15 m. Assume that the array is 12% efficient and that it is tilted 45° toward the south. The location is;
(a) Central California
(b) Washington, D.C.
5. How much sunlight energy (in kWh) strikes the Sahara Desert in 1 year? Assume that the desert has a yearly average of 7 peak SUN-hours per day, that the area of the desert is 3,500,000 square miles, and that there are 365 days in 1 year.

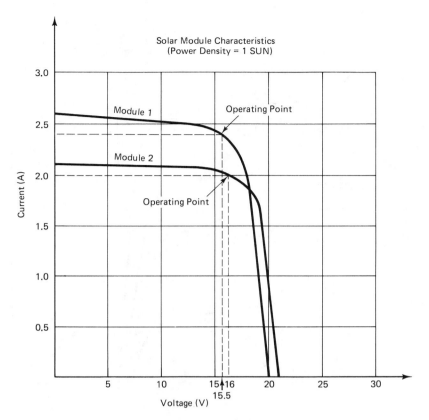

Figure 14-19

6. For solar module number 1, whose characteristic is shown in Figure 14-19, determine for the indicated operating point:
(a) The required load resistance
(b) The power output

7. For solar module number 2, whose characteristic is shown in Figure 14-19, determine for the indicated operating point:
(a) The required load resistance
(b) The power output

8. At what angle should a solar cell be tilted to get the most energy from the sun if it is located in:
(a) Lima, Peru
(b) Sydney, Australia
(c) Johannesburg, South Africa
(d) Honolulu, Hawaii, U.S.A.

9. The solar module whose characteristic is shown in Figure 14-20 is used for a solar generator. A fixed load (R_L) is selected to give the operating point indicated. Determine:
(a) R_L
(b) The power output in 1 SUN

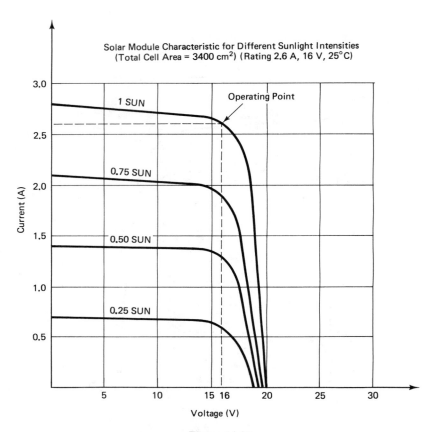

Solar Module Characteristic for Different Sunlight Intensities
(Total Cell Area = 3400 cm^2) (Rating 2.6 A, 16 V, 25°C)

Figure 14-20

(c) The power output in 0.75 SUN
(d) The power output in 0.50 SUN
(e) The power output in 0.25 SUN

10. With reference to Problem 9, determine the value needed for R_L and the new power output if the module voltage is kept constant at 16 V for each of the following intensities:
(a) 0.75 SUN
(b) 0.50 SUN
(c) 0.25 SUN

11. The solar module whose characteristic is shown in Figure 14-20 is used in an application with the indicated operating point (16 V). Find the yearly average energy output per day that could be obtained in each of the following locations. Assume that the module is tilted at 45° toward the south.
(a) Nome, Alaska
(b) Houston, Texas
(c) Phoenix, Arizona

12. Repeat Problem 11 for the 4-week period from December 7 to January 4.

13. For the solar module whose characteristic is shown in Figure 14-20, determine for the indicated operating point at a temperature of 50°C.

(a) The current output
(b) The voltage output
(c) The power output
(The module has a total of 128 cells. There are 4 parallel paths with 32 cells connected in series in each path.)

14. Repeat Problem 13 at a temperature of 75°C.

15. What is the efficiency in 1 SUN at 25°C of the module whose characteristic is shown in Figure 14-20? See Problem 13 for the module construction.

16. Repeat Problem 15 at 65°C.

17. A solar panel is rated 1.3 A, 15.8 V, 25°C, and has a cell area of 275 in^2. What is its efficiency at 25°C?

18. Repeat Problem 17 at 45°C. The new panel output is 1.35A, 14V.

19. A solar module is rated 5.2 V, 35 W. It is used to power a 28-V supply requiring 20 A of current. Find:
(a) The number of modules needed
(b) Their connection (i.e., how many in series and/or parallel)
(c) The required value of load resistance to obtain the rated power

Chapter 15

Wind Energy:
The Wind Generator

Wind energy refers to energy or power that is present and available in the wind. Wind can be defined as the movement of air parallel to the earth. Where does wind come from? The main cause of wind is the source of energy we were introduced to in Chapter 14, the sun. Due to the uneven heating of the earth's surface by the sun, differences in atmospheric pressure are created. Air will move (creating a wind) from a high-pressure area to one of low pressure. The greater the pressure difference, the stronger the wind will be. The actual direction of the wind is further influenced by the earth's rotation.

Like the sun, wind energy depends on geographic location and the time of year. However, the wind is less predictable than the sun. It is possible for many days to pass without having a wind strong enough to turn a generator. On the other hand, we know that the sun will rise every morning producing electrical energy, even if it is cloudy. There are areas, however, where a good strong wind is blowing most of the time. In this case a **wind generator*** would be very practical. Furthermore, it would have the ability to provide power day and night.

15-1   A BRIEF HISTORY

Humankind has been making use of windpower for many centuries. The wind powered the sailboats of the ancient Egyptians, Vikings, and explorers as they traveled from one land to another. It is not known when windpower was first converted to rotary motion; however, it was done as early as the thirteenth century. In this case the rotary power derived was used to grind corn. Sometime later it was used for pumping

---

*A wind generator is an electromechanical system that converts wind energy to electrical energy.

water. In any event, a windmill was something very common which could be seen throughout the world.

Today the windmill is starting to reappear. Now, however, its main function is to drive a device which can convert rotary motion to electricity (generator or alternator). This is nothing new; the principle has been around quite a while. For many years now, utility companies have generated electricity from water power. In this chapter we discuss what a typical **wind energy conversion system** (WECS) involves.

15-2   *THE WIND AND ITS ENERGY*

The wind has a tremendous amount of energy. Unlike the sun, however, it is not too dependable or predictable. The wind changes from day to day and minute to minute for any given location. We shall see in Section 15-3 that predicting the amount of electrical energy that can be obtained over a period of time is extremely difficult. In this section we examine the instantaneous power present in the wind and the equation used to determine it. Figure 15-1 shows a wind with velocity v. The quantity ρ_a is the mass density of the air and is relatively constant. We are actually interested in the power present in this wind for a given cross-sectional area **A**. This is the case since a wind generator will capture only the wind power caught by its blades.

Equations 15-1 give the power in watts present in a wind for a given swept area **A**. The units given in Table 15-1 should be used.

(English)

$$P = \tfrac{1}{2}\rho_a \mathbf{A} v^3 \times \frac{746}{550} \tag{15-1a}$$

or

$$P = 0.678 \rho_a \mathbf{A} v^3$$

(SI)

$$P = \tfrac{1}{2}\rho_a \mathbf{A} v^3 \tag{15-1b}$$

Note that the two equations are actually identical ($\tfrac{1}{2}\rho_a \mathbf{A} v^3$); however, the English version gives power in ft-lb/s. It is therefore divided by the constant 550 ft-lb/hp and multiplied by 746 W/hp to change the units of power into watts.

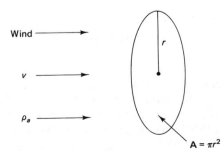

Wind

v

ρ_a

$A = \pi r^2$

Figure 15-1   Diagram of a wind through a circular cross-sectional area, **A.**

Table 15-1

English (Eqs. 15-1a, 15-3a)	SI (Eqs. 15-1b, 15-3b)
$\rho_a = 0.0024$ lb-s^2/ft^4	$\rho_a = 1.24$ kg/m^3
$\mathbf{A}$ ft^2	$\mathbf{A}$ m^2
v ft/s	v m/s
F_W lb	F_W newtons

The following conversion will be needed when working in the English system since wind speed data is usually given in miles per hour (mph).

(English)

$$v(\text{ft/s}) = 1.47 \times v(\text{mph}) \tag{15-2}$$

Example 15-1 (English)

How much power is present in a 10-mph wind striking a windmill whose blades have a radius of 10 ft?

Solution

The area swept by the blades is

$$\mathbf{A} = \pi r^2 = \pi(10 \text{ ft})^2 = 314 \text{ ft}^2$$

Using Eq. 15-2, we have

$$v = 1.47 \times 10 \text{ mph} = 14.7 \text{ ft/s}$$

Substituting into 15-1a gives us

$$P = 0.678(0.0024)(314)(14.7)^3$$
$$= 1623 \text{ W}$$

Example 15-2 (SI)

How much power is present in a 5-m/s wind striking a windmill whose blades have a radius of 3 m?

Solution

The area swept by the blades is

$$\mathbf{A} = \pi r^2 = \pi(3 \text{ m})^2 = 28.3 \text{ m}^2$$

Substituting into Eq. 15-1b gives us

$$P = \tfrac{1}{2}(1.24)(28.3)(5)^3$$
$$= 2193.3 \text{ W}$$

As can be seen from Examples 15-1 and 15-2, there is a great amount of power in the wind. It should be noted, however, that this power is mechanical in its form. By the time it is converted to electrical power, much of it will be lost. We shall see in Section 15-5 that a typical WECS has an efficiency of 20 to 30%. This means that of 2000 W present in a wind, about 500 W will be converted to usable electric power.

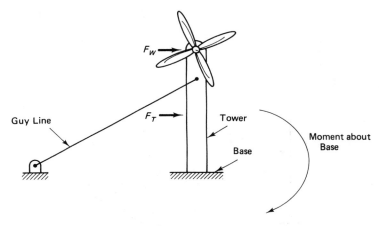

Figure 15-2 Diagram showing forces acting on a wind generator and tower.

15-2.1 Force of a Wind

Although we will not go into the structural considerations in detail, we will mention them here. In any WECS design, the support of the tower on which the wind generator is mounted must be considered.

When a wind blows on a windmill it exerts a force on the blades. The force is given by Eqs. 15-3. The units are the same as those given in Table 15-1.

(English)

$$F_W = 0.44\rho_a \mathbf{A} v^2 \qquad (15\text{-}3a)$$

(SI)

$$F_W = 0.44\rho_a \mathbf{A} v^2 \qquad (15\text{-}3b)$$

In addition to this force, there is a wind force exerted on the tower (F_T) which holds the wind generator. Referring to Figure 15-2, we can see that the effect of these forces is to produce a moment about the tower base in the direction shown. The overturning moment is a function of the wind speed, size of the blades, and the height of the wind generator. Thus large wind machines mounted on high towers must be adequately supported. Many wind machines have an automatic high-wind shutdown feature. This automatically turns the blades so that they are parallel to the wind. In this way the force given by Eqs. 15-3 is greatly reduced.

15-3 PREDICTING THE POTENTIAL OF A SITE

Whenever a wind generator is to be installed, a wind site analysis should be performed. This involves the use of instruments that record wind data continuously. The data should be taken for at least 3 months.

There are many factors that affect wind. Among them are tall buildings, trees, contour of the ground in the surrounding areas, and elevation. Referring to Figure

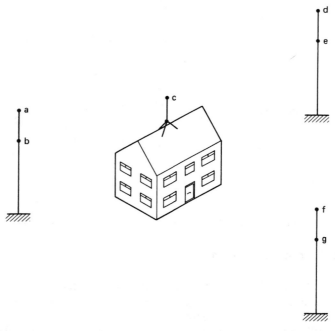

Figure 15-3 Picture showing possible locations for a wind generator (points a to g).

15-3, the average wind speed at each of the indicated points would most likely be different. The best location is the one with the highest average wind speed. If the highest wind speed existed at two or more locations, other factors should be considered. First, the wind machine should be mounted as far as possible from living areas since there is a noticeable sound made by the rotating blades. Furthermore, if it was mounted on a rooftop, a slight vibration would be felt. There is also a cost consideration. If the average wind speed at points a, b, and d were the same, point b would be the choice. At point b a shorter, less expensive tower could be used.

There is also the choice of using a smaller (blade diameter) wind machine at a higher elevation (greater wind) rather than a large machine at a lower elevation and still getting the same power. Referring to Eqs. 15-1, it should be clear that the wind speed (since it is cubed) has a bigger effect on power than does the swept area (blade diameter squared). The following examples will illustrate this.

Example 15-3 (English)

Referring to Figure 15-3, the average wind speed at point a is 10 mph, while that at point b is 8 mph. Find the blade diameter (d) for a machine at each point to capture 1 kW.

Solution

First convert the wind speed to ft/s using Eq. 15-2.

$$v = 10 \times 1.47 = 14.7 \text{ ft/s} \quad \text{(at point } a)$$

$$v = 8 \times 1.47 = 11.76 \text{ ft/s} \quad \text{(at point } b)$$

Rearranging Eq. 15-1a to solve for **A,** we have

$$A = \frac{P}{0.678\rho_a v^3}$$

At point a,

$$A = \frac{1000 \text{ W}}{0.678 \times 0.0024 \times 14.7^3}$$

$$= 193.5 \text{ ft}^2$$

$$A = \pi\frac{d^2}{4}$$

$$d = \sqrt{\frac{4 \times 193.5}{\pi}} = 15.7 \text{ ft}$$

At point b,

$$A = \frac{1000 \text{ W}}{0.678 \times 0.0024 \times 11.76^3}$$

$$= 377.9 \text{ ft}^2$$

$$A = \pi\frac{d^2}{4}$$

$$d = \sqrt{4 \times \frac{377.9}{\pi}} = 21.9 \text{ ft}$$

Note that a smaller cheaper wind machine could be used at point a and provide the same power as a larger machine at point b.

Example 15-4 (SI)

Referring to Figure 15-3, the average wind speed at point a is 5 m/s, while at point b it is 4 m/s. Find the blade diameter (d) for a machine at each point to capture 1 kW.

Solution

Rearranging Eq. 15-1b to solve for **A,**

$$A = \frac{P}{0.5\rho_a v^3}$$

At point a,

$$A = \frac{1000 \text{ W}}{0.5 \times 1.24 \times 5^3}$$

$$= 12.9 \text{ m}^2$$

$$A = \pi\frac{d^2}{4}$$

$$d = \sqrt{4 \times \frac{12.9}{\pi}} = 4.05 \text{ m}$$

At point b,

$$A = \frac{1000 \text{ W}}{0.5 \times 1.24 \times 4^3}$$

$$= 25.2 \text{ m}^2$$

$$A = \pi \frac{d^2}{4}$$

$$d = \sqrt{4 \times \frac{25.2}{\pi}} = 5.66 \text{ m}$$

Note that a smaller cheaper wind machine could be used at point a and provide the same power as a larger machine at point b.

The basic instruments needed to conduct a wind site analysis are a **rotating cup anemometer** and a **wind data compilator.** The anemometer shown in Figure 15-4 works with a **diametral** flow of air. As the wind blows, the anemometer rotates at a speed proportional to the wind speed. Typically, a permanent magnet dc generator is connected to the rotating shaft. A voltage is thus produced which is proportional to the wind speed at every instant of time.

The second instrument necessary is a wind data compilator. It is an electronic instrument that is connected to the anemometer and records the windspeed continuously. After a period of time (2 to 3 months) the data can be retrieved. The average windspeed for any period of time (hour, day, week, etc.) can then be calculated. It should be pointed out at this point that erroneous predictions can be made using valid and correct data. The following examples will illustrate this.

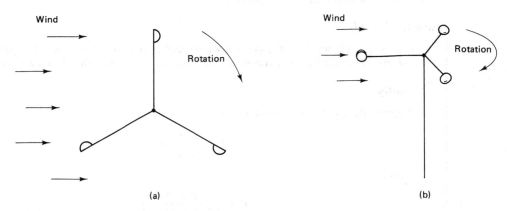

Figure 15-4 Rotating cup anemometer: (a) top view; (b) side view.

Example 15-5 (English)

A wind generator has blades with a 10-ft radius. At the location where it is mounted, data was taken and it was found that the windspeed was 5 mph for 3 h and 15 mph for another 3 h. How much energy was intercepted by the wind machine?

Solution

The problem will be solved two ways. The first method is **incorrect,** however, even though the data used is valid.

Method 1 (Incorrect)   The average windspeed for the 6-h period is 10 mph.

$$\text{average} = \frac{5 \text{ mph} \times 3 \text{ h} + 15 \text{ mph} \times 3 \text{ h}}{6 \text{ h}}$$

4-24-84 AC1

$$10 \text{ mph} \times 1.47 \frac{\text{ft/s}}{\text{mph}} = 14.7 \text{ ft/s}$$

$$\mathbf{A} = \pi (10 \text{ ft})^2 = 314 \text{ ft}^2$$

Using Eq. 15-1a, we have

$$P = 0.678 \times 0.0024 \times 314 \times 14.7^3$$

$$= 1623 \text{ W}$$

$$\text{energy} = P \times t = 1623 \text{ W} \times 6 \text{ h}$$

$$= 9.74 \text{ kWh}$$

Method 2 (Correct)   The energy will be calculated independently for each 3-h period.

1. Average windspeed is

$$5 \text{ mph} \times 1.47 \frac{\text{ft/s}}{\text{mph}} = 7.35 \text{ ft/s}$$

Using Eq. 15-1a gives us

$$P = 0.678 \times 0.0024 \times 314 \times (7.35)^3 = 203 \text{ W}$$

$$\text{energy} = 203 \text{ W} \times 3 \text{ h} = 0.609 \text{ kWh}$$

2. The average windspeed is

$$15 \text{ mph} \times 1.47 \frac{\text{ft/s}}{\text{mph}} = 22 \text{ ft/s}$$

Using Eq. 15-1a, we have

$$P = 0.678 \times 0.0024 \times 314 \times (22)^3 = 5.44 \text{ kW}$$

$$\text{energy} = 5.44 \text{ kW} \times 3 \text{ h} = 16.32 \text{ kWh}$$

The total energy is the sum of the two:

$$\text{energy} = 16.32 \text{ kWh} + 0.609 \text{ kWh}$$

$$= 16.9 \text{ kWh}$$

Note that the second method, which is correct, yields a much larger energy than that obtained with the first method. The reason for the discrepancy is the nonlinear (cube) term in Eq. 15-1a. Note that

$$\left(\frac{a+b}{2}\right)^3 \neq \frac{a^3+b^3}{2}$$

Thus we can see that if the average windspeed for a given location is 10 mph over a 1-month period, it will not give an accurate energy or power prediction. What is needed is a set of data for shorter intervals of time.

Example 15-6 (SI)

A wind generator has blades with a 4-m radius. At the location where it is mounted, data was taken and it was found that the windspeed was 4 m/s for 3 h and 12 m/s for another 3 h. How much energy was intercepted by the wind machine?

Solution

The problem will be solved two ways. The first method is **incorrect,** however, even though the data used is valid.

Method 1 (Incorrect)   The average windspeed for the 6-h period is 8 m/s.

$$\text{average} = \frac{4 \text{ m/s} \times 3 \text{ h} + 12 \text{ m/s} \times 3 \text{ h}}{6 \text{ h}}$$

$$= 8 \text{ m/s}$$

$$\mathbf{A} = \pi(4 \text{ m})^2 = 50.24 \text{ m}^2$$

Using Eq. 15-1b, we have

$$P = 0.5 \times 1.24 \times 50.24 \times 8^3$$

$$= 15,948 \text{ W}$$

$$\text{energy} = P \times t = 15,948 \text{ W} \times 6 \text{ h}$$

$$= 95.7 \text{ kWh}$$

Method 2 (Correct)   The energy will be calculated independently for each 3-h period.

1. Using Eq. 15-1b, we have

$$P = 0.5 \times 1.24 \times 50.24 \times 4^3 = 1.99 \text{ kW}$$

$$\text{energy} = 1.99 \text{ kW} \times 3 \text{ h} = 5.98 \text{ kWh}$$

2. Using Eq. 15-1b gives us

$$P = 0.5 \times 1.24 \times 50.24 \times 12^3 = 53.8 \text{ kW}$$

$$\text{energy} = 53.8 \text{ kW} \times 3 \text{ h} = 161.5 \text{ kWh}$$

The total energy is the sum of the two:

$$\text{energy} = 5.98 \text{ kWh} + 161.5 \text{ kWh}$$

$$= 167.5 \text{ kWh}$$

Note that the second method which is correct yields a much larger energy than that obtained with the first method. The reason for the discrepancy is the nonlinear (cube) term in Eq. 15-lb. Note that

$$\left(\frac{a + b}{2}\right)^3 \neq \frac{a^3 + b^3}{2}$$

Thus we can see that if the average windspeed for a given location is 8 m/s over a 1-month period, it will not give an accurate energy or power prediction. What is needed is a set of data for shorter intervals of time.

The problems and inaccuracies encountered by using average windspeed data can be eliminated with a **wind energy monitor.** This is a special instrument used to determine the energy-producing potential at any given site in kilowatt-hours. Its input comes from an anemometer mounted at the site. The monitor must be programmed at the factory with the characteristics of a specific WECS. This is just data from the manufacturer of power versus windspeed. At the end of any test period the monitor will indicate the number of kilowatt-hours that would have been produced if the WECS had actually been operative.

15-4 MECHANICAL CONSIDERATIONS

There are three basic ways in which the wind can turn a shaft. One of these, called a **diametral** flow of air, is shown in Figure 15-4a. The other two shown in Figure 15-5 work with an **axial** flow and a **tangential** flow. For the generation of electrical power from the wind there are two common methods used; axial flow and diametral flow. In the diametral flow method the axis of rotation is vertical. This is shown in Figure 15-6 in what is called a **Darrieus** design. These machines tend to have excessive vibration.

In the axial flow method the axis of rotation is horizontal. Here there are two basic subdivisions. The windwheel can be either upwind or downwind of the mast. In the upwind case a tail vane is needed to point the machine into the wind. In the downwind case the machine automatically lines up with the wind, thus eliminating the need for a tail vane. The two are shown in Figure 15-7. In many cases an automatic **yaw control system** is part of the wind machine. This is an electromechanical system that uses a motor to point the windwheel into the wind as it changes direction. Note that tracking the wind direction is not necessary in a vertical-axis machine.

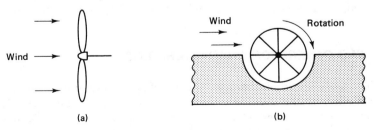

(a)

(b)

Figure 15-5 Two wind-flow mechanisms: (a) axial flow; (b) tangential flow.

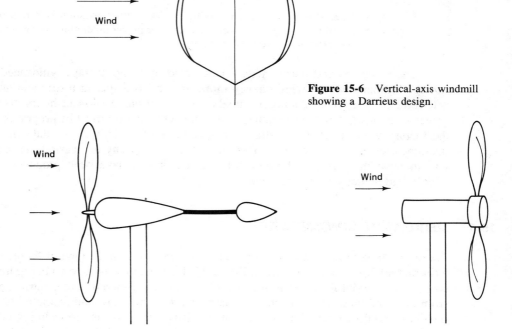

Figure 15-6 Vertical-axis windmill showing a Darrieus design.

(a)

(b)

Figure 15-7 Two horizontal-axis wind machines: (a) windwheel upwind; (b) windwheel downwind.

The number of blades in a horizontal-axis machine is important. If there are many blades, such as on farm windmills, the windwheel will turn at a slow speed, producing high torque. Fewer blades (three or four) will turn at a high speed, producing low torque. This is the condition needed to generate electricity. If two blades are used, the windmachine will tend to jerk as it turns with the changing wind direction.

15-5   *WIND GENERATOR CHARACTERISTICS*

The most important characteristic of a wind generator is its **power curve.** It is a graph supplied by the manufacturer of a particular wind machine. The curve shows the approximate power output as a function of wind speed. Typical power curves are shown in Figures 15-8 and 15-9.

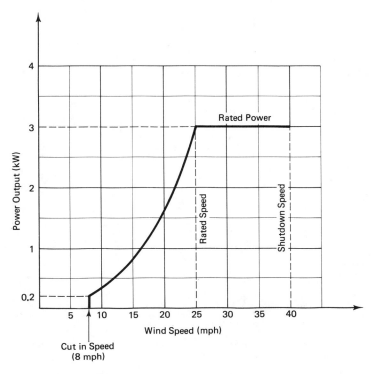

Figure 15-8 Typical wind generator power curve in English units.

There is much information given in a wind generator's power curve. Aside from the power output at any given wind speed, the curve also contains the following information:

1. The **cut in speed,** which is the minimum wind speed needed to start the blades turning and produce a useful output.
2. The **rated power,** which is the maximum power output that the wind machine will produce.
3. The **rated speed,** which is the minimum wind speed required for the wind machine to produce rated power.
4. The **shutdown speed** (also called the **furling speed**) which is the maximum operational speed of the wind machine. Beyond this speed the blades are either folded back or turned to a high pitch position. This is done to prevent damage to the system from high winds.

Note that items 2 and 3 contain the wind machine's ratings. Thus the wind generator whose power curve is shown in Figure 15-8 would be rated 3 kW/25 mph. For the power curve in Figure 15-9 the wind generator's rating would be 15 kW/12 m/s.

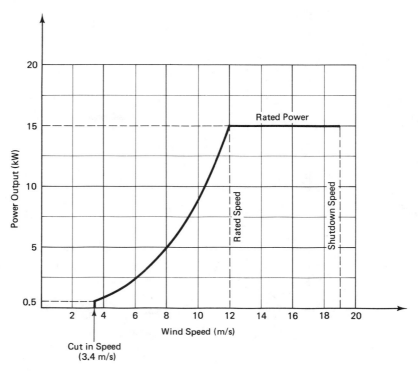

Figure 15-9 Typical wind generator power curve in SI units.

Example 15-7 (English)

The wind generator whose power curve is shown in Figure 15-8 is being used. During an 8-h period the wind had the following average speeds;

$$5 \text{ mph for } 2 \text{ h}$$

$$12 \text{ mph for } 2 \text{ h}$$

$$15 \text{ mph for } 3 \text{ h}$$

$$20 \text{ mph for } 1 \text{ h}$$

Find the electrical energy output for the 8-h period.

Solution

The energy is computed for each of the four windspeeds and time intervals.

1. At 5 mph we are below cut in speed; hence the output is zero.
2. At 12 mph the output from the curve is 0.5 kW.

$$\text{energy} = 0.5 \text{ kW} \times 2 \text{ h} = 1 \text{ kWh}$$

3. At 15 mph the output is 0.8 kW.

$$\text{energy} = 0.8 \text{ kW} \times 3 \text{ h} = 2.4 \text{ kWh}$$

4. At 20 mph the output is 1.6 kW.

$$\text{energy} = 1.6 \text{ kW} \times 1 \text{ h} = 1.6 \text{ kWh}$$

Total energy output for the 8 h is

$$\text{energy} = 0 + 1 \text{ kWh} + 2.4 \text{ kWh} + 1.6 \text{ kWh}$$
$$= 5 \text{ kWh}$$

Example 15-8 (SI)

The wind generator whose power curve is shown in Figure 15-9 is being used. During an 8-h period the wind had the following average speeds;

2.5 m/s for 2 h

5　m/s for 3 h

8　m/s for 2 h

10　m/s for 1 h

Find the electrical energy output for the 8-h period.

Solution

The energy is computed for each of the four windspeeds and time intervals.

1. At 2.5 m/s we are below cut in speed; hence the the output is zero.
2. At 5 m/s the output from the curve is about 1.5 kW.

$$\text{energy} = 1.5 \text{ kW} \times 3 \text{ h} = 4.5 \text{ kWh}$$

3. At 8 m/s the output is 5 kW.

$$\text{energy} = 5 \text{kW} \times 2 \text{ h} = 10 \text{ kWh}$$

4. At 10 m/s the output is about 8.5 kW.

$$\text{energy} = 8.5 \text{ kW} \times 1 \text{ h} = 8.5 \text{ kWh}$$

The total energy output for the 8 h is

$$\text{energy} = 0 + 4.5 \text{ kWh} + 10 \text{ kWh} + 8.5 \text{ kWh}$$
$$= 23 \text{ kWh}$$

15-5.1 WECS Efficiency

Equations 15-1 gave the power present in a wind for a given velocity and swept area. All of this power, however, cannot be collected by a wind generator. The theoretical maximum fraction of available wind power that can be collected by a wind generator is given by the **Betz coefficient.**

The energy in the wind is kinetic energy. To capture this energy, the blades of a wind machine must slow the wind down as it passes through them. Thus after the wind has passed through the wind machine, its velocity (hence kinetic energy) is less than it originally had. The energy it lost has been converted to the kinetic energy of the rotating blades. If after passing through the blades, the wind speed has decreased to one-third of its initial value, the blades will have theoretically captured a maximum fraction of the available wind energy. This maximum fraction is given by

$$\text{Betz coefficient} = 0.5926 \qquad (15\text{-}4)$$

This means that the actual power input for a wind generator will be (at best) 59% of the power given by Eqs. 15-1. The actual blade efficiency is somewhat less than the Betz coefficient. It is a function of a quantity called the **tip speed ratio** (TSR), which is defined by

$$\text{TSR} = \frac{\text{tangential velocity of blade tip}}{\text{wind speed}} \qquad (15\text{-}5)$$

A typical characteristic for each of the wind machines discussed in Section 15-4 is shown in Figure 15-10.

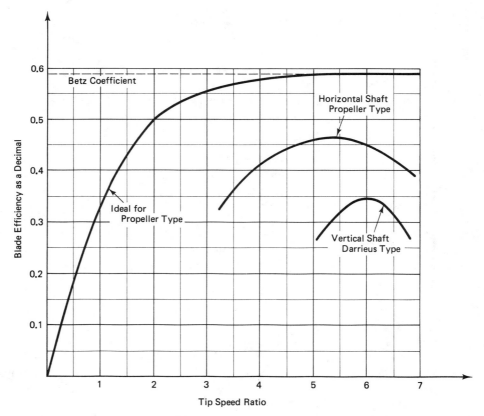

Figure 15-10 Typical efficiency versus tip speed ratio for two different wind machines.

Depending on the generator being used (dc, synchronous, or induction) its efficiency can be anywhere from 60 to 90%. Furthermore, if the generator produces an ac voltage, it may be necessary to turn it at a greater speed than the blades are turning so as to produce 50 or 60 Hz. To do this, a transmission is needed. Its efficiency will be about 95%. In addition, some WECS, depending on their configuration, may include an inverter, rectifier, or batteries. Whatever the component is, its efficiency must also be included in the overall system efficiency. The typical net efficiency for a WECS is 20 to 30%. This assumes that the input power is that which is given by Eqs. 15-1 and the output power is the usable ac electrical power coming out of the WECS.

Example 15-9 (English)

A wind generator whose power curve is shown in Figure 15-8 has a blade diameter of 16 ft. Assume that the power output is 120 V at 60 Hz. Find the net efficiency of the WECS at a windspeed of 12 mph.

Solution

First convert the windspeed from mph to ft/s using Eq. 15-2.

$$v = 1.47 \times 12 \text{ mph} = 17.64 \text{ ft/s}$$

Now calculate the input power using Eq. 15-1a.

$$P_i = 0.678 \times 0.0024 \times \pi \times 8^2 \times 17.64^3$$

$$= 1795 \text{ W}$$

From Figure 15-8 the output power at 12 mph is

$$P_o = 0.5 \text{ kW} = 500 \text{ W}$$

The efficiency can now be calculated.

$$\eta = \frac{P_o}{P_i} \times 100$$

$$= \frac{500 \text{ W}}{1795 \text{ W}} \times 100 = 27.9\%$$

Example 15-10 (SI)

A wind generator whose power curve is shown in Figure 15-9 has a blade diameter of 10 m. Assume that the power output is 230 V at 50 Hz. Find the net efficiency of the WECS at a windspeed of 8 m/s.

Solution

First calculate the input power using Eq. 15-1a,

$$P_i = 0.5 \times 1.24 \times \pi \times 5^2 \times 8^3$$

$$= 24.9 \text{ kW}$$

From Figure 15-9 the output power at 8 m/s is

$$P_o = 5.2 \text{ kW}$$

The efficiency can now be calculated

$$\eta = \frac{P_o}{P_i} \times 100$$

$$= \frac{5.2 \text{ kW}}{24.9 \text{ kW}} \times 100 = 20.9\%$$

Example 15-11 (English)

If the wind generator of Example 15-9 is rotating at 80 rpm, determine the blade efficiency at a windspeed of 12 mph. Use Figure 15-10.

Solution

$$\text{wind speed} = 12 \times 1.47 = 17.64 \text{ ft/s}$$

The circumference that the blade tip traces out is

$$2\pi r = 2\pi \times 8 \text{ ft} = 50.27 \text{ ft}$$

The blade tip speed is

$$50.27 \frac{\text{ft}}{\text{rev}} \times 80 \frac{\text{rev}}{\text{min}} \times \frac{1}{60 \frac{\text{s}}{\text{min}}} = 67 \text{ ft/s}$$

From Eq. 15-5,

$$\text{TSR} = \frac{67 \text{ ft/s}}{17.64 \text{ ft/s}} = 3.8$$

From Figure 15-10 for TSR = 3.8, the blade efficiency is about 40%.

Example 15-12 (SI)

If the wind generator of Example 15-10 is rotating at 9 rad/s, determine the blade efficiency at a windspeed of 8 m/s. Use Figure 15-10.

Solution

The circumference that the blade tip traces out is

$$2\pi r = 2\pi \times 5 \text{ m} = 31.4 \text{ m}$$

The blade tip speed is

$$9 \frac{\text{rad}}{\text{s}} \times \frac{1}{2\pi \frac{\text{rad}}{\text{rev}}} \times 31.4 \frac{\text{m}}{\text{rev}} = 45 \text{ m/s}$$

From Eq. 15-5,

$$\text{TSR} = \frac{45 \text{ m/s}}{8 \text{ m/s}} = 5.6$$

From Figure 15-10 for TSR = 5.6, the blade efficiency is about 46%.

15-6 COMMON WIND GENERATOR CONFIGURATIONS

In this book we have examined three machines that can convert mechanical power (a rotating shaft) into electrical power: the dc generator (Chapter 4), the synchronous alternator (Chapter 8), and the induction generator (Chapter 11). Each of these can be used in a WECS. Depending on which one is used, different components will be needed and the system will have different characteristics. In each case the assumption is made that ultimately we want ac power at a specific line frequency.

15-6.1 WECS Using a DC Generator

The voltage produced by a dc generator is proportional to both flux and speed (see Eqs. 4-1). A typical inverter, which is needed to convert the generated dc to ac, has an allowable input range of 2:1. In other words, for a 120-V ac inverter to operate, the input can vary from 50 to 100 V. Since the windspeed is variable over a wider range, some method of regulation must be employed. By regulating the speed of the generator (wind machine) and/or its field the dc voltage can be maintained within a specified range. Speed regulation is usually accomplished by varying the pitch of the propeller blades. If the dc voltage is sensed, the field strength can be varied accordingly to control the generated voltage. If this is done, we have a voltage control system. A simplified picture of one is shown in Figure 15-11.

With the generator shaft turning at an intermediate speed, the reference voltage (E_R) is adjusted to get a desired output voltage. An increase in shaft speed will now tend to increase the output. This will decrease the field excitation, however, which will restore the output to its desired level. The explanation given is quite simplified. Considerations such as loading, torque variation, and saturation effects have been omitted. These topics would be rightfully covered in a text on feedback control systems.

The overall WECS would look like the one shown in Figure 15-12. A transmission that increases the rotating blade speed to that required for the generator has been included. A wind machine typically rotates in the range 50 to 100 rev/min (5 to 10 rad/s). Depending on the generator, this must be geared up to 1000 to 2000 rev/min (100 to 200 rad/s). The net efficiency would depend on the efficiency of the blades, transmission, generator, regulating circuitry, and inverter.

15-6.2 WECS Using a Synchronous Alternator

We saw in Chapter 8 that an alternator producers an ac voltage whose frequency is proportional to shaft speed. Even with speed regulation there will still be enough of

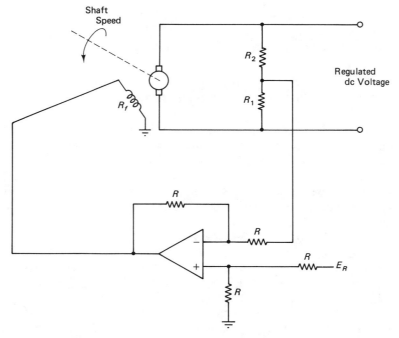

Figure 15-11 Voltage control system which can be used to regulate generator voltage.

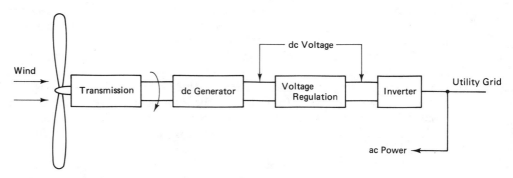

Figure 15-12 Diagram of a WECS using a dc generator.

a variation in frequency and phase to prevent connection of the alternator directly to the utility grid. Thus the alternator is allowed to turn at different speeds, producing a variable-frequency output. The alternator output is then rectified, converting it to dc. The magnitude will be constant since the alternator field is constant. It is usually a permanent-magnet alternator. The dc is now fed to a synchronous inverter, whose line frequency output can be connected directly to the utility grid. With this configuration the need for a transmission is eliminated and the alternator can be connected directly to the windwheel. A diagram of this configuration is shown in Figure 15-13. The net

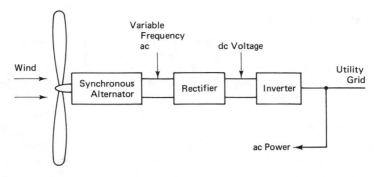

Figure 15-13   Diagram of a WECS using a synchronous alternator.

system efficiency would depend on the efficiency of the blades, alternator, rectifier, and inverter. The transmission's inefficiency and its cost have been eliminated.

15-6.3 WECS Using an Induction Generator

The induction generator is well suited for a wind energy system provided that utility power is available. A review of Section 11-1 indicates that for the induction machine to operate as a generator, a separate source of reactive power is necessary to excite the machine. Furthermore, it must be driven slightly faster than synchronous speed. It is not essential for the speed to be constant, merely to maintain a negative slip. Rated power and peak efficiency are generally obtained at about -3% slip. It is important to remember that when the induction generator is connected directly to the utility grid, the frequency of the generated voltage depends on synchronous speed, not the speed of its rotor (see item 6 in Section 11-1).

The only components necessary for this WECS are a transmission to gear the speed of the blades up to that needed for negative slip and the induction generator. Note that loss of utility power automatically disables the WECS since the field excitation is lost. The net system efficiency depends on the efficiency of the blades, transmission, and generator. A picture of this configuration is shown in Figure 15-14. Some means of speed regulation is needed to maintain the required slip.

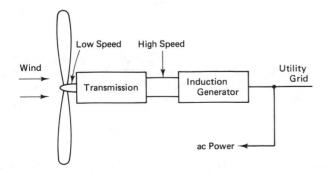

Figure 15-14   Diagram of a WECS using an induction generator.

Example 15-13 (English)

A WECS such as the one shown in Figure 15-14 uses a four-pole three-phase induction machine. The line frequency is 60 Hz and the average windspeed is 12 mph. The blades have a 32-ft diameter and peak efficiency when the tip speed ratio is 6. If the generator efficiency is a maximum at a slip of -2.8%, what should the transmission gear ratio be for peak system efficiency?

Solution

We will find the speeds required for the blades and generator. The transmission will then be selected to match the two speeds. From Eq. 9-7a we can find the required generator speed.

$$S = 120 \times 60 \frac{1 - (-0.028)}{4}$$

$$= 1850.4 \text{ rev/min}$$

The average windspeed is, from Eq. 15-2,

$$12 \times 1.47 = 17.64 \text{ ft/s}$$

From Eq. 15-5 the blade tip speed should be

$$\text{TSR} \times 17.64 \text{ ft/s} = 6 \times 17.64 = 105.84 \text{ ft/s}$$

The circumference traced out by the blade tip is

$$2\pi(16 \text{ ft}) = 100.53 \text{ ft/rev}$$

Thus the blade speed must be

$$\frac{105.84 \text{ ft/s}}{100.53 \text{ ft/rev}} = 1.05 \text{ rev/s}$$

Converting this to rev/min, we get

$$1.05 \text{ rev/s} \times 60 \text{ s/min} = 63.17 \text{ rev/min}$$

The transmission must gear up from 63.17 rev/min to 1850.4 rev/min. Thus the required gear ratio is

$$\text{gear ratio} = \frac{1850.4}{63.17} = 29.3$$

Example 15-14 (SI)

A WECS such as the one shown in Figure 15-14 uses a six-pole three-phase induction machine. The line frequency is 50 Hz and the average windspeed is 5.5 m/s. The blades have a diameter of 9 m and peak efficiency when the tip speed ratio is 5.8. If the generator efficiency is a maximum at a slip of -3%, what should the transmission gear ratio be for peak system efficiency?

Solution

We will find the speeds required for the blades and generator. The transmission will then be selected to match the two speeds. From Eq. 9-7b we can find the required generator speed.

$$\omega = 4\pi \times 50 \, \frac{1 - (-0.03)}{4}$$

$$= 161.8 \, \text{rad/s}$$

From Eq. 15-5 the blade tip speed should be

$$\text{TSR} \times 5.5 \, \text{m/s} = 5.8 \times 5.5 = 31.9 \, \text{m/s}$$

The circumference traced out by the blade tip is

$$2\pi(4.5 \, \text{m}) = 28.27 \, \text{m/rev}$$

Thus the blade speed must be

$$\frac{31.9 \, \text{m/s}}{28.27 \, \text{m/rev}} = 1.13 \, \text{rev/s}$$

Converting this to rad/s, we get

$$1.13 \, \text{rev/s} \times 2\pi \, \text{rad/rev} = 7.09 \, \text{rad/s}$$

The transmission must gear up from 7.09 rad/s to 161.8 rad/s. Thus the required gear ratio is

$$\text{gear ratio} = \frac{161.8}{7.09} = 22.8$$

In this type of configuration, a utility company will usually impose a reactive power charge. Typically, the WECS will operate at a power factor around 0.5. If a capacitor bank is used for power factor correction, this can easily be raised to 0.8. The capacitors, used for power factor correction, supply some (and possibly all) of the reactive power drawn from the utility grid. A significant cost reduction can thus be realized by correcting the power factor.

It should be mentioned, however, that the induction generator can self-excite. That is, in the event of a utility shutdown, the capacitor bank can continue exciting the field. To eliminate the possibility of this potential danger, automatic breakers should be used to disconnect the capacitors from the generator.

SYMBOLS INTRODUCED IN CHAPTER 15

Symbol	Definition	Units	
		English	SI
A	Area swept by the blades of a wind generator	ft^2	m^2
v	Wind velocity	ft/s	m/s
ρ_a	Mass density of air	0.0024 lb-s^2/ft^4	1.24 kg/m^3

Symbol	Definition	Units English	SI
F_W	Force on the blades of a wind generator due to wind	pounds	newtons
F_T	Force on the support tower of a generator due to wind	pounds	newtons
d	Blade diameter	feet	meters
r	Blade radius	feet	meters
TSR	Tip speed ratio	—	—
S	Generator speed in English system	rev/min	—
ω	Generator speed in SI	—	rad/s

QUESTIONS

1. What is wind? What causes it?

2. What is a wind generator?

3. What are the different factors that affect the wind in a given area?

4. Explain the difference between diametral, axial, and tangential flow of air.

5. Define the following terms: anemometer; wind energy monitor; yaw control system.

6. What is the power curve of a wind generator?

7. With reference to a wind generator, define the following terms: cut in speed, rated power, rated speed, furling speed.

8. Define the following terms: Betz coefficient; tip speed ratio.

9. Explain how the power present in the wind is transferred to the wind machine. Do it using the term "kinetic energy."

10. Why is a transmission sometimes needed in a WECS?

11. When would an inverter be used in a WECS?

12. How is a wind generator's speed usually controlled?

PROBLEMS

(English)

1. How much power is present in an 8-mph wind striking a windmill whose blades have a diameter of:

(a) 20 ft?

(b) 28 ft?

(c) 36 ft?

2. How much power is present in a wind striking a wind generator whose blades have a 12-ft radius when the wind speed is:

(a) 6 mph?

(b) 11 mph?

(c) 18 mph?

3. What must the wind velocity be to double the power present in an 8-mph wind?

4. What is the horizontal force exerted by a 12-mph wind on the blades of a wind generator if their diameter is 24 ft?

5. Repeat Problem 4 if the wind velocity is 25 mph.

6. At a given location the average wind speed is 10 mph 50 ft above the ground. At the same location, however, 80 ft above the ground, the average wind speed is 11.5 mph. How big must the blade diameter be at 50 ft to capture the same power as a machine with a blade diameter of 32 ft at the height of 80 ft?

7. With reference to Problem 6, how big must the blade diameter be to capture 2.5 kW of wind power if the wind machine is mounted at the 80-ft height?

8. A wind generator has blades with a 32-ft diameter. While in operation, the data taken indicated that the windspeed was 6.4 mph for 3 h, 9 mph for 3 h, and 12 mph for 2 h. How much energy was captured by the wind machine during the 8-h period?

9. A wind generator has a 25-ft blade diameter. During the course of a typical 24-h day the average windspeeds given in Table 15-2 were recorded. Assuming an overall WECS efficiency of 20%, how much electrical energy is derived during this typical day? The wind generator has a cut-in speed of 6.5 mph.

Table 15-2

Average windspeed (mph)	8.5	0	6	10.5	15
Time (h)	5	3	4	7	5

10. A wind generator, whose power curve is shown in Figure 15-8, was in operation for a full 24-h period. The average windspeed was recorded and is given in Table 15-2. Find the electrical energy output for the full day.

11. A wind generator with a 33-ft blade diameter rotates at 72 rev/min. Find the tip speed ratio in a wind speed of:

(a) 8 mph

(b) 10 mph

(c) 12 mph

12. A propeller-type wind generator whose power curve is given in Figure 15-8 has a blade diameter of 22 ft. The blades turn at a constant speed of 60 rev/min (their characteristic is shown in Figure 15-10). At a windspeed of 13 mph determine:

(a) The blade efficiency

(b) The overall wind generator efficiency

13. A WECS uses an eight-pole 60-Hz three-phase synchronous alternator driven at synchronous speed. The blades have a 28 ft diameter and a peak efficiency when the TSR = 5.4. What should the transmission gear ratio be for peak system efficiency at a windspeed of 11.5 mph?

14. A WECS uses a six-pole 60-Hz three-phase induction generator. It is excited by a three-phase 60-Hz supply line. The blades have a 34-ft diameter and peak efficiency when the TSR = 6.2. If the generator efficiency is a maximum at a slip of −3.5%, what should the transmission gear ratio be for peak system efficiency at a windspeed of 10 mph?

15. How much power is present in a 3.5-m/s wind striking a windmill whose blades have a diameter of:

(a) 7 m?

(b) 9 m?

(c) 12 m?

16. How much power is present in a wind striking a wind generator whose blades have a 4-m radius when the wind speed is:

(a) 3 m/s?

(b) 5 m/s?

(c) 8 m/s?

17. What must the wind velocity be to double the power present in a 4-m/s wind?

18. What is the horizontal force exerted by a 5-m/s wind on the blades of a wind generator if their diameter is 8 m?

19. Repeat Problem 18 if the wind velocity is 12 m/s.

20. At a given location the average windspeed is 4 m/s, 16 m above the ground. At the same location, however, 27 m above the ground, the average windspeed is 5.5 m/s. How big must the blade diameter be at 16 m to capture the same power as a machine with a blade diameter of 11 m at the height of 27 m?

21. With reference to Problem 20, how big must the blade diameter be to capture 3-kW of wind power if the wind machine is mounted at the 27-m height?

22. A wind generator has blades with a 12-m diameter. While in operation, the data taken indicated that the windspeed was 2.6 m/s for 4 h, 4 m/s for 4 h, and 5 m/s for 2 h. How much energy was captured by the wind machine during the 10-h period?

23. A wind generator has an 8-m blade diameter. During the course of a typical 24-h day the average windspeeds given in Table 15-3 were recorded. Assuming an overall WECS efficiency of 17%, how much electrical energy is derived during this typical day? The wind generator has a cut in speed of 3.2 m/s.

Table 15-3

Average windspeed (m/s)	4	0	3	5	8
Time (h)	6	2	3	7	6

24. A wind generator whose power curve is shown in Figure 15-9 was in operation for a full 24-h period. The average windspeed was recorded and is given in Table 15-3. Find the electrical energy output for the full day.

25. A wind generator with a 10-m blade diameter rotates at 12 rad/s. Find the tip speed ratio in a windspeed of:

(a) 4 m/s

(b) 4.5 m/s

(c) 5 m/s

26. A propeller-type wind generator whose power curve is given in Figure 15-9 has a blade diameter of 7 m. The blades turn at a constant speed of 10 rad/s (their characteristic is shown in Figure 15-10). At a windspeed of 6 m/s determine:

(a) The blade efficiency

(b) The overall wind generator efficiency

27. A WECS uses a six-pole 50-Hz three-phase synchronous alternator driven at synchronous speed. The blades have a 9-m diameter and a peak efficiency when the TSR = 5. What should the transmission gear ratio be for peak system efficiency at a windspeed of 5 m/s?

28. A WECS uses an eight-pole 50-Hz three-phase induction generator. It is excited by a three-phase 50-Hz supply line. The blades have a 12-m diameter and peak efficiency when the TSR = 5.8. If the generator efficiency is a maximum at a slip of -3.8%, what should the transmission gear ratio be for peak system efficiency at a windspeed of 4.5 m/s?

Appendix A

Unit Conversions from the English System to SI

The following equalities can be useful when converting from the English system to SI.

Length: 1 in. = 2.54 cm = 0.0254 m
1 ft = 30.5 cm = 0.305 m
1 mile = 1609 m

Area: 1 square mile = 2.59×10^6 m^2
1 in^2 = 0.000645 m^2
1 in^2 = 6.45 cm^2

Volume: 1 ft^3 = 0.0283 m^3

Linear speed: 1 ft/s = 0.305 m/s = 30.5 cm/s
1 mph = 0.447 m/s
1 in./s = 0.0254 m/s = 2.54 cm/s

Rotational speed: 1 rev/min = 0.105 rad/s = 6 deg/s

Force: 1 lb = 4.45 N

Power: 1 hp = 746 W = 0.746 kW

Torque: 1 ft-lb = 1.356 N-m

Magnetic flux: 1 line = 1 maxwell = 10^{-8} Wb
1 kiloline = 1000 maxwells = 10^{-5} Wb

Magnetic flux density: 1 line/in^2 = 15.5×10^{-6} T
100 kilolines/in^2 = 1.55 T = 1.55 Wb/m^2

Magnetomotive force: 1 ampere-turn = 1 A

Magnetic field intensity: $1 \dfrac{\text{A-turn}}{\text{in.}}$ = 39.37 A/m

Appendix B

Unit Conversions from SI to the English System

The following equalities can be useful when converting from SI to the English system.

Length: 1 m = 100 cm = 39.37 in
1 m = 3.28 ft
1 m = 6.22×10^{-4} mile

Area: 1 m² = 0.386×10^{-6} mile
1 m² = 1550 in²
1 cm² = 0.155 in²

Volume: 1 m³ = 35.3 ft³

Linear speed: 1 m/s = 100 cm/s = 3.28 ft/s
1 m/s = 2.237 mph
1 m/s = 39.37 in./s

Rotational speed: 1 rad/s = 9.55 rev/min = 57.3 deg/s

Force: 1 N = 0.225 lb

Power: 1 kW = 1000 W = 1.34 hp

Torque: 1 N-m = 0.737 ft-lb

Magnetic flux: 1 Wb = 10^8 lines = 10^8 maxwells
1 Wb = 10^5 kilolines

Magnetic flux density: 1 T = 6.45×10^4/lines/in²
1 T = 64.5 kilolines/in²
1 T = 1 Wb/m²

Magnetomotive force: 1 A = 1 A-turn

Magnetic field intensity: 1 A/m = $0.0254 \dfrac{\text{A-turn}}{\text{in.}}$

Appendix C

Answers to Odd-Numbered Problems

CHAPTER 1

(1a) 400 kilolines, **(b)** 1256 kilolines, **(c)** 720 kilolines, **(3)** 4.32 A,
(5) 6269 A-turns/in., **(7)** 11.75 A, **(9)** 212 kilolines,
(11) 63.585 kilolines, **(13a)** 196 A-turns, **(b)** 0.98 A,
(15) 641.6 A-turns, **(17)** 1908.3 A-turns, **(19)** 2400 A, 6000 A/m,
(21) 0.0025T, **(23)** 4.77×10^5 A/m, **(25)** 2.48×10^{-3} Wb,
(27) 0.0025 m^2, **(29a)** 566 A, **(b)** 3.14 A, **(31)** 2771.6 A,
(33) 1241.66 A.

CHAPTER 2

(1) 0.4 V, **(3)** 2.127 V, **(5a)** into paper, **(b)** out of paper,
(c) no current flow, **(d)** out of paper, **(e)** out of paper, **(7)** 0.5645 V,
(9) 420 kilolines, **(11)** 309.33 V, **(13)** 78.7 kilolines, **(15a)** up,
(b) right to left, **(c)** left to right, **(d)** up, **(17a)** out of paper,
(b) into paper, **(c)** no current, **(d)** into paper, **(e)** into paper,
(19) 24.21 A, **(21)** 0.38 V, **(23)** 1.99 V, **(25a)** into paper,
(b) out of paper, **(c)** no current, **(d)** out of paper, **(e)** out of paper,
(27) 0.606 V, **(29)** 0.0072 Wb, **(31)** 285.2 V, **(33)** 0.0011 Wb,
(35a) up, **(b)** left to right, **(c)** right to left, **(d)** down, **(37a)** out of

paper, **(b)** into paper, **(c)** no current, **(d)** into paper, **(e)** into paper, **(39)** 21.43 A.

CHAPTER 3

(1a) 0.9 Ω, **(b)** 60 V, **(c)** 80 A, **(d)** 4800 W, **(3a)** 0.66 Ω, **(b)** 84 V, **(c)** 120 A, **(d)** 10.08 kW, **(5a)** 1.125 Ω, **(b)** 89.1 V, **(c)** 96 A, **(d)** 8.55 kW, **(7a)** 2.72 A, **(b)** 1.533 V, **(c)** 0.043 Ω/cond, **(9a)** 2.08 A, **(b)** 0.67 V, **(c)** 0.022 Ω, **(11a)** 42 V, **(b)** 252 V, **(13a)** 4285.7 rev/min, **(b)** 1428.6 rev/min, **(15a)** 150 V, **(b)** 900 V, **(17a)** 0.0085 Wb, **(b)** 0.0028 Wb, **(19)** simplex wave.

CHAPTER 4

(1a) 118.6 V, **(b)** 81.8 V, **(c)** 147.3 V, **(d)** 163.6 V, **(3)** 11.1 %, **(5)** graph, **(7)** 285.7 rad/s, **(9a)** graph, **(b)** 1.74 V/rad/s, **(11)** 15.91 %, **(13a)** 11.11 %, **(b)** −1.67 %, **(c)** 36.96 %, **(d)** 12.5 %, **(15a)** 1250 W, **(b)** 80 %, **(17a)** 2.5 A, **(b)** 2 A, **(c)** 3 A, **(d)** 1.6 A, **(19)** 350 V, **(21)** 57 Ω, **(23a)** 2 A, **(b)** 65.22 A, **(c)** 67.22 A, **(d)** 297.22 V, **(e)** 460 W, **(f)** 4518.3 W, **(g)** 70.5 %, **(25)** 0.22 Ω, **(27)** 10 Ω, **(29a)** 20 A, **(b)** 1.84 A, **(c)** 21.84 A, **(d)** 228.2 V, **(e)** 372.4 W, **(f)** 40 W, **(g)** 572.4 W, **(h)** 77.9 %, **(31a)** 40 A, **(b)** 1.33 A, **(c)** 41.33 A, **(d)** 139.8 V, **(e)** 160 W, **(f)** 136.7 W, **(g)** 683.3 W, **(h)** 80.3 %, **(33a)** 166.7 A, **(b)** 2 A, **(c)** 168.7 A, **(d)** 625.2 V **(e)** 1235 W, **(f)** 1388 W, **(g)** 2846 W, **(h)** 92.7 %, **(35a)** 166.7 A, **(b)** 2 A, **(c)** 168.7 A, **(d)** 625.3 V, **(e)** 1200 W, **(f)** 1422.5 W, **(g)** 2845 W, **(h)** 92.7 %, **(37a)** 100 A, **(b)** 1.24 A, **(c)** 101.24 A, **(d)** 2194 W, **(e)** 1806 W, **(39)** 29.4 kW, **(41)** unit 1: 15.15 A, unit 2: 12.13 A.

CHAPTER 5

(1a) 29.2 lb-ft, **(b)** 21.9 lb-ft, **(c)** 13.1 lb-ft, **(3a)** 5.8 lb-ft, **(b)** 6.3 lb-ft, **(5)** 575 W, **(7a)** 250 lb-ft, **(b)** 125 rev/min, **(c)** motor 2, **(d)** motor 4, **(9a)** 50 %, **(b)** 20 %, **(c)** 650 %, **(d)** 22 %, **(11a)** 230.7 V, **(b)** 3 hp, **(c)** 9.55 lb-ft, **(d)** 66.6 %, **(e)** 2676 W, **(13)** 780 rev/min, **(15a)** 16.6 A, **(b)** 15.5 A,

(c) 13.8 A,   (17a) graph,   (b) 21 A and 12 hp,   (19a) 289 V,
(b) 3 hp,   (c) 75.5 %,   (d) 17.7 lb-ft,   (e) 19.3 lb-ft,
(21a) 342.65 W,   (b) 727.35 W,   (c) 77.7 %,   (23) 12,707.24 rev/min,
(25) 15 kW,   (27) 3.41 N-m,   (29a) 280 N-m,   (b) 6 kW,
(31) 11.11 %,   (33) 75 rad/s,   (35a) 452.8 V,   (b) 2.5 kW,
(c) 30.12 N-m,   (d) 72.5 %,   (e) 2716.8 W,   (37a) 87 %,
(b) 83.3 %,   (c) 75.8 %,   (39) 320 W,   (41a) 212 V,   (b) 1 kW,
(c) 60.6 %,   (d) 16.7 N-m,   (e) 21.9 N-m   (f) 1314 W,
(43a) 524.74 W,   (b) 235 W,   (c) 86.8 %,   (45a) 243 W,
(b) 6657 W,   (c) 6 kW,   (d) 30 N-m,   (e) 87 %,   (47a) 10.8 A,
(b) 135 %,   (49) 29.71 Ω.

CHAPTER 6

(1a) 33.75 V,   (b) 1.32 rad = 75.64 deg,   (c) 104.36 deg,
(3a) 4.86 ms,   (b) 1.83 rad = 104.85 deg,   (c) 54.11 V,
(d) 180 deg,   (5a) 180 Hz,   (b) 2700 Hz,   (c) 6000 Hz,
(7a) 95.5 Hz,   (b) 1,527.9 Hz,   (c) 3,246.8 Hz.

CHAPTER 7

(1a) 18.03 $\underline{/56.3°}$ Ω,   (b) 12.76 A,   (c) 0.55 lagging,   (d) 2,934.8 VA,
2,435.9 var, 1614.1 W,   (e) graph,   (3a) 41.23 $\underline{/-14.04°}$ Ω,   (b) 5.58 A,
(c) 0.97 leading,   (d) 1,283.4 VA, 1,244.9 W,   (e) graph,
(5a) 25k $\underline{/73.7°}$ Ω,   (b) 2 mA,   (c) 0.28 lagging,   (d) 0.1 VA, 0.096 var,
0.028 W,   (e) graph,   (7) 0.88 lagging,   (9a) 3 kW,   (b) 2.3 kW,
(11a) 2500 turns, 0.2,   (b) 1000 turns, 0.5,   (c) 2609 turns, 0.19,
(d) 26 turns, 19.23,   (e) 130 turns, 3.85,   (13a) 43.5 A,   (b) 21.7 A,
(15a) 3,333.3 A,   (b) 272.7 A,   (17a) 5,   (b) 46 V,   (c) 23 A,
(d) 4.6A,   (19a) 0.25,   (b) 5 A,   (c) 184 Ω,   (21) 245 V,
(23) 96 %,   (25a) 0.088 Ω,   (b) 0.176 Ω,   (c) 2.4 Ω,   (d) 4.8 Ω,
(27) 0.02 %,   (29) 0.85 %,   (31a) 3.5 %,   (b) 5.5 %,   (c) 0.82 %,
(33a) 4.25 kW,   (b) 41.7 A,   (35a) 97.1 %,   (b) 96.8 %,
(c) 96.5 %,   (d) 94.8 %,   (e) 96.2 %,   (37a) 0 A,   (b) 0,
(39a) 75 A,   (b) 25 A,   (c) 9 kVA,   (41a) 568.1 A,   (b) 454.5 A,
(c) 250 kVA,   (43a) 115 kVA,   (b) 5 kVA,   (c) 110 kVA,
(45a) 20 kVA,   (b) 1.8 kVA,   (47a) 120 V,   (b) 288.7 A,
(c) 120 V,   (d) 166.7 A,   (49a) 600 V,   (b) 9.62 A,   (c) 600 V,
(d) 5.55 A,   (51a) 62.76 A,   (b) 21.25 kW,   (c) 13.25 kvar,

(53a) 37.5 kVA, **(b)** 54.13 A, **(c)** 22.5 kvar, **(55a)** 20 kVA,
(b) 33, **(c)** 2.62 A, **(d)** 1.52 A, **(e)** 86.6 A, **(f)** 50.16 A,
(57a) 833.33 kVA, **(b)** 109.35 A, **(c)** 3,007 A, **(d)** 3,007 A,
(e) 109.35 A, **(f)** 0.0364, **(59a)** 230.94 kV, **(b)** 505.2 A,
(c) 24 kV, **(d)** 4,861.1 A, **(e)** 9.62, **(61a)** 53.46 A, **(b)** 152.74 A,
(c) 1.65, **(d)** 18.52 kVA, **(63a)** 1,667 kVA, **(b)** 19,245 A,
(c) 86.6 V, **(d)** 1,110 A, **(e)** 2600 V, **(65)** 28.874 kVA,
(67a) 173, **(b)** 38, **(69)** 788.3 lines/sq. in., **(71a)** 0.05×10^{-3} Wb,
(b) 0.39×10^{-3} Wb, **(73a)** 1.26×10^{-3} sq. m, **(b)** 1.6×10^{-4} sq. m,
(75) 0.06 T.

CHAPTER 8

(1a) 1800 rev/min, **(b)** 1200 rev/min, **(c)** 600 rev/min,
(d) 200 rev/min, **(3)** 56.31 kilolines, **(5)** 2.53 kilolines,
(7) 70.33 V, **(9)** 12 poles, **(11)** 0.066×10^{-3} Wb, **(13)** 120 V,
(15) 50.2 A, **(17a)** 0.14 Ω, **(b)** 2.04 Ω, **(c)** 2.04 Ω,
(19a) 6.1 %, **(b)** 18.4 %, **(c)** −8.3 %, **(d)** 636.6 V,
(21a) 139.2 V, **(b)** 141.7 V, **(23a)** 0.03 %, **(b)** −0.15 %,
(c) 0.13 %, **(25a)** 93.4 %, **(b)** 90.9 %, **(c)** 87.3 %,
(27a) 87.6 %, **(b)** 85 %, **(c)** 83.7 %, **(d)** 67 %, **(29a)** 25 A,
(b) 19.6%.

CHAPTER 9

(1a) 3600 rev/min, **(b)** 1800 rev/min, **(c)** 900 rev/min,
(d) 600 rev/min, **(e)** 225 rev/min, **(3a)** 25.4 ft-lb, **(b)** 270 V,
(5a) 19.1 ft-lb, **(b)** 30 ft-lb, **(7a)** 12.2 %, **(b)** 891 rev/min,
(9) 1640 rev/min, **(11)** 564 rev/min, **(13)** 3.65 %, **(15a)** 0.41 Ω,
(b) 0.11 Ω, **(c)** 0.10 Ω, **(17a)** 690 rev/min, **(b)** 705 rev/min,
(19) 68 %, **(21)** 1440 rev/min, **(23a)** 1389 rev/min, **(b)** 771 rev/min,
(c) 1550 rev/min, **(25a)** 85.4 %, **(b)** 0.87, **(27a)** 314 rad/s,
(b) 157.1 rad/s, **(c)** 104.7 rad/s, **(d)** 39.3 rad/s, **(e)** 17.45 rad/s,
(29a) 37.5 N-m, **(b)** 265.6 V, **(31a)** 30 N-m, **(b)** 47.3 N-m,
(33a) 6.4 %, **(b)** 103.65 rad/s, **(35)** 146.9 rad/s, **(37)** 16.14 rad/s,
(39) 5.1 %, **(41a)** 0.36 Ω, **(b)** 0.09 Ω, **(c)** 0.08 Ω,
(43a) 102.35 rad/s, **(b)** 103.5 rad/s, **(45)** 56 %, **(47a)** 23 %,
(b) 80.6 rad/s, **(49a)** 83.8 rad/s, **(b)** 52.4 rad/s, **(c)** 88 rad/s,
(51a) 81.7 %, **(b)** 0.89.

CHAPTER 10

(1) 85.2 deg,   **(3)** graph,   **(5a)** 0.17 leading,   **(b)** 0.94 leading,
(c) 85.3 %,   **(7a)** 0.4 deg,   **(b)** 5 deg,   **(9a)** 1.25 kvar,
(b) 3.83 kvar,   **(c)** 15.31 kvar,   **(11a)** 1939 var,   **(b)** 4312 var,
(c) 6685 var,   **(13)** derivation,   **(15a)** 31.8 kvar,   **(b)** 1,460 μF,
(17a) 104.9 kvar,   **(b)** 655.6 kvar,   **(19a)** 746.3 kVA,   **(b)** 554 kvar,
(c) 500 kVA,   **(d)** 1,131 kvar,   **(21)** 0.85,   **(23)** 34.2 kVA,
(25a) 2.353 kVA,   **(b)** 3.027 kVA,   **(27a)** 0.80 lagging,
(b) 3.756 kVA,   **(29)** 375 rev/min,   **(31a)** 8.75 ft-lb,   **(b)** 1200 rev/min,
(33) 26.2 rad/s,   **(35a)** 57.3 N-m,   **(b)** 78.54 rad/s,   **(37)** 0%.

CHAPTER 11

(1a) 0.67,   **(b)** 87 %,   **(3a)** 0.69,   **(b)** 80.2 %,   **(5a)** 0.75,
(b) 85.4 %,   **(7a)** 0.68,   **(b)** 85.9 %.

CHAPTER 12

(1) 13 A,   **(3a)** 0.55,   **(b)** 1.15 ft-lb,   **(5a)** 90.1 %,   **(b)** 2.37 ft-lb,
(7a) 3.04 ft-lb,   **(b)** 83.2 %,   **(9)** 7.25 A,   **(11a)** 0.58,   **(b)** 2 N-m,
(13a) 83.6 %,   **(b)** 3.1 N-m,   **(15a)** 5 N-m,   **(b)** 87 %.

CHAPTER 14

(1a) 4000 Wh/m^2,   **(b)** 344.23 Langleys,   **(c)** 20 kWh,
(3) 43 minutes,   **(5)** 2.316×10^{16} kWh/year,   **(7a)** 8 Ω,
(b) 32 W,   **(9a)** 6.15 Ω,   **(b)** 41.6 W,   **(c)** 25.8 W,
(d) 12 W,   **(11a)** 124.8 Wh/day,   **(b)** 195.5 Wh/day,
(13a) 2.71 A,   **(b)** 14.32 V,   **(c)** 38.8 W,   **(15)** 12.2 %,
(17) 11.6 %,   **(19a)** 18,   **(b)** 6 in series with 3 parallel paths,   **(c)** 1.5 Ω.

CHAPTER 15

(1a) 831 W,   **(b)** 1630 W,   **(c)** 2694 W,   **(3)** 10.1 mph,
(5) 645.2 lb,   **(7)** 20.13 ft,   **(9)** 4.531 kWh,   **(11a)** 10.6,   **(b)** 8.46,
(c) 7.05,   **(13)** 14.5,   **(15a)** 1023.4 W,   **(b)** 1690.7 W,
(c) 3006.5 W,   **(17)** 5.04 m/s,   **(19)** 3952 N,   **(21)** 6.1 m,
(23) 22.97 kWh,   **(25a)** 15,   **(b)** 13.33,   **(c)** 12,   **(27)** 18.83.

Index

Transformer, three-phase, 251–59
  basic connections, 253–59
  four-wire delta, 259
  open delta (V), 257
TRIAC, 177
  optically coupled driver, 179
  use in speed control, 411–13

Shunt Compound Generator rated 3Kw, 200V. has stray power losses of 120w at full load. R_f=100Ω R_a=0.9Ω R_S=0.2Ω

$I_L = \frac{P}{V} = \frac{3000}{200} = 15A$

Shunt Field Current

$V_f = V_L + V_S = 200 + I_S R_S$
$= 200 + (15)(0.2) = 203$

$I_f = \frac{V_f}{R_f} = \frac{203}{100} = 2.03A$

ARMATURE CURRENT
$I_a = I_f + I_L = 2.03 + 15 = 17.03$

GENERATED VOLTAGE
$V_t = E_g - I_a R_a$
$E_g = V_t + I_a R_a$
$= 200 + (17.03)(0.9) = 218.33$

MECHANICAL POWER CONVERTED TO ELECTRICAL POWER
$P_g = E_g I_a$
$= 218.33 \times 17.03$
$= 3718.2$ w

COPPER LOSSES

$shunt - P_f = R_f I_f^2 = 100 \times (2.03A)^2 = 412w$

$ARMATURE - P_a = R_a I_a^2$

$SERIES - P_S = R_S I_L^2$

$TOTAL - P_f + P_a + P_S = TOTAL COPPER LOSSES = 718w$

Efficiency (η)
$P_i = P + COPPER LOSSES + STRAY POWER LOSS$
$= 3Kw + 718w + 120w$
$= 3838$ $\eta = \frac{P_0}{P_i} \times 100 = \frac{3000}{3838} \times 100 = 78.2\%$

COMPOUND

Short Shunt

SERIES GENERATOR

5.5Kw 220V Shunt Generator R_f=140Ω, R_a=0.5Ω Stray power losses of 95w at rated conditions

FIELD CURRENT
$I_f = \frac{V_t}{R_f} = \frac{220}{140} = 1.57A$

ARMATURE CURRENT
$I_a = I_L + I_f = 25A + 1.57 = 26.57$

LOAD CURRENT
$I_L = \frac{P}{V} = \frac{5.5}{220} = 25A$

GENERATED VOLTAGE
Kirchoffs $E_g = I_a R_a + V_t$
$= 26.57 \times 0.5 + 220$
$P_f = R_f I_f^2 = 140 \times (1.57)^2$ $= 13.3 + 220 = 233.3$
$P_a = R_a I_a^2 = .5 \times (26.57)^2$

LONG SHUNT

LOAD CURRENT
$I_L = \frac{5X}{125} = 40$

FIELD CURRENT
$I_f = \frac{V_t}{R_f} = \frac{125}{125} = 1A$

ARMATURE CURRENT
$I_a = I_L + I_f = 40 + 1 = 41$

A LONG SHUNT COMPOUND RATED 5Kw, 125V HAS AN η OF 80%. R_f=125, R_a=0.2 R_S=0.05

COPPER LOSSES
$shunt - P_f = V_t I_f^2 = 125 \times 1 = 125w$
$ARM - P_a = R_a I_a^2 = 0.2 \times (41)^2 = 336.2w$
$SERIES - P_S = R_S I_S^2 = .05 \times (41)^2 = 84.05w$
$TOTAL - P_f + P_a + P_S = 545.25$

$TOTAL = COPPER + STRAY$

STRAY POWER LOSSES
$\eta = \frac{P_0}{P_i} \times 100$
$80 = \frac{5Kw}{P_i} \times 100$
$P_i = 5Kw \times \frac{100}{80} = 6250w$
$P_i - P_0 = 6250w - 5000w = 1250w$
$TOTAL = 1250 - 545.25 = 704.75w$

ARM loss $P_a = R_a I_a^2 = 1.0 (67.22)^2 = 4518.5w$

15Kw, 230V Shunt Generator, stray power losses of 1300w
R_f=115Ω, R_a=1.0Ω

FIELD CURRENT
$I_f = \frac{V_t}{R_f} = \frac{230}{115} = 2.0A$

LOAD CURRENT
$I_L = \frac{P}{V} = \frac{15Kw}{230} = 65.22$

ARM CURRENT
$I_a = I_L + I_f = 65.22 + 2.0 = 67.22$

SHUNT FIELD LOSS
$P_f = R_f I_f^2 = 115(2.0)^2 = 460w$

EFFICIENCY
$P_i = P_{out} + COPPER + STRAY POWER LOSS$
$= 15K + (460 + 4518.5) + 1300 = 21278.5$
$\eta = \frac{P_0}{P_i} \times 100 = \frac{15000}{21278.5} \times 100 = 70.49\%$

GENERATED VOLTAGE
$= V_t + I_a R_a$
$230 + (67.22 \times 1) = 297.22$

CHAP 5 (BRAKES)   P-129   P-131

$T = K\Phi I_a$

$= \frac{7.04 P(w)}{S}$ (ENG)

$T = \frac{1000 P(Kw)}{\omega}$

STALL TORQUE
LOAD TORQ
NO LOAD SPEED
MOTOR SPEED